AF686445

CARBONATE SEDIMENTOLOGY AND SEQUENCE STRATIGRAPHY

Wolfgang Schlager

Vrije Universiteit/Faculty of Earth and Life Sciences
Amsterdam, Netherlands

Copyright 2005 by
SEPM (Society for Sedimentary Geology)
Laura J. Crossey, Editor of Special Publications
Concepts in Sedimentology and Paleontology No. 8

This book is published with funds from the SEPM Foundation Inc., including the Bruce H. Harlton Publications Fund made possible by Allan P. Bennison

ISBN 1-56576-116-2
©2005 by
SEPM (Society for Sedimentary Geology)
6128 East 38th Street, #308
Tulsa, Oklahoma 74135

Camera-ready by E. van Bentum, Amsterdam
Printed in the United States of America

To
Hanneke Schlager-Vierstraete
wife, mother and silent co-author

Contents

Preface

Sedimentology and stratigraphy are neighbors yet distinctly separate entities within the earth sciences. Put in a nutshell, sedimentology searches for the common traits of sedimentary rocks regardless of age as it reconstructs environments and processes of deposition and erosion from the sediment record. Stratigraphy, by contrast, concentrates on changes with time, on measuring time and correlating coeval events. Sequence stratigraphy straddles the boundary between the two fields. It is a sedimentologic concept as it uses depositional anatomy to reconstruct environments and lateral facies change, and it is part of stratigraphy as it studies the vertical succession of sedimentary rocks and their succession and correlation.

This expose, dedicated to carbonate rocks, approaches sequence stratigraphy from its sedimentologic background. Carbonate sedimentation, in contrast to siliciclastic sedimentation, is largely governed by chemistry and biota of the ocean and thus intimately tied to the ocean environment. Therefore, the presentation starts with essentials of physical and chemical oceanography and biology. It then proceeds to principles of marine carbonate production (and erosion) and the geometry of carbonate accumulations, using the concept of carbonate production systems, or factories, to illustrate the variations among carbonate rocks. Armed with the knowledge on production and accumulation, the text turns to carbonate facies; the sedimentologic part closes with an overview of the rhythms and events governing carbonate deposition in time and space. Chapters 6 through 8 deal with sequence stratigraphy. This part starts with an overview of the standard model of sequence stratigraphy and then develops carbonate sequence stratigraphy on the basis of processes and principles presented in the sedimentologic part and using the three major carbonate factories as a template for discussion.

The book attempts to make progress by combining different specialties and different lines of reasoning, and by searching for principles underlying the bewildering diversity of carbonate rocks. I think the expose provides enough general background, in introductory chapters and appendices, to be easily digestible for sedimentologists and stratigraphers as well as earth scientists at large. The text does appeal to the reader's willingness to engage in scientific discussion, however. It is not a cook book presenting recipes.

The book bears the name of a single author but many have supported me in the endeavor. Gregor Eberli (University of Miami), Mitch Harris (Chevron, San Ramon), Adrian Immenhauser (Vrije Universiteit, Amsterdam) and Rick Sarg (Exxon, Houston) reviewed the entire manuscript, Eberli and Harris did so as official reviewers for SEPM. Advice on special topics was offered by Giovanni Bertotti (Vrije Universiteit, Amsterdam) on tectonics, Bruce Fouke (University of Illinois, Urbana) on sea cliffs, Henk Kooi (Vrije Universiteit, Amsterdam) on slope stability, and Georg Warrlich (Shell, Rijswijk) on computer modeling.

Important data were contributed by Flavio Anselmetti (ETH, Zürich), Hemmo Bosscher (Shell, Rijswijk), Annette George (University of Western Australia, Crawley), Robert Ginsburg (University of Miami), Mitch Harris (Chevron, San Ramon), Lisa Hinnov (Johns Hopkins University, Baltimore), David Hunt (Norsk Hydro, Bergen), Adrian Immenhauser (Vrije Universiteit, Amsterdam), Alexandra Isern (National Science Foundation, Washington), Noel P. James (Queens University, Kingston), Bernd Kaufmann (University of Tübingen), Lorenz Keim (Geological Survey, Bolzano), Jeroen Kenter (Vrije Universiteit, Amsterdam), Don McNeill (University of Miami), Sam Purkis (Nova University, Dania) Franco Russo (University of Cosenza), Orson van de Plassche (Vrije Universiteit, Amsterdam), Valentina Zampetti (Vrije Universiteit, Amsterdam).

Many of the ideas expressed in the book developed in the stimulating environments of Miami's Rosenstiel School of Marine and Atmospheric Science, particularly the Comparative Sedimentology Laboratory founded by Robert Ginsburg, and the Sedimentology Section of the Earth Sciences Faculty at the Vrije Universiteit Amsterdam. At both institutions, Industrial Associates Programs offered valuable feedback from geologists in industry. Another crucial sounding board were the participants of short courses I taught on the subject in North America, Europe and Southeast Asia between 1991 and 2003.

I was very fortunate with the publication process. Elisabeth van Bentum was my partner in the production of the manuscript. Her energy and skills in drafting and camera-ready editing made for smooth sailing all along. SEPM, with Howard Harper at the helm, Laura Crossey as Editor-in-Chief and Kris Farnsworth as publications coordinator supported and encouraged me throughout.

Wolfgang Schlager, Amsterdam

CHAPTER 1

Essentials of neighboring disciplines

INTRODUCTION

Unraveling Earth history is a core business of geology. However, understanding cause and effect of past events requires input from other disciplines that can study processes directly and do not have to reconstruct them from incomplete historic records. This introductory chapter summarizes a very limited number of concepts from neighboring disciplines that are relevant in this respect. The list is woefully incomplete. I did not try to simply cover the most important concepts; rather, I selected those that are highly relevant for the topic of this book yet not sufficiently important for geology at large to be routinely covered by introductory texts or courses in geology.

SOME PRINCIPLES OF OCEANOGRAPHY

Layering of the oceanic water column

The density of ocean water varies as a function of temperature and salinity. Under the influence of the Earth's gravity field denser water sinks and lighter water rises and this creates a density-stratified, layered ocean (Weyl, 1970; Open University, 1989a, 1989b). In first approximation, three layers may be distinguished: (Fig. 1.1)

- A surface layer, where waves and currents are sufficiently strong to preclude the formation of permanent density gradients; the surface layer is thus well mixed and in equilibrium with the atmosphere with regard to oxygen and other chemical agents.
- The thermocline, where density increases steadily downwards and vertical mixing is greatly inhibited.
- The oceanic deep water, where density slowly increases downward but the vertical gradient is so low that it is easily disturbed by horizontal flow.

Water in all layers circulates. Circulation in the surface layer is driven by winds in the atmosphere. The deep layer presently forms one world-wide circulation system that moves slowly under the influence of minor density gradients.

Surface layer and deep layers are generally well separated but they communicate in certain, well defined areas. In the northern North Atlantic and around Antarctica, surface water becomes so dense that it sinks and joins the deep-water body. Conversely, along the west-facing coastlines of Africa and the Americas the surface layer is driven offshore by winds and the Coriolis force such that thermocline water and deep water well up. Deep water also rises to the surface in the Antarctic Current because of the current's extremely low density gradient. More gentle upwelling occurs in the equatorial Pacific where the opposing directions of the Coriolis force of the northern and southern hemisphere drive the surface layer away from the equator. See Fig. 1.2 for cartoons of upwelling mechanisms.

An important principle in all interactions between atmosphere and ocean is Ekman transport: wind blowing over the ocean induces a current in the surface layer approximately at right angles to the wind. This current moves to the right of the wind in the northern hemisphere and to the left in the southern hemisphere.

When using the modern oceans as models for the geologic past, we should keep in mind that currently temperature has a greater effect on the density stratification than salinity. The main reason for the dominance of temperature are the ice caps of the poles. At times in the past when little ice was present, salinity seems to have been the dominant factor in density stratification (e.g. Hay, 1988).

Present surface circulation of the oceans

The surface circulation greatly influences the distribution of temperature, salinity and nutrients in the surface layer and those properties, in turn, are important controls on carbonate precipitation and deposition. We shall examine the present circulation to derive guidelines for interpreting the past. For more detail, see Weyl (1970), Broecker and Peng (1982), Open University (1989a, 1989b), Emiliani (1992, p. 254-309).

The chain of cause and effect between atmosphere and surface ocean starts with the Sun. It heats the equatorial belt of the Earth much more than the polar regions. As a consequence, the air over a stationary Earth would rise over the equator, flow poleward at high altitude, cool, sink and flow back towards the equator near the Earth's surface. This simple circulation model immediately calls for two essential modifications:

- the Coriolis force of the rotating Earth deflects the flow of air. Air starting to move from pole to equator will be deflected to move from east to west as the Coriolis effect turns it to the right in the northern and to the left in the southern hemisphere. Conversely, air starting to flow poleward from the equator will turn to flow eastward.

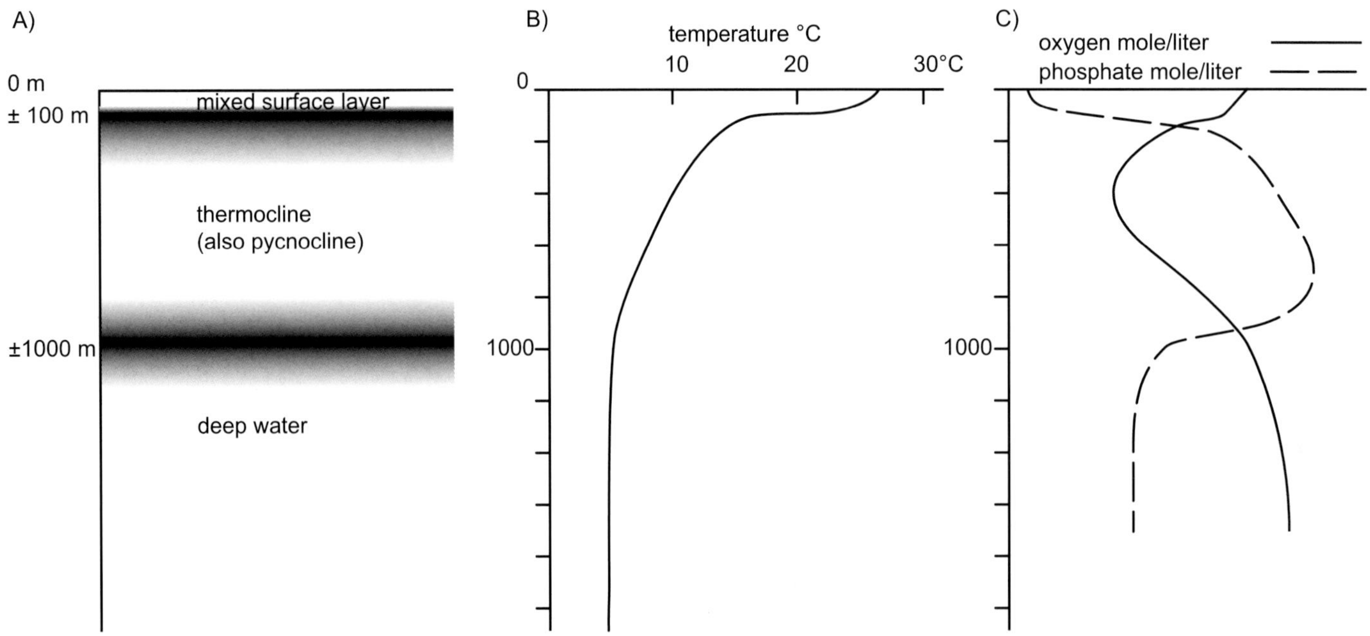

Fig. 1.1.— Upper water column of the tropical and temperate parts of the ocean. A) Principal layering of the water column. B) Temperature profile (solid line) indicates low gradients in the mixed layer, extremely high gradients in the uppermost thermocline, high gradients in the lower thermocline and low gradients in the oceanic deep water. C) Characteristic profiles of the concentration of dissolved oxgen and phosphate (a nutrient). These profiles reflect the interplay of the density structure of the water, gas exchange with the atmosphere and organic growth and decay. Oxygen and nutrient levels are crucial for subdividing the carbonate producing environments. See text for details.

- a single circulation cell per hemisphere cannot be sustained because of rapid cooling in the upper atmosphere and the speed of the Earth's rotation. The single cell breaks up into three cells where air rises and sinks as shown in Fig. 1.3A. The three cells induce three major windfields per hemisphere – easterly trade winds in the low latitudes, westerlies in the temperate belt and easterlies again in the subpolar regions.

The three wind fields between pole and equator induce surface currents in the ocean. If the Earth were entirely covered by water, we would see globe-circling currents flowing from east to west near the poles, from west to east in the temperate zone of westerly winds and from east to west again in the subtropical belt of trade winds. Around the equator where winds are weak and irregular, surface currents will also be weak. The present distribution of land masses severely obstructs the development of globe-circling currents. Pacific and Atlantic both extend from the Arctic to the Antarctic but are bounded in the west and east by continents or archipelagos. The Indian Ocean extends over a little more than one hemisphere in the same fashion. Thus, is behooves us to consider a model ocean stretching from pole to pole but bounded in the east and west by land (Fig. 1.3B; Weyl, 1970). In this model, the latitudinal currents induced by the wind fields are deflected to form circular currents called gyres. The largest ones, the subtropical gyres, develop in both hemispheres as water flows westward under the trades and returns to the east under the westerlies. The smaller subpolar gyres circle counterclockwise under the combined effect of westerlies in the temperate zone and easterlies in the subpolar region. In the equatorial region, our model ocean shows two narrow, counterclockwise gyres as part of the water piled up against the western border of the ocean returns eastward in the equatorial calm zone, the doldrums.

Fig. 1.4 depicts the real surface circulation of the world ocean. It shows most features predicted Weyl's (1970) model ocean. The most significant departure is around Antarctica where north-south land barriers are absent so that a globe-circling current, the Antarctic Current, could develop. Furthermore, the narrow equatorial gyres and the counter current separating them are weakly developed in the Atlantic where the equatorial system points WNW-ESE because of the peculiar shape of the continental borders, and in the Indian Ocean, where India deforms the northern subtropical gyre and monsoonal circulation disturbs the trade-wind system.

Monsoons are winds that reverse direction in opposite seasons because of the different heating and cooling of the air over land and sea. Air over a large land mass at temperate latitudes becomes cool and dense in winter; a high-pressure cell forms and wind streams out from it. Conversely, air in the summer is hot and light; a cell of low pressure develops and sucks in air from adjacent oceans where

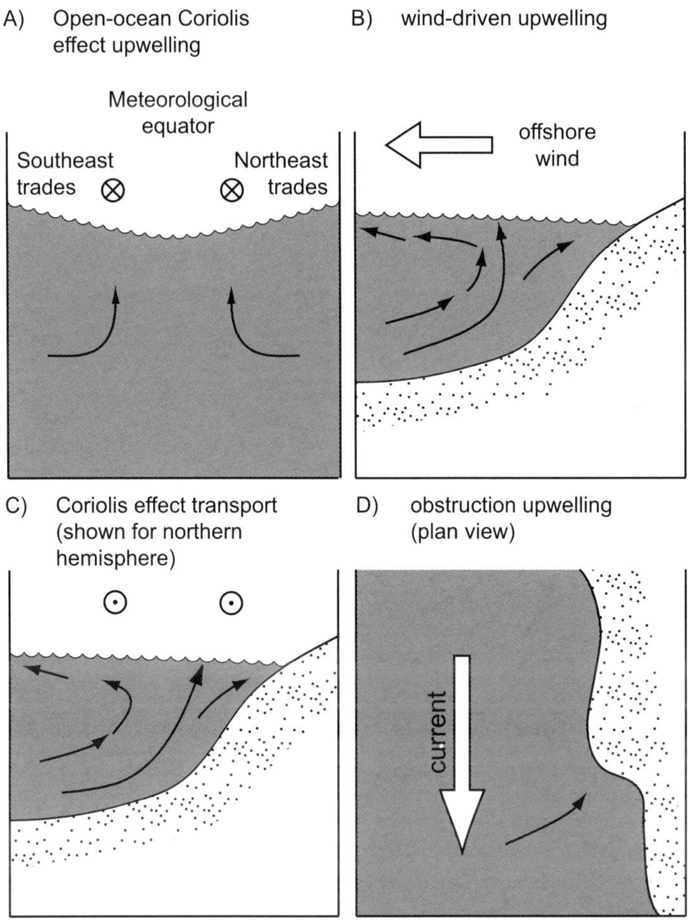

Fig. 1.2.— Most efficient processes causing upwelling. A) Ekman transport along the equator drives the surface layers away in opposite directions. B) Wind blowing offshore drives the surface layer away from the coastal zone and exposes deeper water layers. (Ekman transport starts at high angles to the wind but quickly turns into wind-parallel motion because of space constraints). C) Ekman transport induced by coast-parallel winds or surface currents drives the surface layer offshore and exposes deeper layers (situation shown is for the northern hemisphere). D) Local upwelling where shore-parallel current is impeded by a promontory in the coastline. After Pipkin et al. (1987), modified.

the reversing monsoon winds may induce reversing surface currents. Currently, the strongest monsoons develop between southern Asia and the Indian Ocean.

Fig. 1.4 shows that the present world ocean shows features of Weyl's laterally bounded "gyre ocean" as well as a globe-circling current. Paleoceanographic studies indicate that the changing positions of continents, archipelagos and island arcs of the geologic past also changed the coeval circulation patterns. It seems, however, that with proper adjustments, the concepts of oceanic gyres and globe-circling currents are extremely useful models for the reconstruction of the past, including the reconstruction of shoal-water carbonate deposition.

Present deep circulation of the ocean

The principal driver of present deep-water circulation is the formation of dense cold water in the northern North Atlantic and on the shelves of Antarctica. This water sinks and provides the deep and bottom water for all major ocean basins. The first-order pattern is that most deep water forms in the Atlantic and flows through the Indian Ocean to the North Pacific whence it returns to the source areas via upwelling and surface circulation. (Fig. 1.5; Broecker and Peng, 1982; Open University, 1989b).

This flow pattern forces the three principal oceans to constantly exchange surface water and deep water at a large scale. The effects of this exchange on the three basins are very different, though, and this leads to differences in water chemistry that are relevant for carbonate sedimentation. The Atlantic donates deep water and receives surface water, the Pacific receives deep water and donates surface water, and the situation of the Indian Ocean is intermediate. As nutrient concentrations are generally low in surface water and high in deep water, the Atlantic is depleted in nutrients compared to the Pacific. Fig. 1.6 depicts this principle of "basin-basin fractionation" of dissolved substances.

Another side effect of deep circulation concerns carbonate dissolution at depth. The deep water in the Atlantic is relatively young, thus still rich in oxygen (from the time spent at the sea surface) and low in carbon dioxide because the water has not had time to oxidize much organic matter. Conversely, Pacific deep water is old, low in oxygen and high in carbon dioxide because of the long time spent oxidizing organic matter without an opportunity to take up oxygen from the atmosphere. The low CO_2 content of the deep Atlantic leads to high carbonate saturation and a low position of the carbonate compensation depth (CCD) – the level where the rate of carbonate sedimentation equals the rate of carbonate dissolution. The situation is reversed in the Pacific. There, the deep water is old, rich in CO_2 and thus largely undersaturated with respect to calcium carbonate. This increases the rate of dissolution and raises the carbonate compensation depth. The world map in Fig. 1.7 clearly shows the differences between Atlantic and Pacific.

Distribution of temperature, salinity, nutrients and light in the surface waters

Temperature, salinity and *nutrients* in the surface layer of the sea are important controls on carbonate production. Their distribution is mainly governed by the latitudinal differences of solar radiation and by the patterns of surface circulation. The first-order trend is a decrease in temperature with increasing latitude and salinity maxima in the horse latitudes that correspond to the desert belts on land. From the horse latitudes, salinity decreases pole-ward and towards the equator where high rainfall dilutes the surface water. The effects of gyre circulation are more varied as outlined below.

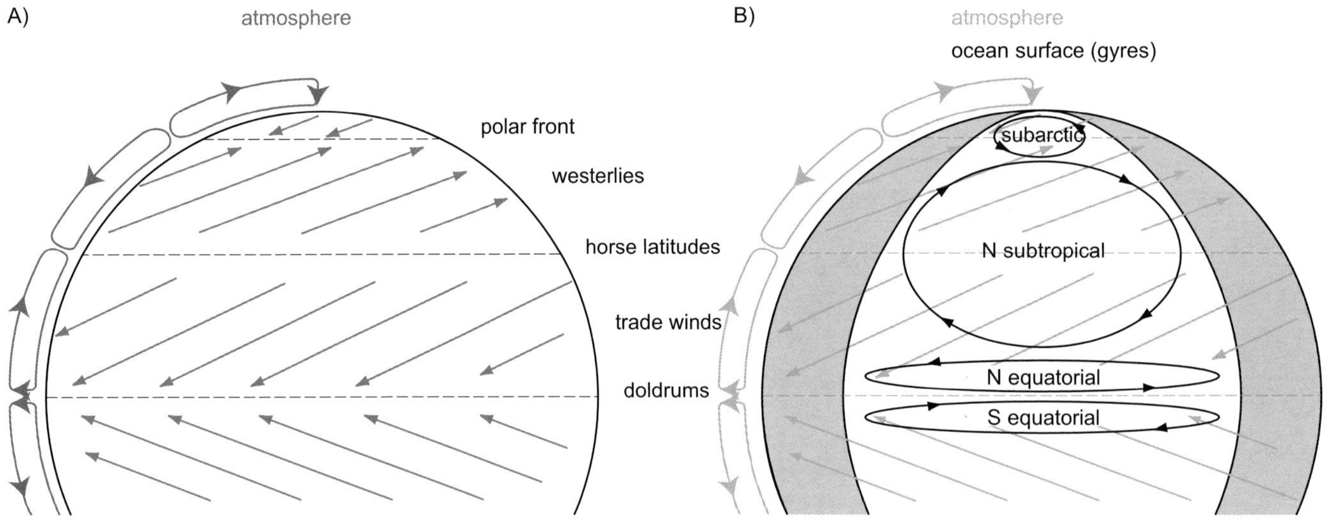

Fig. 1.3.— Weyl's (1970) model of surface circulation in the ocean. A) The atmosphere has three circulation cells per hemisphere. In the northern hemisphere of a non-rotating Earth, these cells would induce winds at the surface of the northern hemisphere that blew northward between equator and horse latitudes, southward between horse latitudes and polar front, and northward again in the polar region. The Coriolis force deflects these winds to the right, thus creating the wind fields we currently observe. B) The wind fields depicted in Fig. A shear over the surface layer of the ocean and induce currents there. The surface currents deviate to the right of the wind vector in the northern hemisphere because of Ekman transport caused by the Coriolis force. The results are west-flowing currents under the trade winds, east-flowing currents under the westerlies, and west-flowing currents in the polar regions. In modern ocean basins, N-S running coasts block most of these currents and deflect them to form large closed circulation loops called gyres.

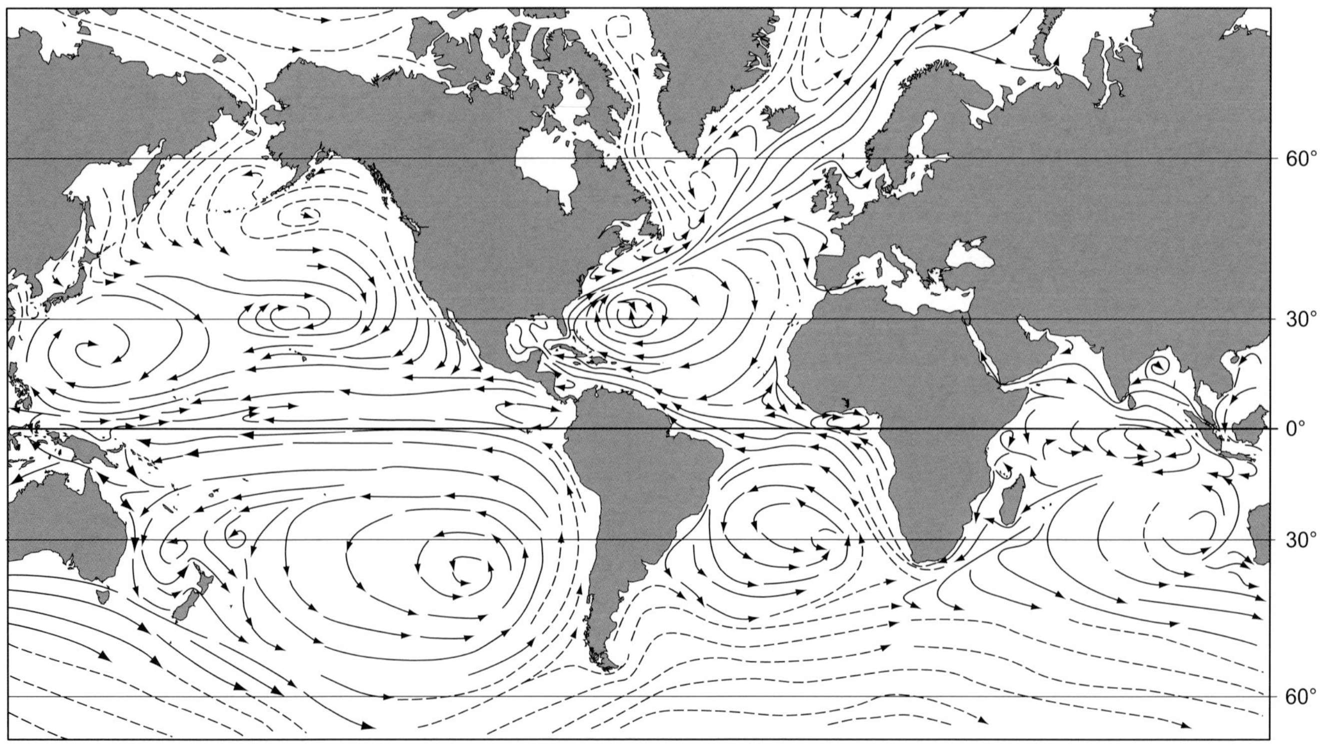

Fig. 1.4.— Surface circulation in the modern oceans. After Strahler (1971), modified. The principal features of Weyl's model are present, albeit with some distortions due to the shape of the ocean basins. In the south, around Antarctica, the gyre-circulation is replaced by the globe-circling Antarctic Current.

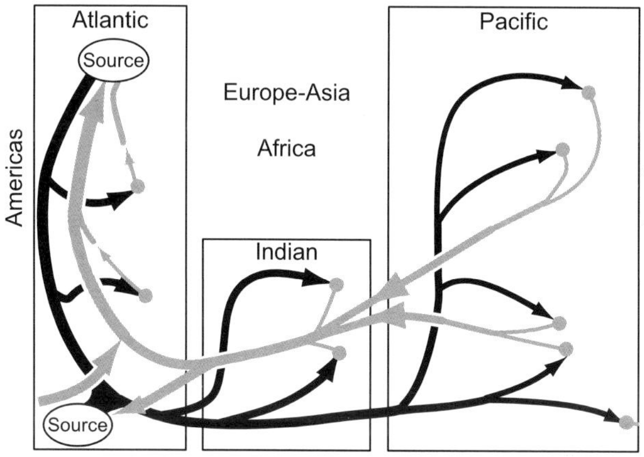

Fig. 1.5.— Present deep-water circulation in the ocean. After Broecker and Peng (1982, Fig. 1-12), modified. Deep-water (black arrows) originates in the northern North Atlantic and on the shelves of Antarctica and flows through all three oceans. Return flow occurs by surface circulation (gray). All pathways are extremely simplified. Deep-water rises to the surface in upwelling areas, mainly on the west sides of continents (dots) and in the Antarctic current.

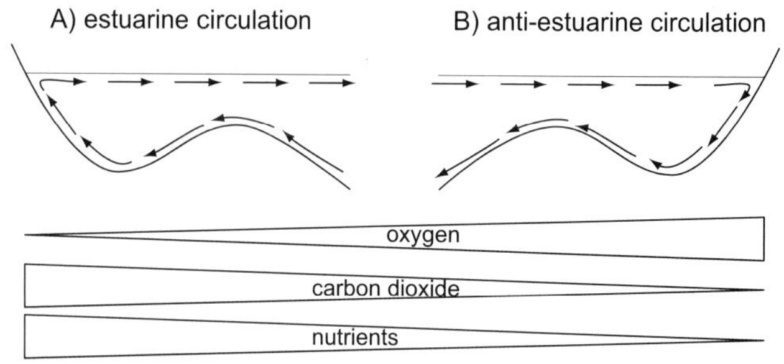

Fig. 1.6.— Basin-basin fractionation of nutrients, oxygen and carbon dioxide (based on Berger and Winterer, 1974; Broecker and Peng, 1982). A) Estuarine circulation. This ocean basin circulates like a river estuary. It donates surface water to, and receives deep water from the world ocean. Consequently, its deep water is old, rich in carbon dioxide and nutrients, and low in oxygen. Modern example: Pacific.
B) Anti-estuarine circulation. This ocean basin circulates like a hypersaline lagoon. It donates deep water and receives surface water. Its deep water is young, rich in oxygen and depleted in carbon dioxide and nutrients. Modern example: Atlantic.

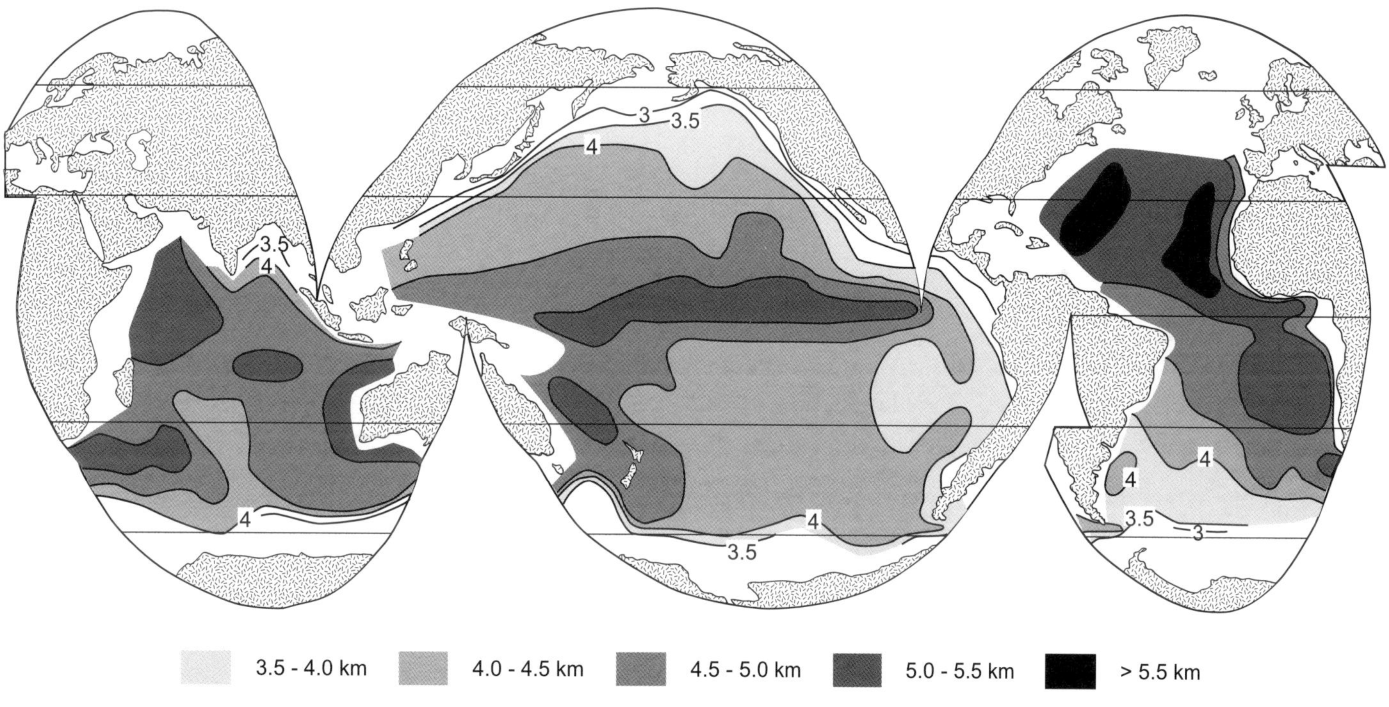

Fig. 1.7.— Carbonate compensation depth (CCD) in the recent oceans, determined from the carbonate content of pelagic sediments. The CCD forms a surface with considerable relief. This surface is relatively deep in the Atlantic; relatively shallow in the Pacific and lies at intermediate depths in the Indian Ocean. The difference among ocean basins is caused by basin-basin fractionation. In all three oceans the CCD shoals towards the ocean margins as a result of high organic productivity and concomitant production of CO_2. Gentle equatorial upwelling in the Pacific increases planktonic carbonate production and thus causes a depression of the CCD. After Berger and Winterer (1974), modified.

The subtropical gyres are of special importance for carbonate sedimentation because of their effect on nutrient concentrations and surface productivity (Fig. 1.8). The basic rule is that areas of upwelling are productivity maxima, areas with old surface waters are productivity minima. The rotation in the gyres is such that the Coriolis force deflects the water towards the center of the gyre, creating a zone of convergence, filled with old, nutrient-depleted water of

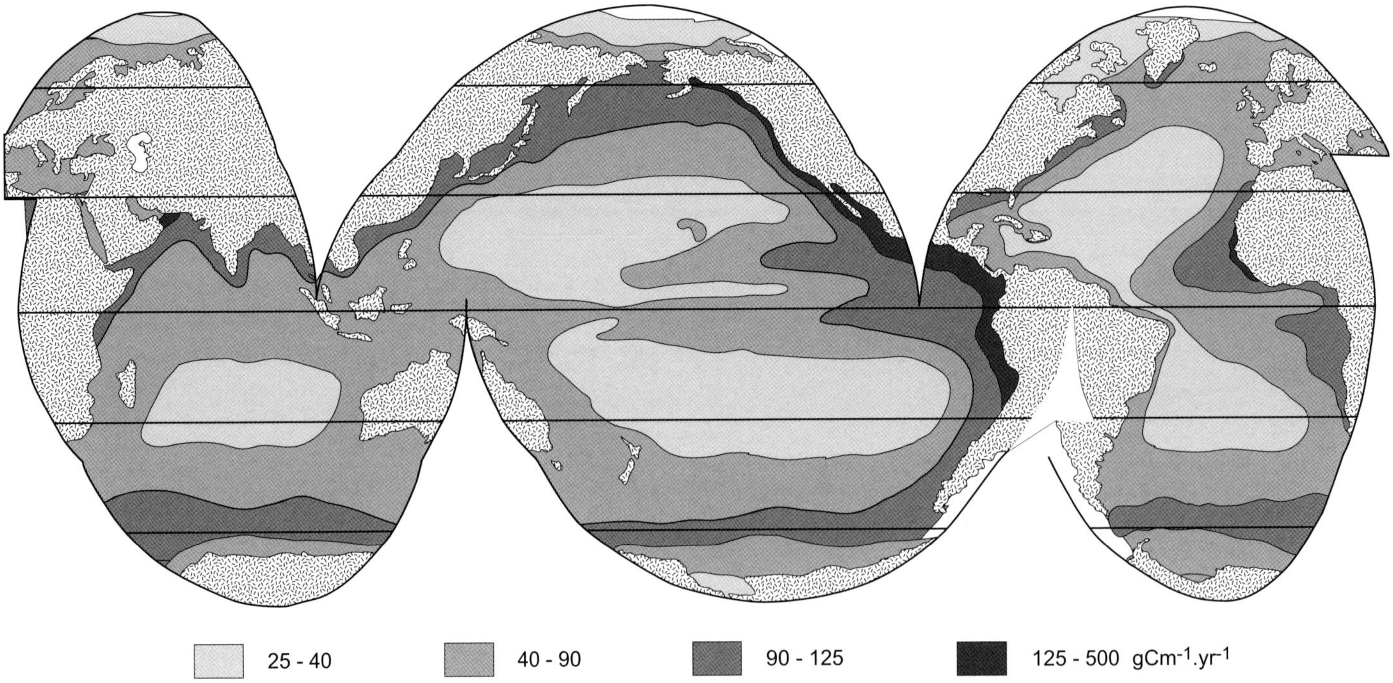

Fig. 1.8.— Organic productivity in the surface layer of the ocean, calculated from phosphate distribution, latitude and distance from shore. Typical are minima ("submarine deserts" in the subtropical gyres and maxima in the coastal upwelling zones, particularly along the west-facing coasts. High productivity also occurs in the globe-circling Antarctic current where water of all depths mixes easily because density gradients are very low. Carbonate production is strongly influenced by nutrient concentration (see chapter 2). After Berger (1989), modified.

high salinity because the gyre centers lie in the arid horse latitudes. The centers of the gyres represent warm, nutrient-depleted deserts of the ocean.

The eastern peripheries of the subtropical gyres represent the other extreme. The flow pattern is such that wind and Coriolis force drive the surface water away from the continent. Upwelling of cool, nutrient-rich water is the result. At the western periphery of the gyres, upwelling is weaker as wind drift and Coriolis force oppose one another.

In the equatorial belt, the opposed direction of the Coriolis force in the two hemispheres leads to surface divergence and gentle upwelling with moderately high nutrient levels. Upwelling and high rainfall lower the salinities.

Light is the basis for photosynthesis, i.e. organic growth that relies on dissolved nutrients in the water and energy from the Sun. Organic growth by photosynthesis is the begin of the food chain of the ocean. Moreover, much carbonate sediment is formed as a byproduct of photosynthesis.

The decrease of light with water depth follows a simple exponential function

$$I_z = I_0 . e^{-kz}$$

where I_0 and I_z are the irradiance at sea level and depth z, respectively, and k is an attentuation coefficient that depends on the turbidity of the water. For instance, k is large in areas of high suspended sediment load or high plankton productivity; k is small in clear, low-productivity waters such as at the centers of the subtropical gyres (Fig. 1.9).

ESSENTIALS OF CARBONATE MINERALOGY AND CHEMISTRY

The material for carbonate sedimentation is extracted from the dissolved load of the sea; the volume of carbonate rocks derived from erosion of older rocks is very small. The precipitation reactions can be summarized as

$$Ca^{2+} + 2HCO_3^- \rightarrow CO_2 + H_2O + CaCO_3$$

Aquatic precipitation proceeds along biotic and abiotic pathways that are discussed further in chapter 2.

Precipitation, preservation and alteration of the carbonate rocks are strongly influenced by their mineralogy. Three minerals appear in significant amounts: aragonite, calcite and dolomite (Fig. 1.10). In practice, calcite is further subdivided into rather pure calcite (also called "low-magnesium calcite") and magnesian calcite (or "high-magnesium calcite"). Magnesian calcites are generally defined as calcites with more than 4 mol% of $CaCO_3$ replaced by $MgCO_3$ (J.A.D. Dickson in Tucker and Wright, 1990). This boundary is justified by the observed changes in solubility (Fig. 1.11). Up to 4mol% $MgCO_3$, magnesium does not seem to significantly influence calcite solubility. Therefore it is practical to draw the boundary at the 4% level. One should be aware, however, that there is a continuous range of magnesium contents in calcite from 0 to over 30 mol% $MgCO_3$.

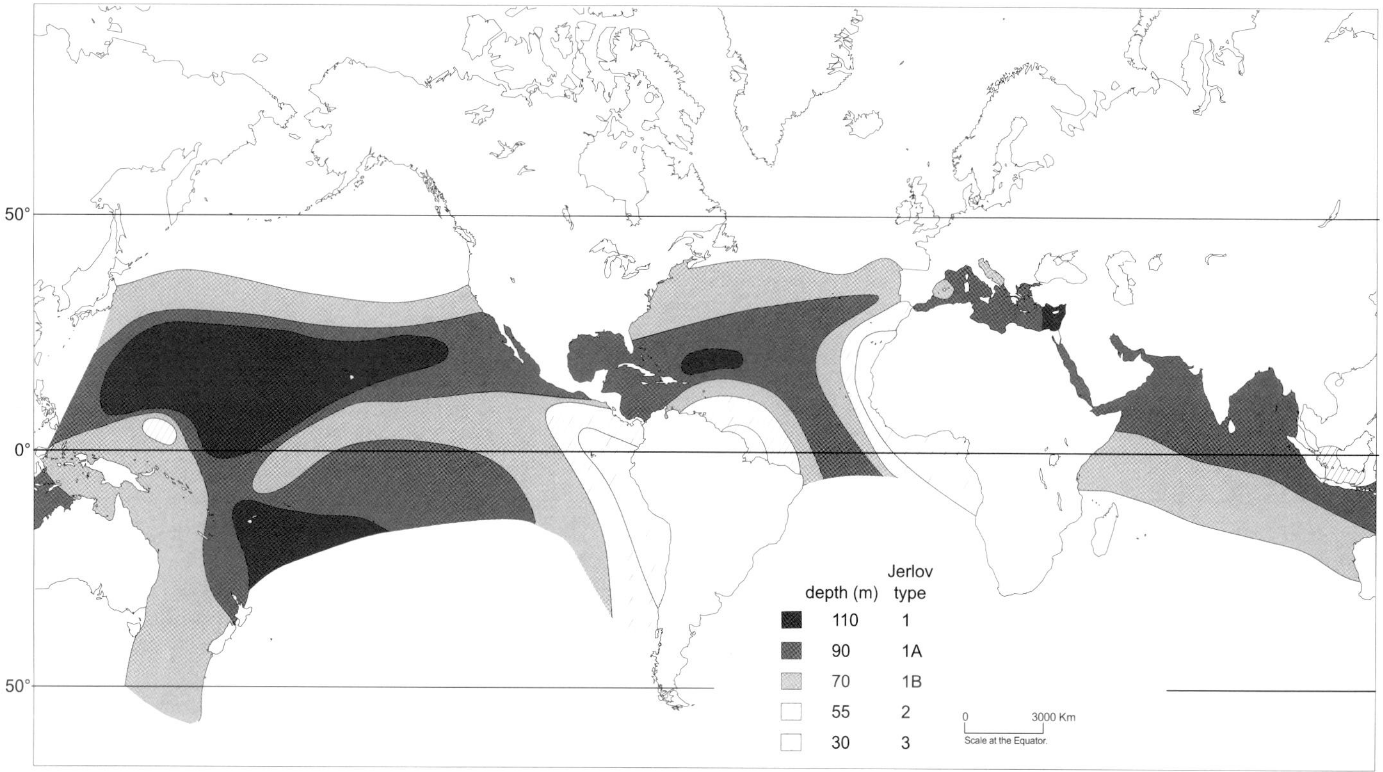

Fig. 1.9.— Crude estimate of the thickness of the euphotic zone based on the optical water types of Jerlov (1976). The euphotic zone is thickest in the subtropical gyres and thins towards higher latitudes. The euphotic zone also thins in the zone of equatorial upwelling and near tropical river discharge (Amazone, Indonesia). Modified after F. Vijn and H. Bosscher (written comm.)

Calcite, aragonite and dolomite differ considerably in solubility and the differences are sedimentologically very important. In sea water (and also in many pore waters) the ranking in terms of solubility is aragonite > calcite > dolomite. The solubility of magnesian calcite depends on the magnesium content as explained above. From 4 mol% onwards, solubility steadily increases with Mg content (Fig. 1.11). Magnesian calcite of >12mol% $MgCO_3$ is more soluble than aragonite, magnesian calcite of lower $MgCO_3$ content is less soluble than aragonite (Morse and Mackenzie, 1990; Morse, 2004).

In view of this solubility ranking, it is surprising to see that carbonate sediments in the tropical marine environment consist largely of aragonite and magnesian calcite with minor fractions of calcite; primary dolomite forms only in special environments. Cool-water carbonates are richer in calcite but still contain mostly aragonite and magnesian calcite. The reason for this paradox is that most precipitation is controlled or induced by organisms and most abiotic reactions are inhibited by reaction kinetics. Thus, thermodynamics is a poor predictor of carbonate precipitation reactions in marine environments. Nearly all marine surface waters are supersaturated with respect to calcite and dolomite but the appropriate precipitation reactions are blocked in various ways. The abiotic reactions that ultimately do occur, such as the formation of fibrous cements in marine environments, produce aragonite or magnesian calcite rather than the thermodynamically expected minerals calcite and dolomite.

mineral			isomorphic substitution	density $g.cm^{-3}$
Aragonite	$Ca\,CO_3$	rhombic	Sr, Na	2.94
Calcite	$Ca\,CO_3$	rhombohedral	Mg, Sr, Na	2.72
Magnesian Calcite	$Ca\,CO_3$ with 5 - 44% Ca replaced by Mg	rhombohedral		
Dolomite	$Ca\,Mg\,(CO_3)_2$	rhombohedral	Sr, Na, Fe	2.89

Fig. 1.10.— Common carbonate minerals, their chemical composition, crystallography, and density.

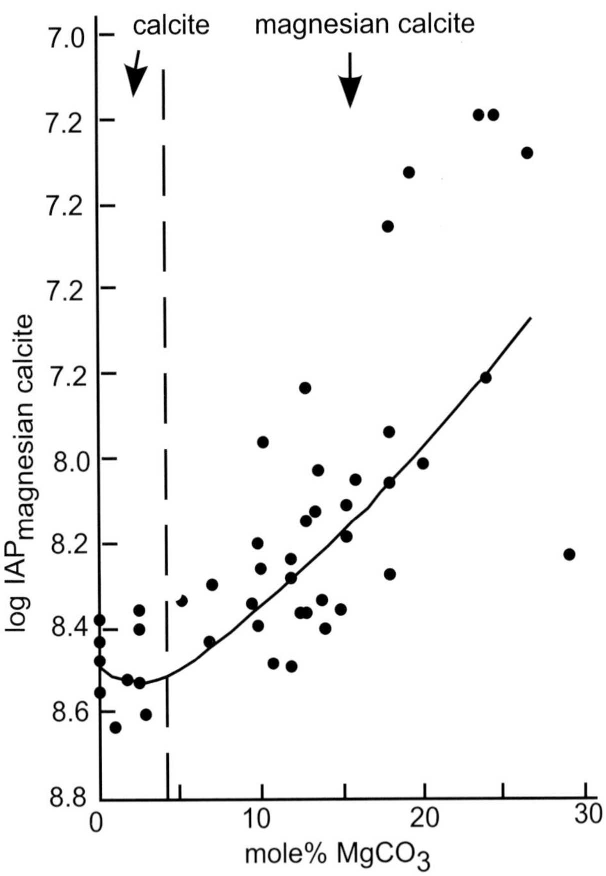

Fig. 1.11.— Solubility of calcites as a function of $MgCO_3$ content. Curve represents the general trend of several experiments indicated by different symbols. Solubility, expressed as ion activity product, increases steadily from about 4 mol% $MgCO_3$ upward. Dashed line: generally accepted boundary of calcite and magnesian calcite. After Mackenzie et al. (1983), modified.

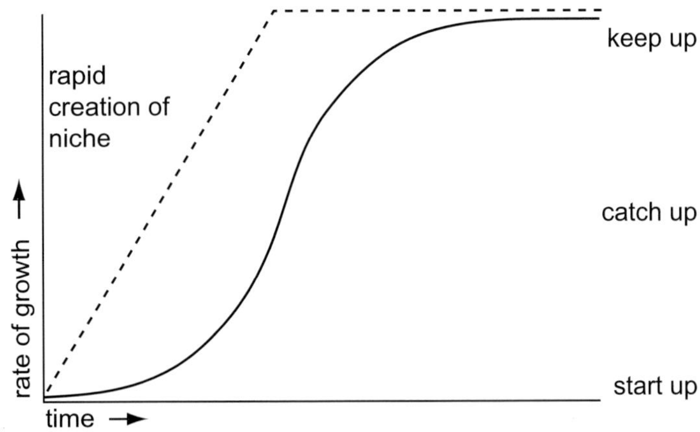

Fig. 1.12.— Sigmoidal growth curves – a common pattern in carbonate production. Populations of organisms respond to the opening up of new living space in three steps: first, growth lags behind the creation of living space, second, population growth exceeds the rate of change in space, finally, population growth is limited by the rate of growth in living space. A sigmoidal growth curve results from this three-phase growth. Most carbonate systems follow this pattern and the terms on the right are widely used for the three stages of sigmoidal carbonate growth, (e.g. Neumann and Macintyre, 1985.)

Aragonite and magnesian calcite are metastable in most marine environments, i.e. they precipitate and exist for considerable time because of the effects of reaction kinetics mentioned above. However, diagenesis replaces aragonite and magnesian calcite by calcite and dolomite on geologic time scales. Replacement proceeds by dissolution-precipitation reactions rather than solid-state inversion. This implies that porosity may be created or destroyed and chemical signals reset in different diagenetic environments.

Most magnesian calcite and aragonite disappear in less than a million years and are replaced by calcite and dolomite. Whether there is also a (much slower) conversion from calcite to dolomite remains debatable because of the likely overprint by the evolution of ocean chemistry (Morse and Mackenzie, 1990, p. 548; Veizer and Mackenzie, 2004). The big bodies of dolomite in the stratigraphic record seem to have replaced limestones relatively early during burial diagenesis, i.e. millions to tens of millions of years after deposition. The amount of dolomite formed in the depositional environment is minor; where dolomite is formed in these settings, the process is often induced by microbes (see Machel, 2004, for review).

ECOLOGY

Marine carbonate precipitation and deposition are closely related to life in the ocean. Ecology, the study of the relationship of organisms and their environment, provides a number of very useful concepts for carbonate sedimentologists. A selection of them is presented below.

Law of sigmoidal growth

Growth of a population of organisms follows a sigmoidal curve consisting of a slow starting phase, a phase of rapid growth and a final phase of slow down where the population approaches a steady state in equilibrium with the carrying capacity of the living space (Fig. 1.12).

The logistic equation is a mathematical expression that produces sigmoidal growth curves (Fig. 1.13). The equation describes the interaction between the intrinsic growth rate of a population and the limits set by the size of the living space. At early stages of growth, living space is virtually unlimited and the growth rate is given by

$$dN/dt = rN$$

where N is the number of individuals in the population, t is time, and r is the intrinsic reproduction rate of an individual. This relationship leads to exponential growth. The

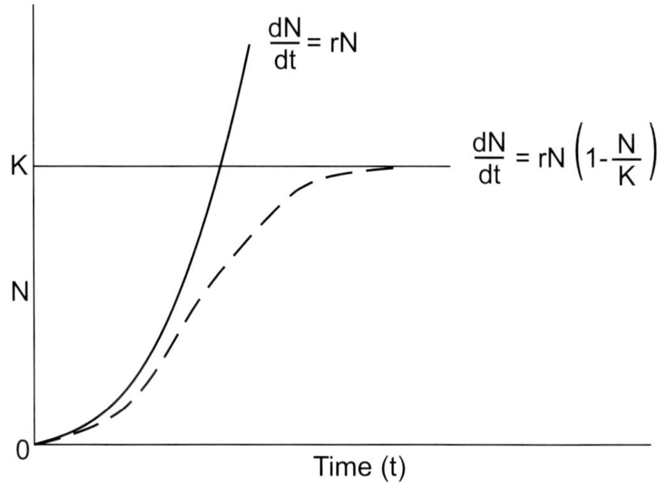

Fig. 1.13.— Exponential growth (bold) and growth governed by the logistic equation (dashed). In natural systems, an initial phase of exponential growth usually gives way to decelerating growth that goes to zero as population size approaches the limits of space or other resources. The logistic equation produces sigmoidal growth by applying a rapidly increasing correction to the exponential growth term. This correction consists of the ratio of population size, N, and the carrying capacity of the environment, K. After Townsend et al. (2003), modified.

limitations of growth imposed by the finite size of the living space can be added to this equation by introducing the term N/K, the ratio of population size, N, and carrying capacity of the living space, K. This yields the logistic equation

$$dN/dt = rN[1 - (N/K)]$$

Solutions to this equation are sigmoidal curves as in Fig. 1.12. Initial growth rates are very nearly exponential because N/K, the growth reduction imposed by the limited size of the living space, is negligibly small. As N/K increases, growth rates progressively deviate from the exponential trend; they approach zero, i.e. constant population size, as N/K approaches unity.

The logistic equation is one of the simplest equations producing sigmoidal curves. It was invented in the 19th century for modelling the growth of populations of living beings but it applies to abiotic systems, too, if there is some sort of competition among the individual components of the population. In carbonate sedimentology, the equation is useful for describing growth of carbonate production systems as well as purely mechanical accumulations.

Life strategies of organisms

The logistic equation and its sigmoidal growth curve lead to another important topic in biotic carbonate production – the different life strategies of organisms. As a rule, energy and other resources available to an organism are limited and need to be partitioned between growth and reproduction. In other words: organisms need to decide to either grow more and reproduce less or vice versa (Townsend et al., 2003). In a very influential book, MacArthur and Wilson (1967) introduced two extreme life strategies that correspond to these alternatives. There are organisms, such as weeds, that quickly appear in newly opened habitats, multiply rapidly, are short-lived and relatively small. They are called r-strategists because they spend most of their life in the r-dominated, nearly exponential part of the population growth curve in Fig. 1.13. At the other end of the spectrum are organisms, such as large trees in a forest, that live in environments of intense competition for limited resources. These organisms grow fast, live long, and reproduce slowly and relatively late in their life; they are called K-strategists as most of their life is spent in the K-dominated part of the growth curve.

Food chains, nutrient levels

Wherever life exists on Earth, the organisms have developed a network of interactions, connecting plant with grazer, predator with prey etc. (Townsend et al., 2003). All interactions in a given ecosystem constitute the food chain. This chain normally begins with photosynthesis, the growth of organic tissue in green plants using energy from the Sun and dissolved inorganic chemical substances. Alternatively, the food chain may start with chemosynthesis, where bacteria grow by oxidizing methane, hydrogen sulfide or other substances. Marine carbonate production nearly always depends on photosynthesis as a starting point.

The organisms at the starting point of the food chain are called autotrophs (literally: self-feeders); organisms further down the food chain depend on other organisms for food and are called heterotrophs.

The rate of photosynthetic production, the primary productivity, in the marine environment depends on the light intensity and the concentration of dissolved nutrients, such as phosphorous, nitrogen or carbon. Fig. 1.8 shows the first-order patterns of nutrient concentration in the world ocean. High-productivity zones rim the major ocean basins, there is an equatorial belt of elevated productivity and another globe-circling belt around Antarctica. Finally, production minima characterize the central parts of the ocean basins, particularly in the subtropics. Comparison of Figs 1.8 and 1.4 reveals that the productivity patterns largely reflect the surface circulation pattern of the ocean – high productivity occurs where upwelling of deeper water brings nutrients to the surface, production minima occur where the water has remained at the surface for a long time. More specifically, the high-productivity rings around the ocean basins coincide with zones of coastal upwelling, the productive equatorial belt marks the gentle upwelling generated by the opposing effects of the Coriolis force in the northern and southern hemisphere, and the belt around Antarctica is caused by upwelling within the Antarctic Current. Finally, the productivity minima lie in the subtropical gyres where old, nutrient-depleted, saline surface water is swept together by wind and

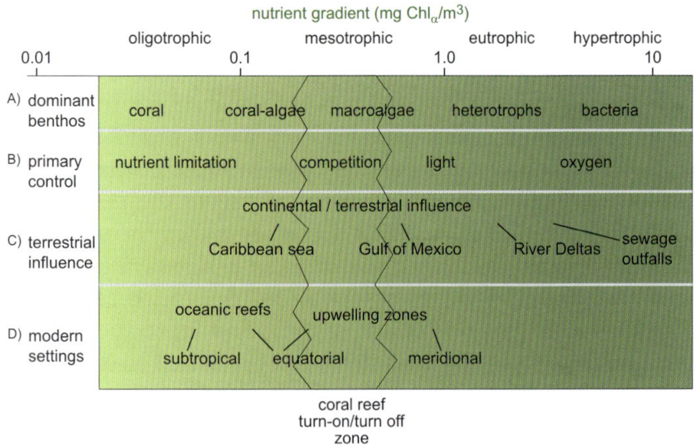

Fig. 1.14.— Nutrient concentration in surface waters of the ocean and its effect on marine biota. Nutrient levels are expressed in milligrams of chlorophyll per cubic meter of seawater on the horizontal axis. On the vertical, various changes induced by changing nutrient levels: A) Dominant benthic organisms. B) Type of limitation imposed on benthic communities. C) Level of nutrient influx from land (with characteristic examples). D) Modern oceanographic settings. After Hallock (2001), modified.

Coriolis effects. In addition to marine nutrients, input of organic matter from land may cause high productivity in the nearshore zone.

Ecologists have subdivided the continuum of nutrient levels in the ocean in discrete categories. Fig. 1.14 presents a classification (Hallock, 2001; Mutti and Hallock, 2003) that subdivides the trophic continuum in four categories. Three of them – oligotrophic, mesotrophic and eutrophic environments – occur on a regional scale in the ocean; they correspond to subtropical gyres, equatorial divergence and coastal upwelling zones respectively. The hypertrophic category represents an extreme situation that may occur locally in the ocean.

Benthic organisms and carbonate sediments secreted by benthos are sensitive indicators of marine nutrient levels as shown in Fig. 1.14. One important message is that the preferred habitats of carbonate-secreting benthos are in oligotrophic and mesotrophic environments. Coral reefs, in particular, thrive in submarine deserts and are easily killed by excess nutrients.

Life and light

Light is a precious commodity for life in the ocean because, most marine food chains start with photosynthesis. Sunlight is progressively absorbed as it travels through the water column. The sunlit part of the ocean, the photic zone, therefore occupies the uppermost part of the water column. The transition to the aphotic part of the ocean is very gradual but marine ecologists have defined two light levels of special significance in the continuum from light to dark – the base of the euphotic zone and the base of the zone of light saturation (Fig. 1.15).

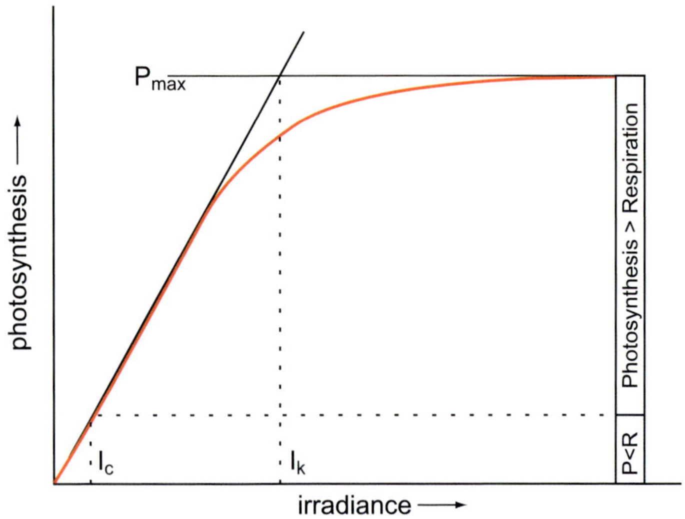

Fig. 1.15.— Irradiance (light energy flux) and rate of photosynthesis (red). Two important levels of irradiance have been defined: I_c is the level at which oxygen consumption by respiration equals oxygen production by photosynthesis; I_c defines the lower limit of the euphotic zone. I_k defines the lower limit of the zone of light saturation. Below this level light is a growth-limiting factor. P_{max} = maximum photosynthetic rate of the system. After F. Vijn and H. Boscher (written comm.)

The zone of light saturation is the uppermost water layer where light is so abundant that it is not a growth-limiting factor. The thickness of this zone is different for different organisms. The euphotic zone is the water layer where the rate of oxygen production by photosynthesis is larger or equal to the rate of oxygen consumption by respiration. The euphotic zone is the site of most benthic carbonate production and therefore of prime importance for carbonate sedimentology. In modern oceans, the base of the euphotic zone lies between 30 m and over 150 m. Fig. 1.9 shows that the thickness of the euphotic zone varies in response to ocean productivity, river discharge and latitude. (For application of these concepts to carbonate sedimentology see chapter 2).

Species diversity and environment

The number of species in an ecosystem is a fundamental question that lacks a simple answer and an evaluation is beyond the scope of this book (see Townsend et al., 2003 for a succinct overview). Here we shall discuss only one example of diversity variation – diversity as a function of salinity. It is particularly relevant for carbonate sedimentology.

Fig. 1.16 shows that species diversity has two maxima, one at normal-marine salinities, the other in fresh water. Diversity is low in brackish and hypersaline environments. It

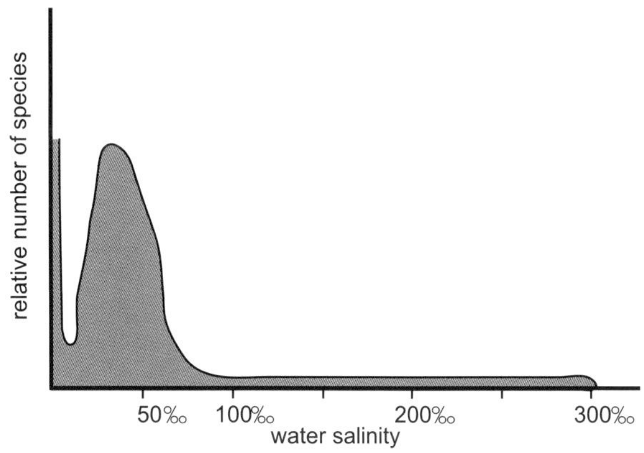

Fig. 1.16.— Salinity effects. Diversity of aquatic fauna and flora plotted against water salinity. Diversity peaks in the two most stable environments – fresh water and open-ocean sea water. Restricted marine environments are characterized by large variability of salinity. This reduces species diversity. Temperature variations have similar effects and tend to occur together with salinity variations. After Remane and Schlieper (1971), modified.

is probably more than a coincidence that the diversity maxima correspond to two salinity ranges that are commonly available at the Earth's surface: fresh water is continuously supplied by condensation of water vapor from the atmosphere and normal sea water is present in large quantities in the ocean. Sea-water composition changes through geochemical cycling but the rate of change is very slow compared to the rates of biotic speciation and extinction (see chapter 5). Regardless of the reason for the salinity-diversity relationship in Fig. 1.16, the pattern provides an excellent tool for determining the degree of restriction of carbonate depositional environments, i.e. the extent to which the environment was cut off from exchange with the open sea (see chapter 2).

Paleontologists have collected large data sets on the number of fossil species in sedimentary rocks and their change through geologic time. After correction for preservation effects, these data provide a basis for estimating biodiversity of past environments.

From oceanography , chemistry and biology to geology

The oceanographic, chemical and biological concepts presented above were derived from the study of the modern world, typically based on observations at time scales of seconds to hundreds of years. Study of the geologic record requires expanding the scope to scales of thousands of years to hundreds of millions of years. The principles of this chapter remain valid but new processes come into view that operate at these expanded time scales. The most important ones are biotic evolution, plate tectonics, and geochemical cycling, i.e. the chemical fractionation and recombination of materials as a result of mantle convection and plate tectonics. The kind of interaction between these long-term processes and the principles just described may be illustrated by a few examples.

Ecologic concepts such as logistic growth and r-K life histories were certainly valid in the past. The difficulty is to properly apply them to organisms that differ from their modern counterparts because of evolutionary change. For instance, the question of which invertebrate carbonate producers of the past had photosynthetic symbionts is a matter of intense debate among paleontologists. The answers are highly relevant for modeling carbonate production of the past (see chapter 2).

A similar situation holds for oceanography. We have no reason to doubt that general principles of oceanic layering and circulation, such as mixed layer, thermocline, Ekman transport etc. were valid in the past. It is also virtually inevitable that surface circulation was always largely governed by wind shear and deep circulation by temperature-salinity related density differences. This notwithstanding, the circulation patterns of past oceans differed considerably from the modern situation. One reason is that plate tectonics continually changed the shapes and positions of ocean basins and continents. For instance, the globe-circling current around Antarctica developed in the Oligocene when tectonic movements opened critical seaways. On the other hand, Cenozoic closure of the seaways between North and South America, and Europe and Africa blocked the globe-circling equatorial current that had existed in the Late Jurassic and Cretaceous. Another reason for major changes in oceanic circulation may be the conditions at the poles. For instance, Hay et al. (2004) questioned that the large gyres in the surface ocean could form under reduced temperature gradients such as in the Mid-Cretaceous. They suggested an alternative model of surface circulation by ephemeral, migrating eddies.

Finally, ocean chemistry as well as the mineralogy of carbonate skeletons varied in the past, mainly because of changing rates of chemical cycling through the crust and the mantle and evolutionary changes of biota at the Earth's surface. Chapter 5 discusses examples of such oscillations that directly affected carbonate sedimentation.

CHAPTER 2

Principles of carbonate production

Three rules capture the peculiar nature of carbonate depositional systems – carbonate sediments are largely of organic origin, they can form wave-resistant structures and they are easily altered by diagenesis because the original minerals are metastable. The implications of these rules are pervasive. We will encounter them throughout the chapters of this book, starting with the present review of principles that govern the production of sediments and the growth of reefs.

MODES OF MARINE CARBONATE PRECIPITATION

Precipitation of solid matter from the dissolved load of the sea occurs either abiotically, governed by inorganic thermodynamics and reaction kinetics, or biotically as a consequence of the metabolism of plants and animals. Precipitation of marine evaporites is an example of an abiotic process, precipitation of marine opal by diatoms or radiolaria an example of a biotic one. Marine carbonate precipitation proceeds along abiotic and biotic pathways and this makes it particularly diverse and complex. The interplay of abiotic and biotic processes during four billion years of organic evolution and environmental change has led to a stunning diversity of precipitation mechanisms that are far beyond the scope of this book. However, even if one focuses on the practically relevant aspects of carbonate precipitation, the subdivision into abiotic and biotic precipitation is inadequate. Work in the past two decades in particular has shown that it is sedimentologically very advantageous to further subdivide the biotic category. I follow Lowenstam and Weiner (1989) who recognized three degrees of biotic influence on precipitation in general and on carbonates in particular: (Fig. 2.1)

a. *Abiotic* (or quasi-abiotic) precipitates where biotic effects are negligible.
b. *Biotically induced* precipitates where the organism sets the precipitation process in motion but organic influence on its course is marginal or absent. The reaction takes place outside the cell and the product is very similar, often indistinguishable, from abiotic precipitates.
c. *Biotically controlled* precipitates where the organism determines location, beginninng and end of the process, and commonly also composition and crystallography of the mineral. All skeletal carbonate falls in this category. From an environmental perspective, it is important to further subdivide skeletal carbonates into:
   c.1 controlled precipitates by photo-autotrophic organisms that generate organic matter from dissolved substances and sunlight, and
   c.2 controlled precipitates by heterotrophic organisms that are independent of light but require particulate organic matter for food.

Currently, organisms have the first hand on precipitation in most carbonate settings and abiotic precipitation will kick in if biotic fixation is insufficient. Thus, abiotic precipitation is a sort of "default setting" in the carbonate system of modern oceans (term credit to Ron Perkins).

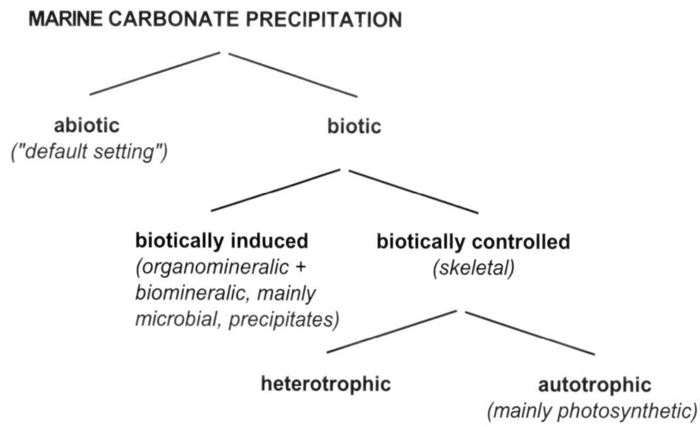

Fig. 2.1.— Pathways of carbonate precipitation in aquatic environments – a cascade of options governed by the degree of biotic influence. After Schlager (2000), modified.

Abiotic marine carbonate precipitation

Organisms and organic matter are so common in depositional environments and have so many ways of influencing carbonate precipitation that it is virtually impossible to demonstrate that a particular carbonate precipitate from a natural environment is totally abiotic. However, there are carbonate materials for which the organic influence, if present at all, is very subtle. Their texture and mineralogy can be reproduced abiotically in the laboratory and their natural occurrence is governed by first-order trends in ocean chemistry. These materials are included here.

The most conspicuous abiotic precipitate is cement formed in the pore space during the early stages of diagenesis when the deposit was still in the depositional environment. Burial cements are excluded from the abiotic carbonate factory because they are not derived from sea water but largely from remobilized sedimentary material. The case for abiotic origin is particularly strong for acicular aragonite cements. Acicular magnesian calcites may be biotically influenced (Morse and Mackenzie, 1990).

Associated with tropical skeletal carbonate we find abiotic precipitates in the form of ooids. Ooids form in high-energy environments by stepwise accretion on a nucleus. Field observations and laboratory experiments indicate a growth history of alternating phases of accretion and rest. The degree of organic influence on the precipitation process remains a matter of debate. However, two arguments tip the balance in favor of abiotic precipitation: (1) growth of aragonitic, Bahama-type, ooids in the laboratory where the precipitation was essentially abiotic and organic matter had only a modulating effect (Davies et al., 1978) and (2) the similarity of ooids and cements in terms of mineralogy and chemical signature (Morse and Mackenzie, 1990). It seems that the role of organisms and organic matter in the formation of ooids is not sufficient to significantly alter the abiotic controls (Morse and Mackenzie, 1990; see Reitner et al., 1997 for contrasting view on Great-Salt-Lake ooids).

The origin of carbonate in whitings, clouds of carbonate suspended in sea water, is a much debated issue. Morse and Mackenzie (1990) conclude that abiotic precipitation (probably on nuclei of suspended sediment) is very likely for Bahamian whitings. However, in-situ experiments by Yates and Robbins (1999) strongly suggest that blooms of unicellular algae trigger the first precipitation, probably followed by extended abiotic growth of the original, biotically induced precipitates (Yates and Robbins, 1999, p. 135). I consider the carbonate mud from whitings a mixture of biotically induced and abiotic precipitation.

The occurrence of acicular cements, ooids and whitings in modern oceans indicates strong control by inorganic marine chemistry of the sea water. In first approximation, oolites, aragonitic sea-floor cements and whitings occur in the zone of highest carbonate supersaturation in the ocean – the mixed layer of the tropical seas. Ooids and whitings are restricted to this zone, acicular cements are most abundant there. In the temperate latitudes, sea-floor cementation is rare and destructive sea-floor diagenesis tends to dominate. The correlation of needle cements, ooids and whitings with sea-water chemistry is very helpful in predicting at least first-order trends in the geologic record and warrants their classification as principally abiotic precipitates in spite of evidence of some biotic influence.

Biotically controlled precipitation

The majority of carbonate material in modern oceans is precipitated as highly structured skeletons of organisms. Precipitation is primarily controlled by the biochemistry of the respective organisms (such as algae, foraminifera or corals); the organisms, in turn, are influenced by the conditions of the sea they live in, particularly light, temperature and water chemistry (for instance the degree of carbonate saturation of the sea water). To appreciate the effects of various environmental factors we need to recall the two fundamental types of metabolism introduced on p. 9. Autotrophic organisms nourish themselves by utilizing inorganic materials to synthesize their own living matter; heterotrophic organisms have to rely on organic material to do so. Autotrophic organisms among the carbonate producers are almost exclusively photo-autotrophic: they perform photosynthesis and thus depend on light for their livelihood. Some carbonate-secreting organisms are themselves heterotrophs but live in symbiosis with autotrophic algae. As a result, the system of host plus symbiont becomes autotrophic. Fig. 2.2 gives an overview of important autotrophic and heterotrophic carbonate producers. Note that the metabolism of extinct groups can only be deduced from circumstantial evidence.

Autotrophic producers	**Heterotrophic producers**
Cyanobacteria (only biotically induced precipitates) Coccolithophorid algae (Haptophyceae) Green algae (such as dasycladaceans, codiaceans) Red algae **Autotrophic production via symbionts** Many larger foraminifers Hermatypic corals (Scleractinia) Certain bivalves (Tridacnids, rudists?)	Foraminifera Archaeocyathans Sponges (e.g. pharetronids, stromatoporoids, chaetetids) Ahermatypic corals (Scleractinia) Most bivalves Gastropods Cephalopods Arthropods (e.g. trilobites, ostracodes, barnacles) Brachiopods Bryozoans Echinoderms

Fig. 2.2.— Important autotrophic and heterotrophic carbonate producers.

Light is arguably the most important control on skeletal carbonate precipitation because of the dominance of photo-autotrophic organisms in carbonate production – at least in the Cenozoic. Photosynthesis is a complex, and only partly understood process. The basic reaction may be simplified as

$$CO_2 + H_2O + \text{solar energy} \rightarrow HCHO + O_2$$

where HCHO represents a simple summary formula for organic matter. The formula clearly illustrates the link be-

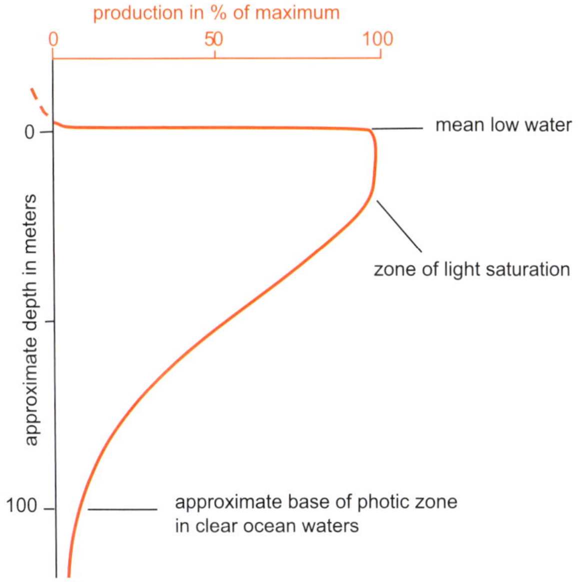

Fig. 2.3.— The profile of carbonate production (red) in a tropical setting from terrestrial elevation to subphotic depth. In most terrestrial environments, production is negative as carbonate rocks are being dissolved by rainwater and acidic soils. Maximum production is in the upper part of the photic zone (zone of light saturation), from where it decreases approximatly exponentially with depth.

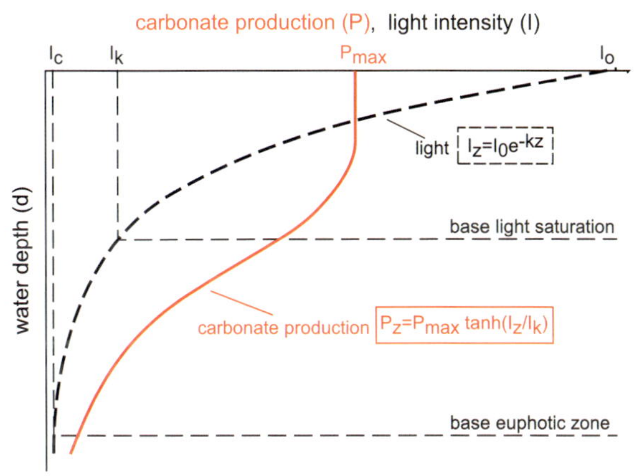

Fig. 2.4.— Change of light intensity and tropical carbonate production with water depth. Light displays a simple exponential decrease with water depth (black curve and equation). The production of organic matter can be related via a hyperbolic-tangent function to light intensity (red curve and equation). Production shows a shallow zone of light saturation, where light is not a growth-limiting factor, followed by rapid decrease of organic growth with water depth (definitions in Fig. 1.15). In the tropical carbonate factory, organic production can be taken as a good estimate of carbonate production. In tropical environments, the zone of light saturation reaches to about 20 m for corals, the euphotic zone to about 100 m. I_z = light intensity at depth z, I_s = light intensity at base of saturated zone, P = organic production (and also a proxy of carbonate production), z = water depth, k = extinction coefficient of light. After Bosscher and Schlager (1992), modified.

tween photosynthesis and carbonate chemistry. Photosynthesis extracts CO_2 from the sea water, thus increasing its carbonate saturation and facilitating precipitation of carbonate minerals. For the organisms themselves, precipitation of $CaCO_3$ has the added advantage that potentially deleterious Ca^{2+} ions can be removed from the system and a protective skeleton can be constructed.

The link between skeletal carbonate fixation, photosynthesis and light explains the decrease of skeletal carbonate production with water depth in tropical environments. Above sea level, carbonate production rapidly drops to zero in the supratidal zone and becomes negative in most terrestrial environments as carbonate material dissolves in rain water and acidic soils. The typical pattern is shown in Fig. 2.3 and Fig. 2.4 whereas Figs. 2.5 and 2.6 show spe-

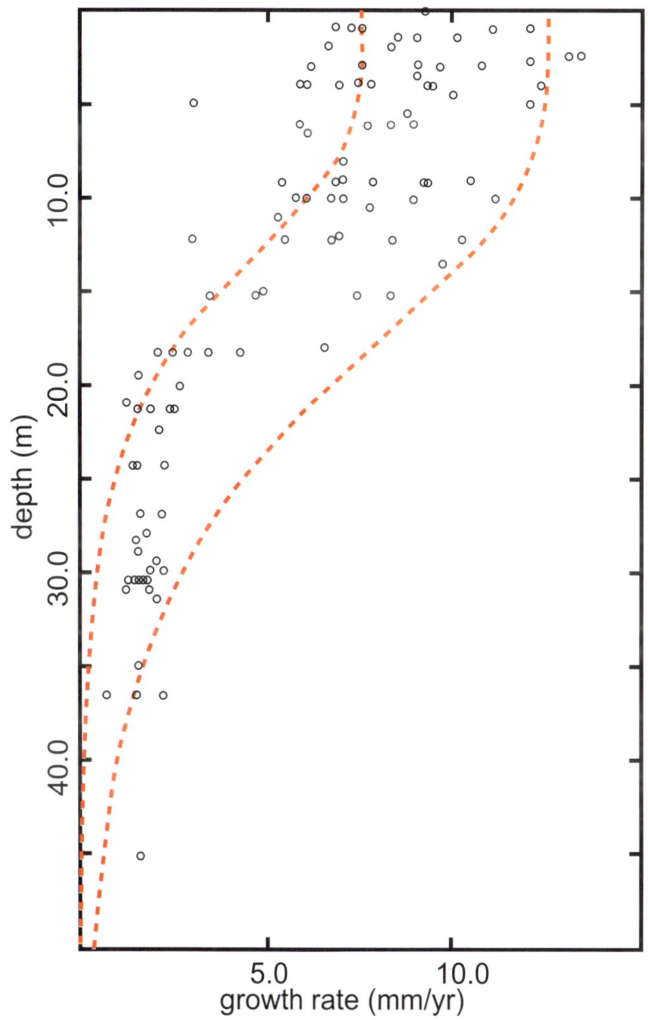

Fig. 2.5.— Predicted and observed values of coral growth vs. depth. Circles: measured growth rates of Caribbean reef coral *Montastrea annularis*; red curves: growth rates predicted by the light-growth equation of Fig. 2.4 for common values of water turbidity in the Caribbean. After Bosscher and Schlager (1992), modified.

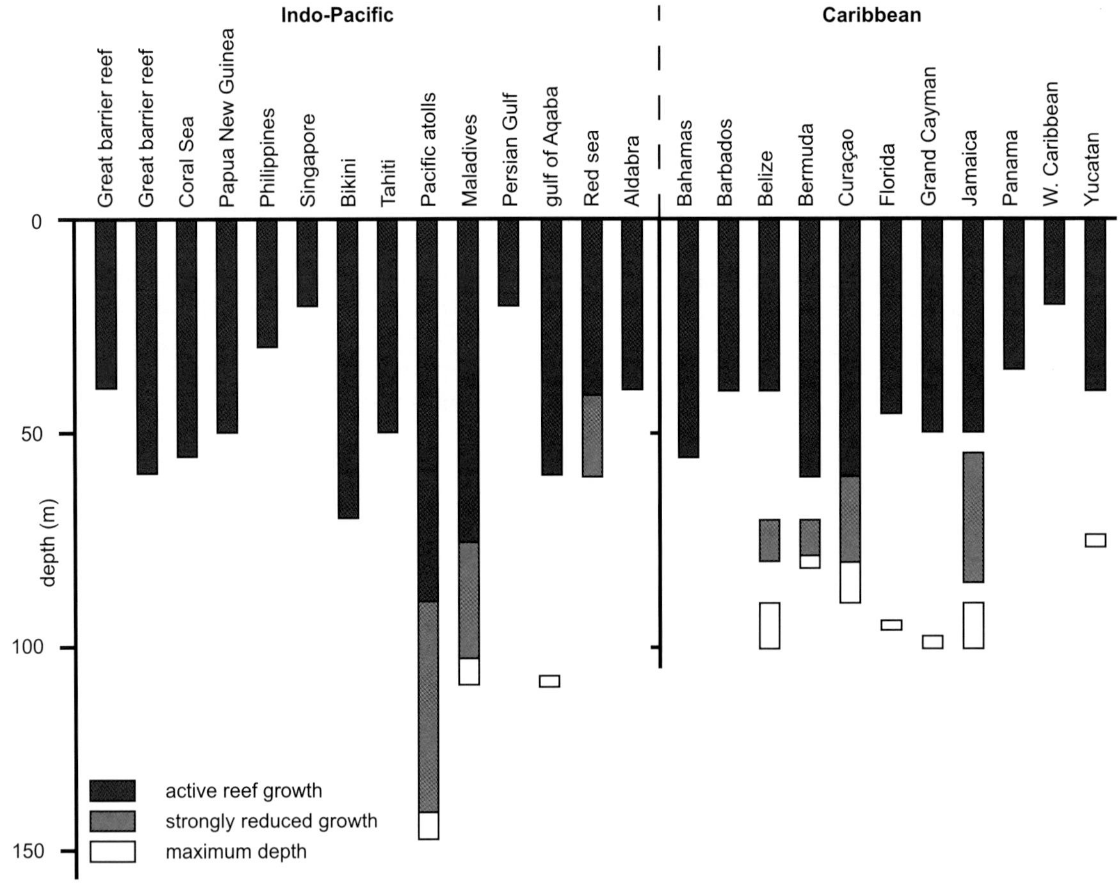

Fig. 2.6.— Depth of the euphotic zone in the Indo-Pacific and the Caribbean, constrained by the limits of reef growth. The base is gradational and varies regionally by tens of meters. After Vijn and Bosscher (written communication).

cific data, mainly from reef environments. The growth-depth curve displays a shallow zone of light saturation followed by a zone of rapid decrease and a third, deep zone where growth asymptotically approaches zero. This growth curve can be derived, via a hyperbolic function, from the well-established exponential decrease of light intensity with depth(Fig. 2.4). The growth-depth patterns of most other carbonate-producing organisms are less well known but seem to follow similar trends.

Characterization of growth-depth curves requires definition of two parameters – the *light saturation zone* and the *euphotic zone*. For both exist stringent biological definitions (Fig. 1.15). Geologists are normally not able to measure the required variables and have to resort to proxy indicators. The euphotic zone is defined in the geologic record as the interval where abundant growth of photosynthetic, carbonate-secreting benthos is possible. The zone of light saturation has not been defined in geological terms. Loosely speaking, it is the interval where light has no recognizable control on the rates of growth and calcification of organisms. The growth forms of corals indicate the zone of severe light limitation by a change from massive to platy colonies (Fig. 2.7). In the light-saturated zone, corals indicate the presence of a very turbulent surface layer of the sea by dominance of branching growth forms.

Temperature rivals light in its effect on skeletal carbonate production. Generally, warmer is better, but there exist upper temperature limits for the various carbonate-secreting organisms. Thus, the temperature window of calcifying benthos is different for different organisms. Most hermatypic (i.e. symbiotic) corals function in the range of 20-30 °C. The upper temperature boundary sets important limits to carbonate production, particularly in restricted lagoons where temperatures frequently exceed 30°C. The most important effect of temperature, however, is the global zonation of carbonate deposits by latitude (Figs 2.8, 2.9). In spite of what has just been said about the importance of light, the boundary northern and southern limit of coral reefs, and thus the boundary of tropical and cool-water carbonates, in the modern oceans seems to be controlled by winter temperature rather than radiation. This indicates that as one moves poleward in the modern oceans, the temperature limit for hermatypic coral growth is reached before the light limit. In the past, this need not always have been the case. The temperature limit and the light limit may have shifted relative to one another during geologic history. The fairly stable position of 30-35° latitude for the boundary of Phanerozoic tropical carbonates may reflect the joint control by temperature and light.

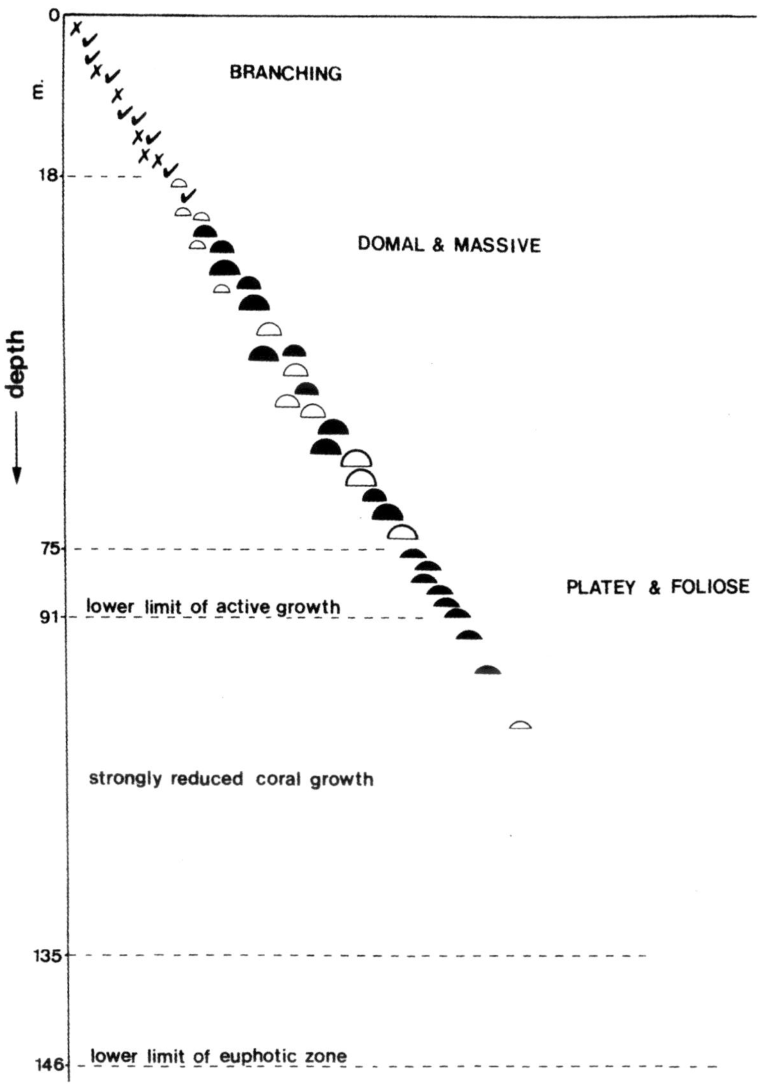

Fig. 2.7.— Growth forms of corals reflect the environmental changes related to water depth. Example is from the Caribbean. Branching forms dominate the uppermost, high-energy layer of the sea. Domal and massive forms occupy intermediate depths. Below the zone of light saturation, corals become platy and foliose in order to capture a maximum amount of light. At the biologically defined lower limit of the euphotic zone, coral growth in this region is already negligible. Vijn and Bosscher, written communication.

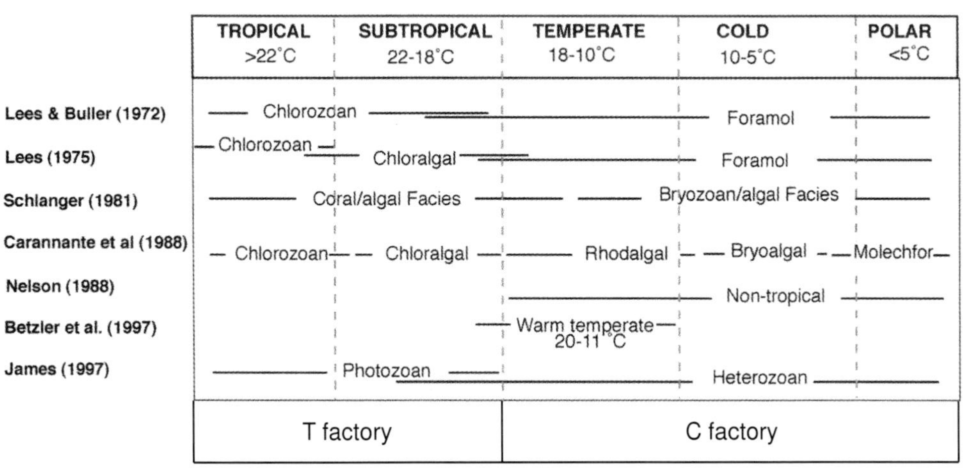

Fig. 2.8.— Comparison of terminologies proposed for carbonate sediments of different climatic zones. Most terms are created by combining parts of the names of the most common groups of organisms. Boundary between T and C factories used in this book approximately coincides with the subtropical/temperate boundary of most authors. Based on Mutti and Hallock (2001) and Schlager (2003).

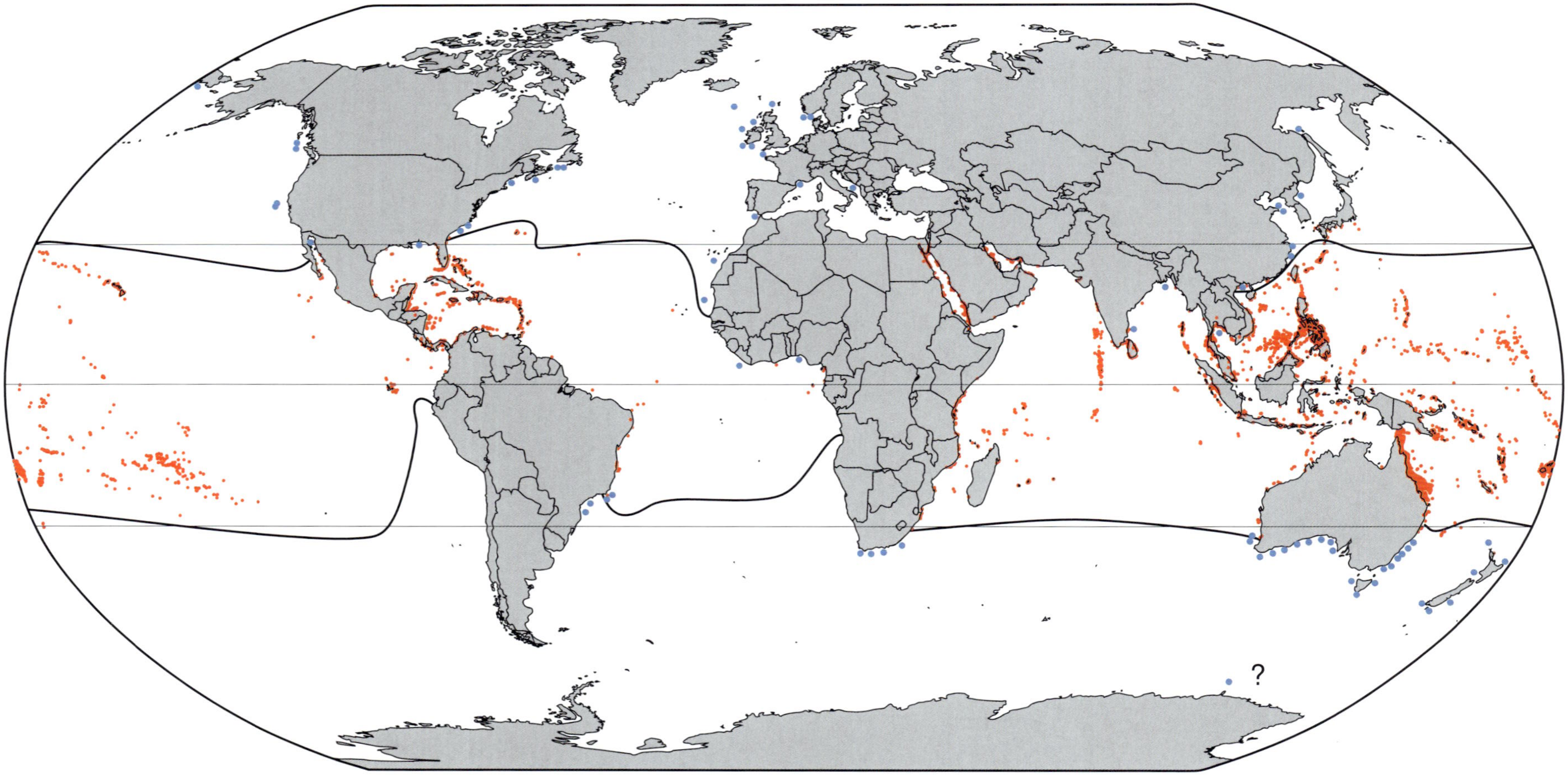

Fig. 2.9.— Temperature control on reefs. Recent tropical reefs (red) are limited in the north and south by the position of the 20°C-isotherm for the coldest winter month, shown here as a bold line. Cool-water bioherms (blue) occur almost exclusively pole-ward of this line. The 20°C-isotherm follows the 30° latitude line only approximately. In the eastern parts of the Atlantic and Pacific Oceans the isotherms bend towards the equator because cold water is upwelling there. If radiation were the dominant control on the distribution of reefs, the northern and southern limits of the reef belt should parallel latitude much more closely. Compiled from ReefBase (www.reefbase.org), James (1997) and Van Loon (1984).

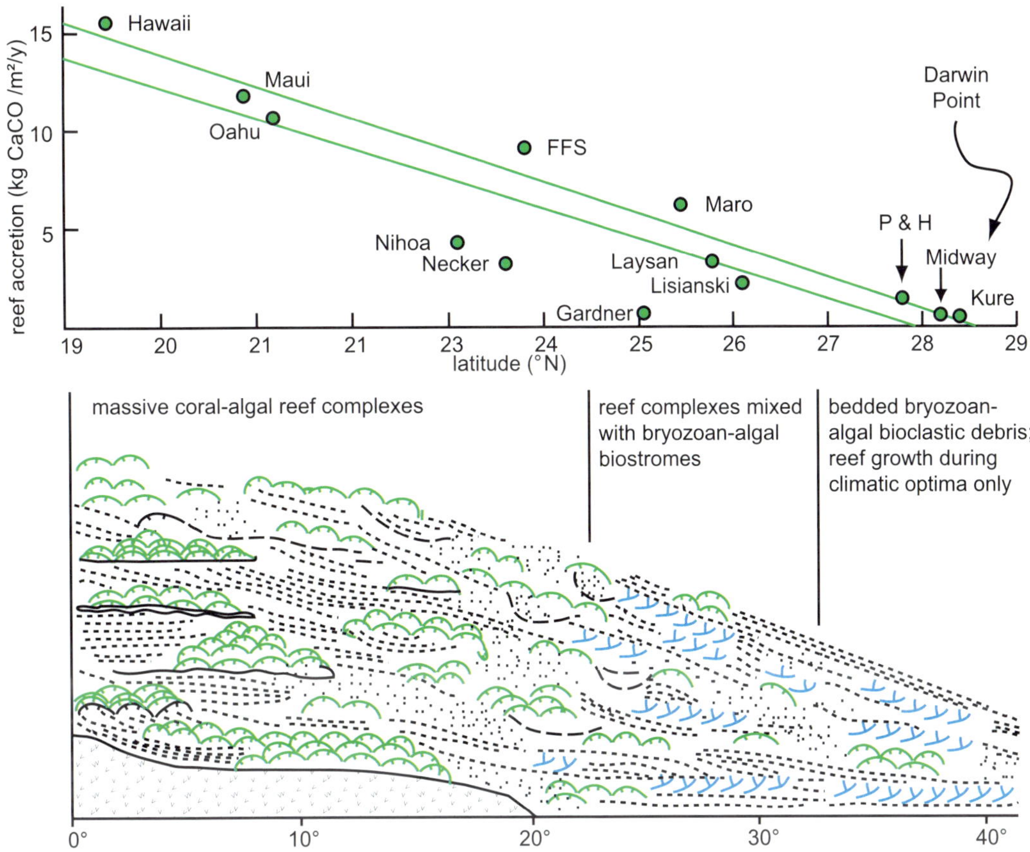

Fig. 2.10.— Northward decrease of reef growth and change to cool-water carbonate deposition in the North Pacific. Upper panel: decrease of the rate of reef growth with latitude. Darwin Point marks the northern limit of reef growth. Lower panel: latitudinal change from tropical to cool-water carbonate facies observed on the Hawaii-Emperor chain of islands and seamounts. Black dots and dashes – carbonate debris, green – coral reefs, blue – bryozoan biostromes. Compiled using Schlanger (1981) and Grigg (1982).

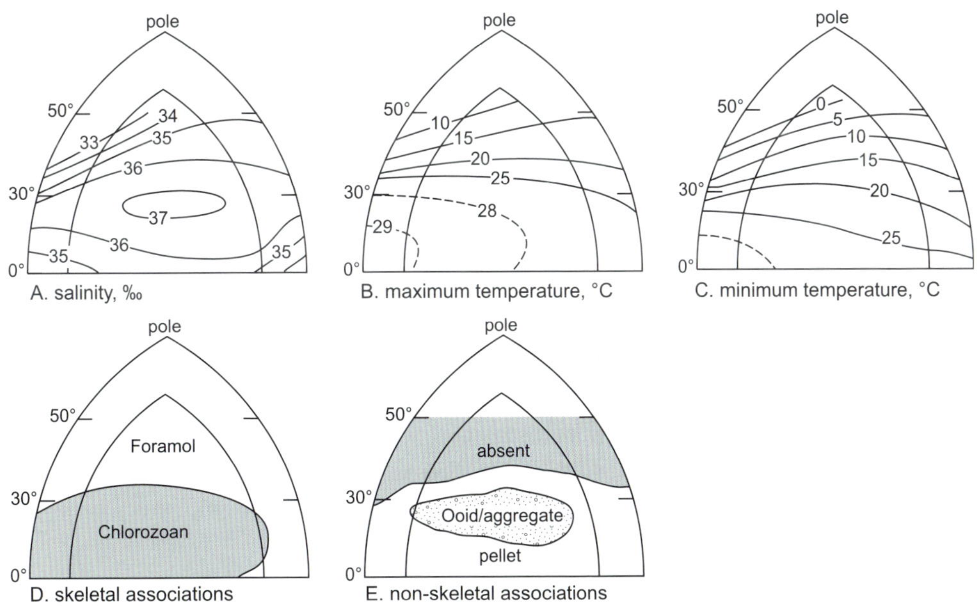

Fig. 2.11.— Carbonates in cool latitudes and tropical latitudes – a comparison. Figs A, B, C illustrate changes in environmental conditions; Fig. D illustrates difference in skeletal carbonate – temperate (= cool-water) carbonates are dominated by benthic foraminifers and molluscs ("foramol" association), tropical latitudes by green algae and corals ("chlorozoan" association); Fig. E shows that non-skeletal grains (ooids, peloids) are virtually absent in temperate-water carbonates. After Lees (1975), modified.

Latitudinal zonation of skeletal production. Skeletal carbonate production changes very significantly with latitude. The differentiation into tropical and cool-water carbonates is widely applied and often further subdivided (Figs 2.8, 2.10, 2.11; Lees, 1975; Tucker and Wright, 1990; James and Kendall, 1992; James, 1997). Tropical carbonates are dominated by photosynthetic organisms and usually include metazoan reefs, abundant green algae and larger foraminifera. Cool-water carbonates lack these deposits and consist mainly of skeletal sand and gravel derived from molluscs, bryozoans, smaller foraminers and red algae. The contribution of photo-autotrophs to cool-water carbonate production is limited to red algae that are normally not the dominant component. Consequently, the depth window of cool-water carbonate production is much wider.

It should be noted that the zone of "tropical carbonates" reaches to 30-35° of latitude and thus extends from the humid tropics to the desert belt of the horse latitudes (Fig. 2.9). The cool-water realm extends over several climate zones, reaching from the northern limit of the desert belt to the polar regions (Fig. 2.11). The differences of the tropical and the cool-water realm are not restricted to the skeletal material. Cool-water carbonates also are distinct by the absence of mud, shallow reefs and oolitic sand shoals with early cementation. The lack of reefs and cemented shoals has fundamental implications for the depositional anatomy.

Nutrients. Contrary to common expectations, high-nutrient environments are unfavorable for many carbonate systems. Nutrients, to be sure, are essential for all organic growth, including that of carbonate-secreting benthos. However, the carbonate communities dominated by autotrophs, such as reefs, are adapted to life in submarine deserts. They can produce their organic tissue with the aid of sunlight from sea water with very low nutrient levels and are very efficient in recycling nutrients within the system. In high-nutrient settings, the carbonate producers are outpaced by soft-bodied competitors such as fleshy algae, soft corals or sponges. Furthermore, the destruction of reef framework by bio-erosion increases with increasing nutrient supply.

Salinity varies relatively little in the open-marine environment. The effects of these subtle variations on carbonate production are not well known. Where access to the open ocean is restricted, salinity varies greatly and significantly affects the diversity of the biota (Fig. 1.16). The combined effects of salinity and temperature variations allow one to subdivide carbonate environments (Fig. 2.12).

Biotically induced precipitation

In the last two decades, it has been demonstrated that a subdivision of shoal-water carbonates into abiotic and biotically controlled (skeletal) material is inadequate. A significant portion of the non-skeletal carbonate material has been precipitated under the influence of organisms and thus cannot be classified as abiotic. Commonly precipitation is induced by micro-organisms, mostly bacteria and cyanobacteria. Micrite is a major, often dominant component of these deposits. The term "(mud) mound" is commonly used as a field-geologic term (Wilson, 1975; James and Bourque, 1992). For the deposits themselves, the term "microbialite" is widely used. The drawback is that the word has a strong genetic connotation. If one wishes to avoid this explicit statement on genesis, the term "automicrite" is recommended. It stands for autochthonous micrite as opposed to allochthonous micrite that was transported and deposited as fine-grained sediment (Wolf, 1965). Whether the micrite formed as a rigid precipitate can often be deduced from thin sections or polished slabs.

The past decade brought enormous progress on the orgin of mud mounds and other automicrite deposits. The combination of detailed field work, petrography and collaboration with biologists and organic chemists has led to detailed insight in a geologically very important carbonate precipitation mode that differs significantly from the more conspicuous skeletal mode (Monty et al., 1995; Reitner et al., 1995a; 1995b; Neuweiler et al., 2003).

The *environmental controls* on microbial precipitation are less well known than those of skeletal precipitation. An important property of the microbial mode of precipitation is its near-independence of light. Microbial precipitates may form in the photic zone or below, certainly to depths of 400 meters. On modern reefs, the microbial deposits are best developed in the forereef environment. However, stromatolites in the uppermost photic zone (e.g. Reid et al. 2000) and automicrite in the interstices of coral framework (Camoin et al. 1999) demonstrate that the microbial mode of carbonate fixation finds its niches even in the prime domains of skeletal production.

An important chemical requirement is supply of alkalinity in the form of the anions HCO_3^- and CO_3^{2-}. A likely source of alkalinity is sulfate reduction combined with decay of organic matter in oxygen-deficient layers of the ocean such as the oxygen minimum of the thermocline. The estimated water depth and organic-rich ambient sediments of many mud mounds support this assumption.

Whether temperature sets practically relevant limits for microbial carbonate precipitation is unclear. Mud mounds seem to be best developed in low latitudes. However, the paleo-latitude of many Paleozoic mounds is not well constrained and narrow latitudinal restriction is not to be expected with a production system that demonstrably functions at low light levels and in intermediate water depths, i.e. at temperatures significantly below tropical surface temperatures.

Precipitation modes in comparison

The boundaries of the three precipitation modes are gradational. The degree of biotic influence in the induced and controlled categories varies considerably and even the

PROTECTION AND RESTRICTION IN CARBONATE ENVIRONMENTS

PROTECTION \ RESTRICTION	NO RESTRICTION	BIOTIC RESTRICTION	EVAPORITES PRESENT
AGITATED (no mud)	e.g. skeletal sand shoal "White Bank" in Florida Reef Tract	e.g. sand shoals in Hamelin Basin (Shark Bay)	e.g. rubble and sand "sub-littoral sheet" Hamelin Basin (Shark Bay)
CALM OR EPISODICALLY AGITATED (muddy sediments)	e.g. deep muddy lagoons of Pacific Atolls (Enewetak)	e.g. muddy sands of Florida Bay or Bahamas, W of Andros	e.g. muddy sands of tidal flats Hamelin Basin (Shark Bay)

Fig. 2.12.— Protection against water turbulence and restriction of exchange with the open sea are two independent variables that can be used to classify carbonate depositional environments in a protection-restriction matrix. This version is rather simple and meant for immediate use in the field, refinement for specific case studies is easily possible.

abiotic category is not always free of subtle biotic influences (e.g. Webb, 2001). "Quasi-abiotic" may be an appropriate term for those who find the term abiotic too stringent. Another point merits mention. The bio-induced category includes carbonates precipitated by the action of living cells as well as material precipitated under the influence of non-living organic matter. Trichet and Defargue (1995) distinguished these two categories as "bio-mineralic" and "organo-mineralic" respectively. The reason for lumping them here lies in the difficulty of distinguishing them at the scale of outcrops, formations or sequences.

Lowenstam and Weiner (1989) proposed the above classification to distinguish basic modes of mineral precipitation. It is comforting to see that the three classes can be recognized in most instances by petrographic analysis. Nearly all biotically controlled carbonate precipitates are organic skeletons, endowed with characteristic shape, crystal fabric, mineralogy and, commonly, chemical signature. The stunning success of carbonate microfacies analysis in the 1960's and 70's was largely due to the study of skeletal particles. Abiotic precipitates in the form of cements were also recognized early by petrographic techniques and have contributed important information on environmental conditions in depositional and diagenetic environments. The third group, the bio-induced precipitates, became clearly identifiable when techniques of organic chemistry were combined with standard petrography during the past two decades (e.g. Reitner et al., 1995b; Trichet and Defargue, 1995; Reid et al., 2000). Petrographically, bio-induced precipitates commonly occur as fine-grained carbonate (i.e. micrite) in the form of pellets, lumps or masses with variable, often concentric lamination. Distinction from depositional mud is sometimes difficult but surprisingly often one can establish that the material was hard or firm upon formation and did not accumulate as soft, muddy sediment. Thus, the classification as automicrite (see above) can often be backed by textural evidence.

FROM PRECIPITATION MODES TO CARBONATE FACTORIES

If one increases the scale of observation from samples and thin-sections to mappable formations and beyond, it turns out that all large accumulations are mixtures of the three precipitation modes described above (Fig. 2.13). These mixtures are not random. They cluster into three preferred production systems, or factories, that differ in dominant precipitation mode (Fig. 2.14), mineral composition (Fig. 2.15), depth range of production (Fig. 2.16) as well as growth potential (see p. 24f). I introduced the classification by using the terms tropical, cool-water and mud-mound factory (Schlager, 2000, 2003). I think the terms T factory, C factory and M factory are preferable because there exist several definitions for each of the three key words and each letter stands for at least two important properties of the respective factory (see below).

T factory

T stands for "tropical" and "top-of-the-water-column". Biotically controlled precipitates dominate. Most abundant among them are photo-autotrophic organisms, for instance algae and animals with photosynthetic symbiotic algae, such as hermatypic corals, certain foraminifers and certain molluscans. The other characteristic products are abiotic precipitates in the form of marine cements and ooids. Clay-size precipitates, the "whitings", are probably mixtures of abiotic or biotically induced precipitates (Morse and Mackenzie, 1990; Yates and Robbins, 1999; Thompson, 2001). Heterotrophs devoid of photo-symbionts are common, but non-diagnostic contributors. Construction of wave-resistant structures by organic framebuilding or rapid marine cementation is common, particularly at the shelf-slope break.

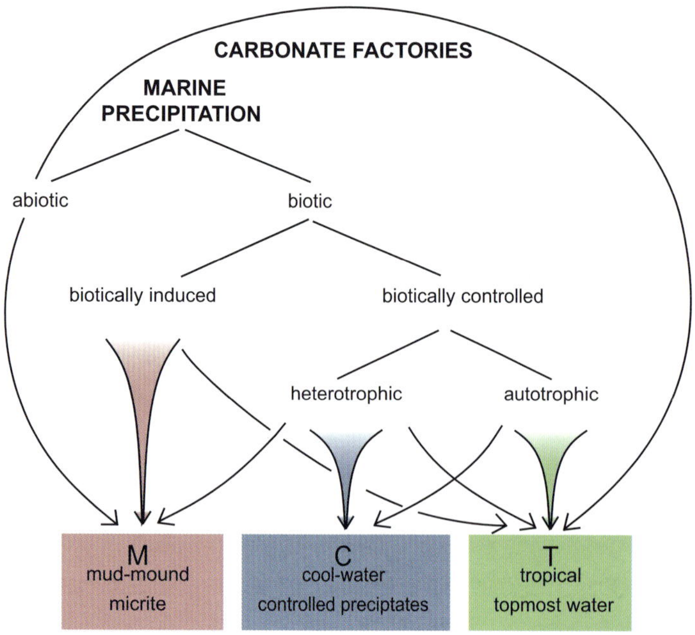

Fig. 2.13.— From precipitation modes to carbonate factories. At the scale of geological formations, the pathways of precipitation of Fig. 2.1 combine in characteristic ways to form carbonate factories. The typical material of the T factory is biotically controlled precipitate from tropical autotrophic organisms (or heterotrophic organisms with autotrophic symbionts); the C factory is dominated by heterotrophic organisms and the M factory by biotically induced precipitates, mostly micrite. After Schlager (2003), modified.

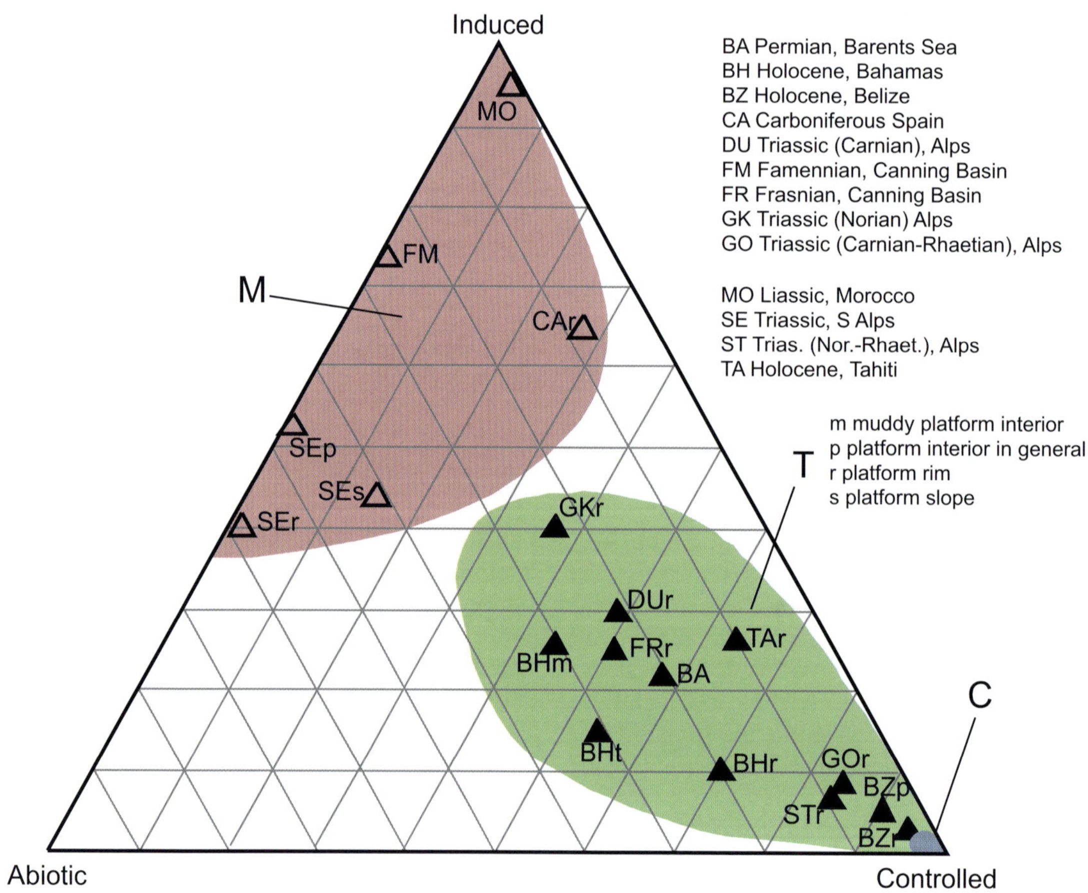

Fig. 2.14.— Proportions of abiotic, biotically induced and biotically controlled material in factory output estimated from the composition of some well-known Phanerozoic carbonate formations. C factory consists almost entirely of one category. M and T factories are mixtures of all three categories and grade into one another with the M factory producing mainly bio-induced and abiotic material, T factory producing biotically controlled material. After Schlager (2003), modified.

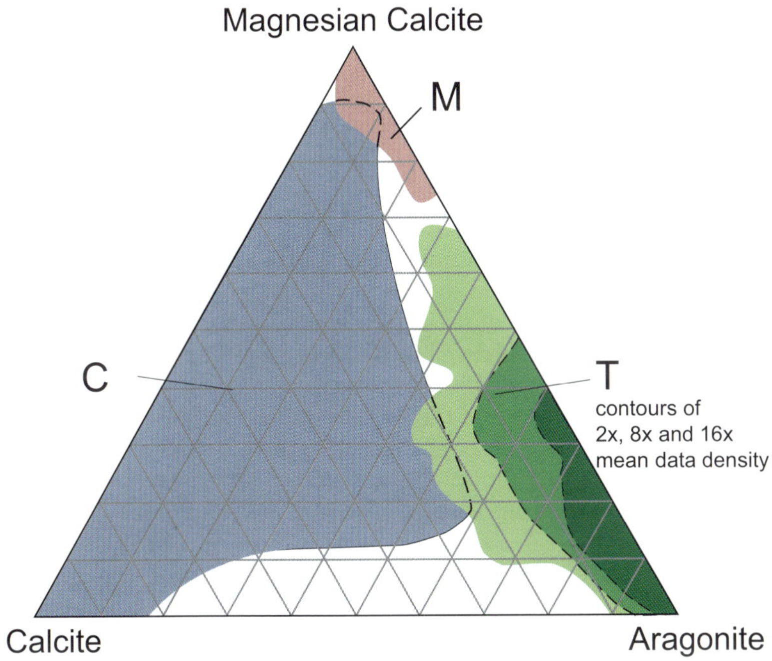

Fig. 2.15.— Mineralogic composition of sediments of the three factories. C and T factory based on X-ray analyses of recent sediments based on Bathurst (1971), Milliman (1974), Morse and Mackenzie (1990). M factory based on point-counting of thin-section micrographs in Russo et al. (1997), Triassic, and Neuweiler (1995), Cretaceous. Assumptions: Automicrite = magnesian calcite; fibrous cements = 1:1 mixture of aragonite and magnesian calcite.

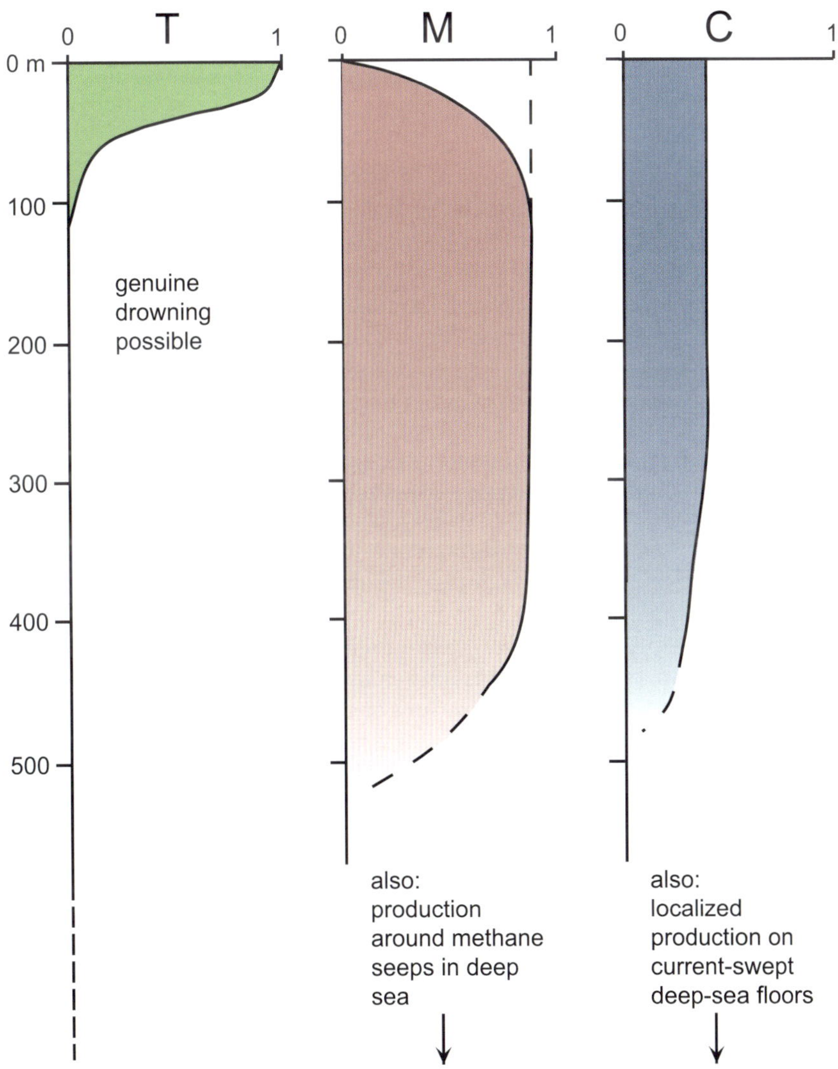

Fig. 2.16.— Production rates and depth window of production of carbonate factories. Width of shaded bars represents estimated production rate at a given depth as a fraction of the tropical standard. Dominance of photo-autotrophic (i.e. light-dependent) organisms in the tropical factory leads to very high production rates but only in a narrow depth window. Production of the other factories is largely independent of light, the depth windows extend over hundreds of meters and their lower limits are poorly known. In modern oceans, production by the M factory is low at shallow depths, probably because of competition by the T factory . After Schlager (2003), modified.

The T factory is restricted to the warm, sunlit waters of the ocean that are high in oxygen because of constant equilibration with the atmosphere and low in nutrients because of intensive competition. In modern oceans, the characteristic settings are the surface waters between 30° N and S of the equator. The northern and southern limit of the T factory closely follows the line where the mean temperature of the coldest month is about 20 °C (Fig. 2.9). The T factory may also pass into the cool-water factory downward in the water column, for instance at the boundary between the warm surface layer of the ocean and the thermocline. Furthermore, transitions from T to C factory occur in shallow tropical upwelling areas where cool, nutrient-rich water comes to the surface (Lees and Buller, 1972; Pope and Read, 1997, p. 423; Brandley and Krause, 1997, p. 365; James, 1997).

C factory

The letter C is derived from cool-water and controlled precipitation. The products are almost exclusively biotically-controlled precipitates. Heterotrophic organisms dominate. The contribution of photo-autotrophic organisms in the form of red algae and symbiotic larger foraminifers is sometimes significant (Lees and Buller, 1972; Nelson, 1988; Henrich et al., 1997; James, 1997). The sediment typically consists of skeletal hash of sand-to-granule size. Cool-water carbonates lack shoal-water reefs and oolites; carbonate mud and abiotic marine cements are scarce (Figs 2.10, 2.11).

The cool-water factory extends poleward from the limit of the tropical factory (at about 30°) to polar latitudes (Fig. 2.9). The transition to the T factory is very gradual (Betzler et al., 1997). The C factory also occurs at low latitudes in the thermocline below the warm surface waters and in upwelling areas.

The oceanic environments of the C factory are photic or aphotic waters that are cool enough to exclude competition by the T factory and sufficiently winnowed to prevent burial by terrigenous fines. Nutrient levels are generally higher than in the tropical factory. These constraints set a wide depth window for the cool-water factory from upper neritic to bathyal and even abyssal depths. The most common setting is the outer neritic, current-swept part of continental shelves. The transition to the T factory normally extends over more than thousand kilometers (Fig. 2.10; Schlanger, 1981; Collins et al., 1997).

M factory

The letter M alludes to mud-mound, micrite and microbes. Intensive work in the past 15 years established the significance of this carbonate factory in the Phanerozoic (Lees and Miller, 1985; James and Bourque, 1992; Monty, 1995; Lees and Miller, 1995; Pratt, 1995; Reitner et al., 1995a; Webb, 1996, 2001). The characteristic component is fine-grained carbonate that precipitated *in situ* and was firm or hard upon formation. A number of detailed case studies suggests that precipitation of this fine-grained carbonate was caused by a complex interplay of biotic and abiotic reactions with microbes and decaying organic tissue playing a pivotal role (Reitner et al., 1995b; Monty, 1995; Neuweiler et al., 1999; Neuweiler et al., 2000; Reitner et al., 2000). The term "automicrite" for micrite precipitated in-situ (Wolf, 1965; Reitner et al., 1995a) is very useful in instances where the microbial origin is uncertain and the term "microbialite" not justified, for instance if precipitation is driven by non-living organic matter as in the concept of organomineralization (Trichet and Defargue, 1995; Neuweiler et al., 2003). Abiotic marine cement is the second most important product of this factory. It forms typically in vugs (such as *Stromatactis*) within the rigid framework of automicrite (Fig. 2.17.) Biotically controlled (skeletal) carbonate may occur but is not characteristic.

The typical setting of the M factory in the Phanerozoic are dysphotic or aphotic, nutrient-rich waters that are low in oxygen but not anoxic (Leinfelder et al., 1993; Neuweiler et al., 1999; Stanton et al., 2000; Boulvain, 2001; Neuweiler et al., 2001). These conditions often prevail in the thermocline, i.e. at intermediate depths below the mixed layer of the sea (Fig. 2.18). However, in the Proterozoic and after severe extinctions in the Phanerozoic, a carbonate production system dominated by biotically induced micrite and abiotic marine cements also extended into the shallow environments normally occupied by the T factory. The products are suffciently similar to the classical M factory to include them here. Textures and structures of the carbonate products do not indicate that the involvement of phototrophic microbes – as opposed to aphotic ones – fundamentally changes the precipitation process of the biotically induced carbonates. At present it is very difficult to distinguish between photically and aphotically formed automicrite unless associated sessile skeletal organisms provide the necessary information.

SEDIMENTATION RATES AND GROWTH POTENTIAL OF THE THREE FACTORIES

Siliciclastic systems depend on outside sediment supply. For carbonate factories, the ability to grow upward and produce sediment is an intrinsic property of the system. Conceptually, one can distinguish between the ability to build up vertically and track sea level, the aggradation potential, and the ability to produce and export sediment, the production potential. In most instances, however, it is only possible to quantify the lower limit of the growth potential by determining aggradation rates.

Fig. 2.19 presents rates calculated from thickness and stratigraphic ages of ancient deposits. The upper limit of the observed rates was interpreted as a crude estimate of the growth potential. Rates of all three factories were found to decrease as the length of the time interval increases. Different tests have shown that this trend has real, physical meaning and is not just a consequence of the fact that geologists calculate sedimentation rates by dividing thickness by time such that one variable, time, appears on both axes (Gardner et al., 1987; Schlager et al., 1998). The reason for the scaling of sedimentation rates lies in the distribution of hiatuses (Sadler, 1981). Sedimentation and erosion are episodic or pulsating processes and the record is riddled with hiatuses of highly variable duration. The fractal Cantor set is a good mathematical model of this situation (Fig. 2.20; Plotnick, 1986).

On log-log plots, the regression lines of the rate-time plots show slopes of approximately -0.5. A slope of -0.5 is typical of random noise, for instance a trend that is generated by the superposition of many unrelated effects. This is indeed what one may expect considering the many factors that can affect sedimentation rates. Sadler (1981, 1999) examined very large data sets and found regression slopes of approximately -0.5 in the domain of $10^3 - 10^8$ yr, but also significantly higher slopes in certain time windows, such as the frequency band of the Earth's orbital perturbations (chapter 5).

The upper limit of the observed rates, the estimated growth potential, also scales with a factor of -0.5. In the geologically particularly relevant interval of $10^6 - 10^7$ years the T-factory rates are highest, decreasing from 250 to 100 μ/y. C rates are about 25% of the tropical rates. Rates of the M factory are about the same as the T rates. However, field observations indicate that mud-mounds shed far less sediment into the adjacent basins. I therefore estimate the growth potential of the M factory to be only 80-90 % of that of the T factory.

The growth potential also varies with time. Particularly important is that during transgressions the factories need to be started up again after exposure and this start-up phase is coupled with slow growth followed by rapid growth (Fig. 1.12). Population dynamics as described by the logistic equation (Fig. 1.12) is the reason for the sigmoidal growth curve. In carbonate sedimentology, this pattern is known as the start-up, catch-up and keep-up stages of growth and reef response to the Holocene sea-level rise is a typical example of this rule (Fig. 2.21; Neumann and Macintyre, 1985). However, loose-sediment accumulations such as oolite shoals or lagoonal muds, also display this pattern. The law of sigmoidal growth implies that the growth potential of the system is significantly lower in the start-up phase. In the Holocene, the effect lasts only 2,000 to 5,000 years. Lag effects of millions of years have been postulated for growth in the wake of mass extinctions (Hottinger, 1989).

The growth potentials derived from Fig. 2.20 should be viewed as very crude estimates. They are based on limited data and they consider only vertical aggradation which is a rather imperfect substitute for sediment production by volume or mass. However, data on volumetric sediment production of the distant geologic past are very rare and hampered by the fact that carbonate factories are open systems that export much sediment to the surrounding ocean where it dissolves or becomes highly diluted and unrecognizable.

The ability to grow upward, the aggradation potential, is an important parameter in its own right. In the T factory, vertical growth determines the system's ability to keep up with relative sea-level rise and remain in the photic zone. Usually, the aggradation potential is significantly higher at the margins of shoal-water platforms than in the platform interior (Fig. 2.22) and this may lead to a growth morphology of raised rims and deep lagoons (chapter 3). In the M factory, the vertical growth potential is critical for staying above the adjacent sea floor and avoid being buried. Where the M factory builds platforms that rise to sea level, one occasionally also observes raised rims and deep lagoons (e.g. Adams et al., 2004).

CARBONATE-SPECIFIC ASPECTS OF DEPOSITION AND EROSION

Carbonate sedimentation obeys the same mechanical processes that control sedimentation in siliciclastics or other sediment accumulations. A discussion of these principles is beyond the scope of this book. There are, however, some aspects of deposition and erosion in carbonate environments that deserve special attention.

Deposition – source and sink

In siliciclastic systems, source and sink, i.e. erosional hinterland and place of deposition, are normally well separated. In shoal-water carbonate systems, the two roughly coincide. However, the spatial coincidence is only approximate. The action of waves and tides in shoal-water carbonate environments is normally so intense that redistribution of sediment is common and may move sediment by tens of kilometers (Fig. 2.23). Large-scale redistribution of the output of carbonate platforms is also apparent from the record of the deep-water sediments surrounding the shoal-water sources. Periplatform muds, turbidites, and debris flows attest to the fact that a platform such as the Bahama Bank exports material not just from the rims but also the platform interior (Neumann and Land, 1975; Droxler& Schlager, 1985; Haak and Schlager, 1989; Roth and Reijmer, 2004). Determining extent and pattern of sediment redistribution requires knowledge of the platform facies outlined in chapter 4.

Another noteworthy aspect of carbonate sedimentation is the ability to form rigid structures upon deposition. This is done by organisms building skeletal framework, by bio-induced precipitation of micrite or by abiotic precipitation of cement. In reefs, organic framebuilding and abiotic cementation combine to create structures that can resist the force of all but the most turbulent marine environments.

Erosion

Erosion of the surface of the solid Earth is, of course, pervasive and often dominant. In the context of this book, we will only consider erosion that occurs within the depositional environment or so close to it in space or time that it influences subsequent deposition.

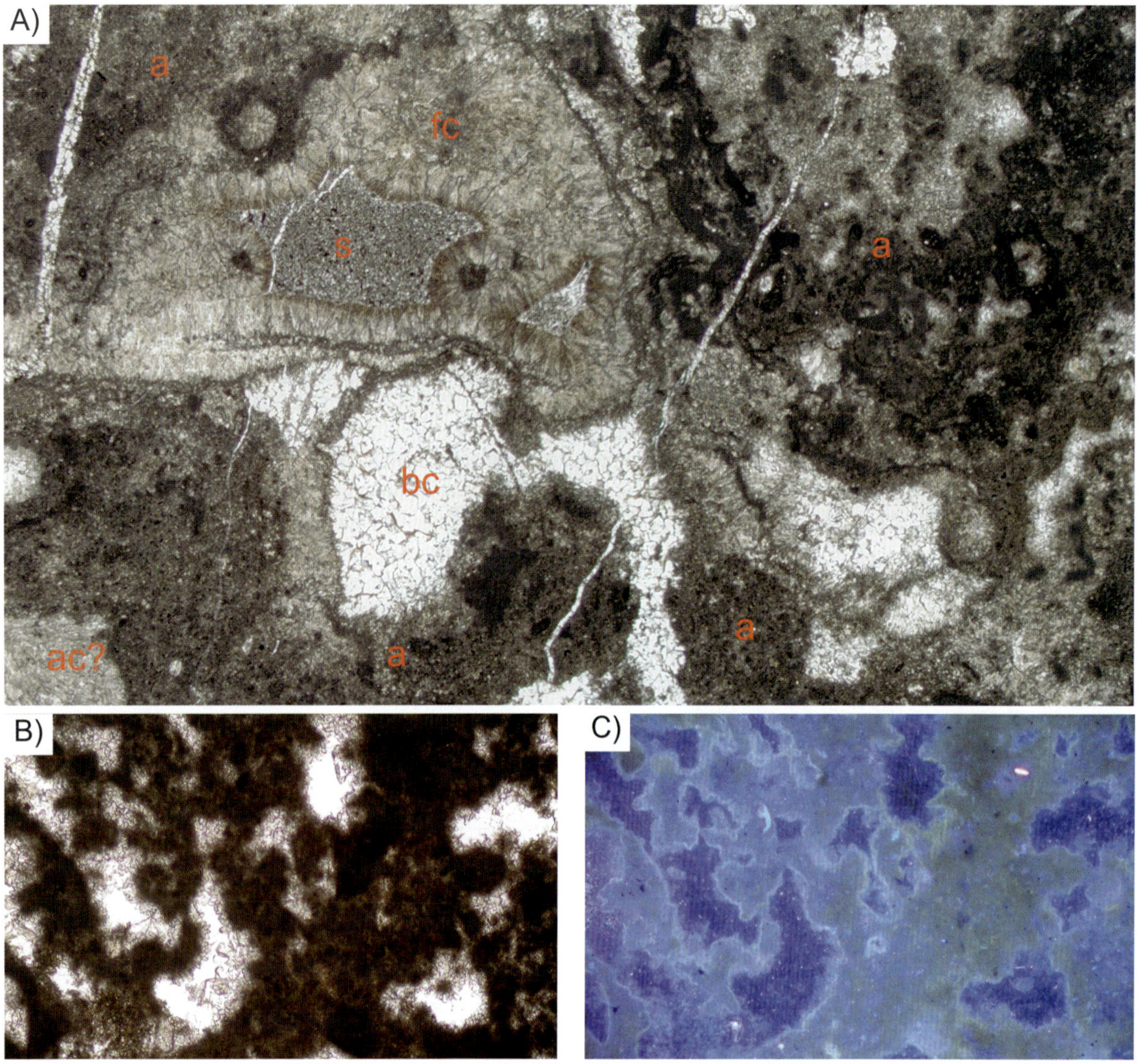

Fig. 2.17.— A) Thin-section of automicrite facies (long side = 1.5 mm). Dark automicrite, (a), with clotted or pelleted texture provides a basic framework and supports a large cavity filled by inclusion-rich, brownish fibrous calcite (fc) and muddy sediment, (s). Other parts of the cavity are filled by (late) blocky calcite (bc) and some questionable fibrous aragonite, (ac?). Close intergrowth of automicrite and fibrous calcite in upper left and lower right. Triassic, Southern Alps, Italy. After Russo et al. (1997), modified. B) Thin section of clotted and pelleted automicrite with primary pores filled by late blocky calcite. Plane-polarized light. C) Epifluorescence image of thin section in B.. Fluorescence is induced by organic matter. Automicrite fluoresces considerably, blocky cement remains dark. High content of (primary) organic matter is characteristic of automicrite. Triassic, Molignon Hut, Southern Alps, Italy. Images B and C by L. Keim, outcrop documentation in Brandner et al. (1991).

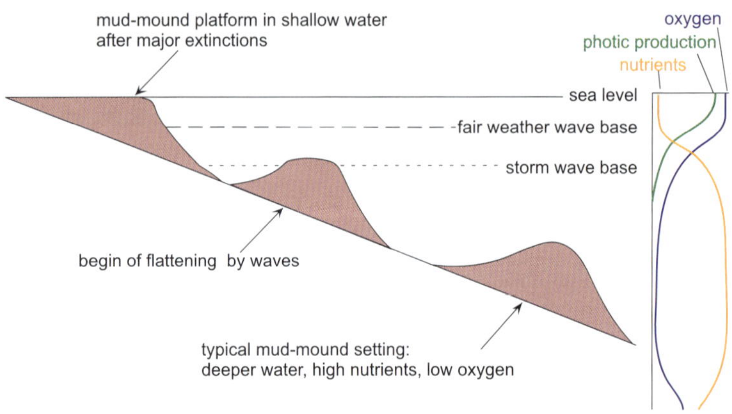

Fig. 2.18.— Environmental setting of the marine M factory. The factory is best developed in the nutrient-rich waters of the thermocline but may extend into the zone of wave action if the T factory is weak. Below the zone of wave action, accumulations usually are mounds, in the zone of wave action they become flat-topped platforms. Mud-mounds connected to cold seeps or hot vents may deviate from these environmental requirements. After Schlager (2003), modified.

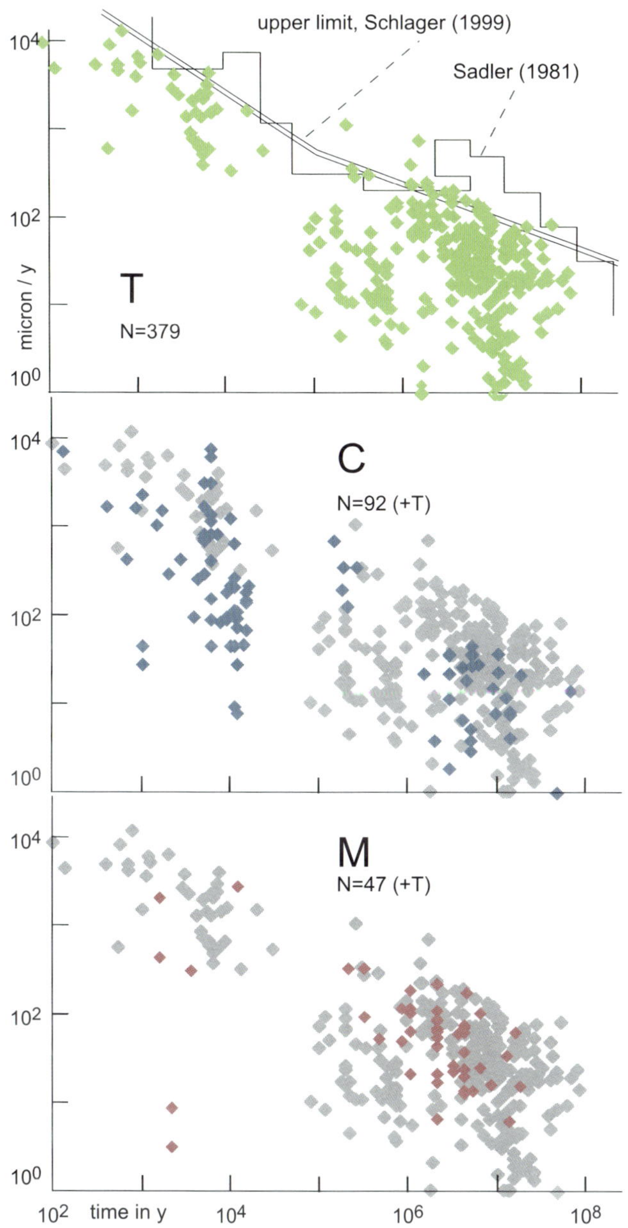

Fig. 2.19.— Sedimentation rates of the T, C, and M factories plotted against the length of the time interval of observation. Rates of all three factories decrease with increasing length of time – a general pattern of sedimentation rates caused by the occurrence of hiatuses on all scales in the record. Rates of T factory are highest, M factory is similar but overall production is lower as the factory exports less sediment laterally. Rates of C factory are about 25% of the T-factory rates in the million-year domain; high rates in the thousand-year domain are attributed to extensive reworking of the slowly lithifying accumulations. After Schlager (1999a, 2003), modified.

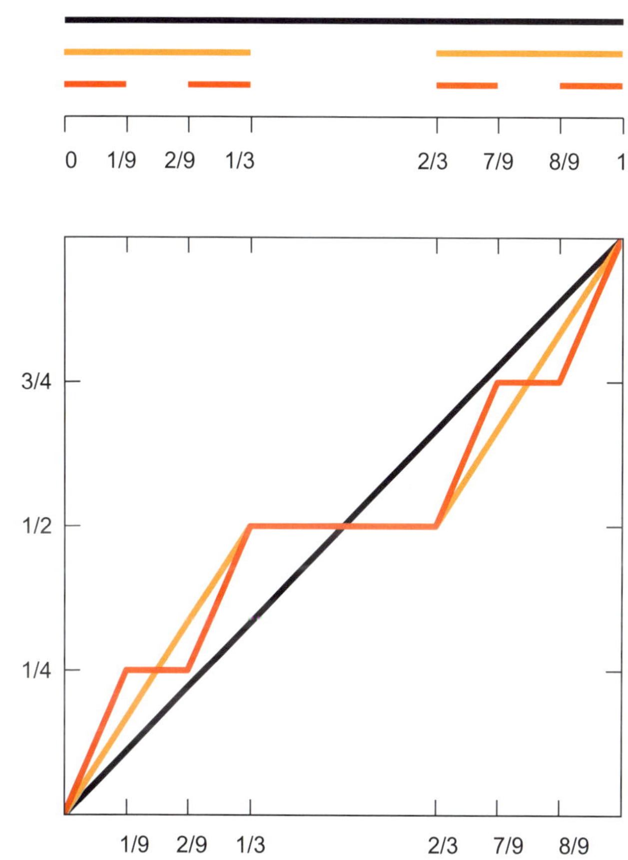

Fig. 2.20.— Cantor set and devil's staircase provide a mathematical model of the decrease of sedimentation rates with increasing time. Upper panel: Construction of the Cantor set starts with a straight-line segment of unit length and proceeds by erasing the middle third of the remaining segments. The set has a fractal dimension of 0.63 as compared to dimension 1 for an Euclidean line; stratigraphic sections probably resemble random Cantor sets with variable dimensions and patterns. Lower panel: The middle-third erasing Cantor set can be transposed into the devil's staircase by plotting, as a function of x in the unit interval, the relative weight y of the set that lies to the left of x. Horizontal segments represent gaps in the set. The devil's staircase can be viewed as a geological accumulation curve with plateaus representing hiatuses. Note that with increasing number of recognized hiatuses, the inclination of the remaining oblique segments steepens, i.e. sedimentation rates increase as shorter and shorter time intervals are considered. The model correctly predicts the trend observed with real sedimentation rates. Compiled from Plotnick, (1986) and Schroeder (1991).

As in siliciclastic systems, mechanical erosion may have a profound effect on sediment accumulations in carbonate environments. In addition, carbonates may be modified by chemical dissolution and bioerosion (Fig. 2.24). The different types of erosion may act in consort. Furthermore, sea-floor lithification and mechanical erosion are often coupled because lithification is a relatively slow process that operates most effectively where currents sweep the seafloor.

Mechanical erosion

Mechanical erosion is distributed very unevenly over carbonate depositional environments. It is most intensive at platform margins where the construction of reefs and lithified sand shoals drastically alters the equilibrium profile between the sediment and the wave energy in the water column. Siliciclastic systems adjust easily to the ocean's energy regime and develop shelf breaks on average around 100 m water depth. In contrast, reefs or carbonate sand shoals can build to sea level at the same position. In fact, modern coral reefs are able to withstand all but the seas and swells of ocean-facing coasts in the trade wind belt – the most energetic wave regime in the tropics. It is obvious that mechanical erosion is intensive in these settings and constant frame-building and cementation are required to repair the damage. It should be noted that sea-floor lithification is common and geologically coeval with deposition. This greatly reduces the rate of mechanical erosion and makes piles of carbonate sediment more resistant to lateral displacement.

Drowned platforms in open-ocean settings are affected by particularly intensive mechanical erosion. (Figs 2.25, 2.26, 2.27). Like submarine volcanoes, oceanic platforms disturb the normally sluggish oceanic tidal waves by inducing eddies that circle the platform. Velocities in these "topography-trapped waves" may become an order of magnitude higher than the orbital velocity of the original tidal wave (Fig. 2.26), leading to erosion and sediment redistribution. On living platforms, this erosion is restricted to the rim and upper slope. On drowned platforms, however, the currents may sweep freely across the top. As a result, the edges of drowned oceanic platforms are often bare and the pelagic sediment cap on the inner platform is thin, lens-shaped and replete with hiatuses. Hiatuses in this setting easily span tens of millions of years (Figs 2.25, 2.27). The interplay of mechanical erosion and sea-floor lithification on current-swept, drowned platforms creates very irregular morphology that may resemble subaerial karst.

Interplay of mechanical erosion and lithification is also observed on the flanks of modern carbonate platforms. Contour currents and turbidity currents are the eroding agents and the high amount of metastable minerals shed from the platform factory drives sea-floor lithification in areas that are swept clean by the currents (Schlager and James, 1978). As on seamounts, bizarre micromorphology reminiscent of karst or desert weathering is the result.

Chemical erosion

At present, the surface waters of the ocean are nearly everywhere saturated with respect to calcite. Dissolution in shoal-water carbonate settings is restricted to the most soluble mineral phases, i.e. magnesian calcites with very high magnesium contents (>18 mol%) and aragonite grains of very small size (1 micron or less). In the deep sea, however, sea-floor dissolution is intensive. There is no fixed depth level that marks the onset of dissolution. Currently, the tropical Atlantic is undersaturated with respect to calcite below about 4 km; the eqivalent level in the Pacific lies at approximately 1 km. The analogous levels for aragonite are 1-3 km for the tropical Atlantic and 0.2 km for the tropical Pacific. The difference between the two ocean basins is the result of deep-sea circulation and basin-basin fractionation (p. 3). We may assume that similar ranges in dissolution levels occurred in the geologic past. Carbonate dissolution need not be restricted to the deep sea. There is a tendency to develop a shallower dissolution maximum in the thermocline, between about 0.1 and 1.5 km in the water column that may affect epeiric seas and intracratonic basins.

Terrestrial dissolution of limestones or dolomites is conveniently summarized under the term *karst* . The dominant factor in karst erosion is dissolution by rain water and associated goundwater. Mechanical erosion plays a secondary role. Dissolution occurs at the surface and within the carbonate rocks, using matrix pores and fractures as avenues. A zone of intensive internal dissolution is around the water table and in the mixing zone of fresh water and sea water (Moore, 2001, p. 210). Karst erosion goes hand in hand with cementation and is normally described as diagenetic alteration if it occurs within a carbonate formation.

The rates of karst erosion increase with net precipitation and CO_2 content of the water, but with decreasing temperature. The rates of surface lowering ("surface denudation") are particularly important for sedimentologists as they influence the morphology of the subsequent phase of marine deposition. Surface denudation is often less dramatic than intuitively assumed. Fig. 2.28 shows a depositional topography of sand bars and channels that survived 120 ky of karst erosion in a humid tropical setting. Rates of karst surface denudation are low compared to rates of carbonate deposition. Fig. 2.29 presents rates of surface denudation as well as rates calculated with an empirical formula including the principal parameters mentioned above (Dreybrodt, 1988). Rates determined at time scales of $10^{-1} - 10^{1}$ y by chemical techniques are in the range of 10-100 μ/y (Fig. 2.29). Rates determined from the Neogene and Quaternary geologic record for time scales of $10^{3} - 10^{5}$ y and reported by Purdy and Winterer (2001) are similar or higher. The absence of a scaling trend analogous to the trend for sedimentation rates (Fig. 2.19) is somewhat surprising but the datasets are very small. In any case, the rates of karst denudation are lower than the carbonate sedimentation rates reported for similar time scales.

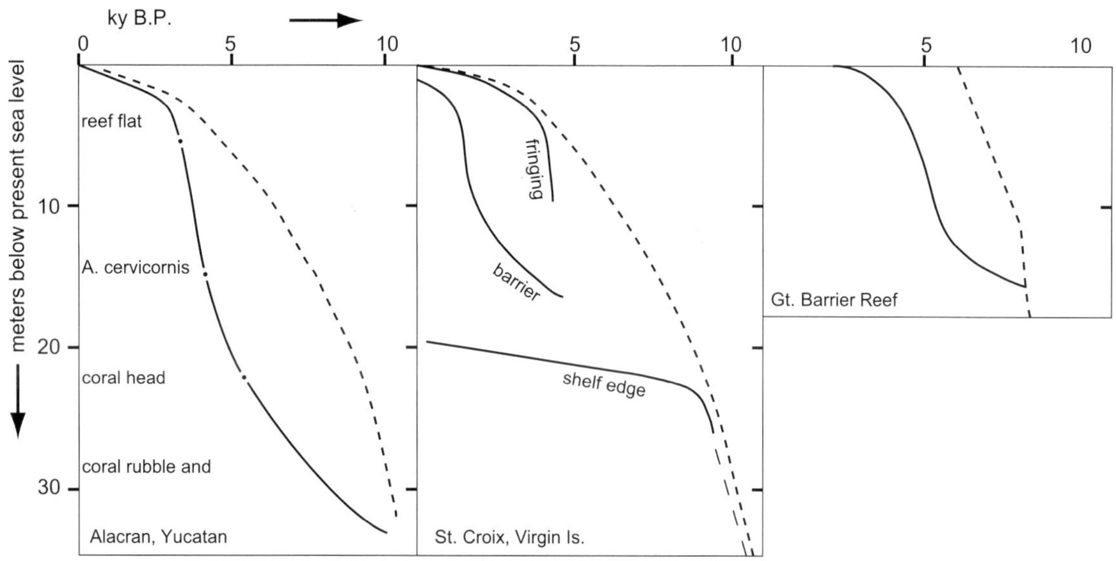

Fig. 2.21.— Growth history of Holocene reefs. Bold curves – growth, dashed – sea level. Note sigmoidal growth curves following the logistic equation (Fig. 1.13). Compiled from Schlager (1981), Neumann and Macintyre (1985).

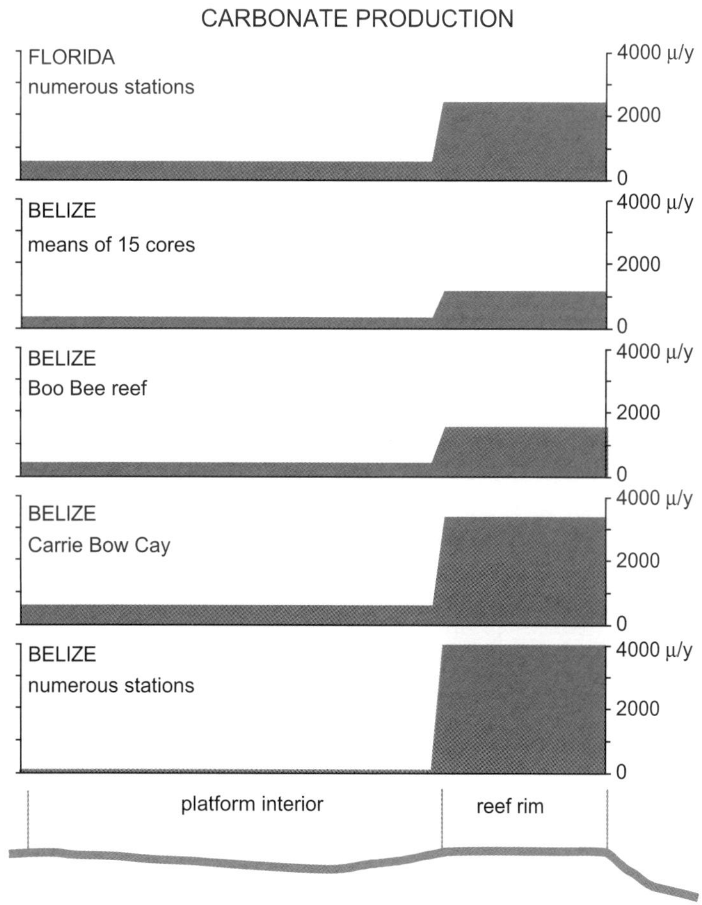

Fig. 2.22.— Sedimentation rates of platform margins and lagoons in Belize and Florida. Rates of reef rims are 3 to 35 times higher than the lagoon rates. The difference must reflect higher growth potential of the rim because the lagoons are deep and have unused accommodation. After Schlager (2003), modified .

Fig. 2.23.— Satellite image of Great Bahama Bank. Note that the sea floor is visible over most of the bank as water depth is less than 10 m in most areas. The entire bank is part of the carbonate factory and produces sediment. However, much of this material is reworked and redistributed by waves and tidal currents. Hydrodynamic shape of the sediment bodies is particularly obvious on the oolite sand shoals south of Tongue of the Ocean (down center) and north of Exuma Sound (upper right). NASA Johnson Space Center, photo STS029-90-9.

Bioerosion

Erosion by organisms is particularly intensive in limestones and dolomites because they are highly soluble and consist of relatively soft minerals that are easily abraded. Furthermore, most carbonate skeletons contain abundant food in the form of organic matter between the carbonate minerals and this is an added incentive for organisms to attack the grains.

The intensity of bioerosion varies enormously. It seems to be most intensive in the intertidal zone of the tropics where notches cut by bioerosion advance at rates of 2500 μ/y averaged over 10^3 y (Neumann and Hearty, 1996). The peak of bioerosion in the intertidal zone is largely responsible for the formation of the spectacular limestone cliffs because the rapidly advancing notch undercuts the cliff like a saw; from time to time the cliff collapses, putting more material in the intertidal zone where it is rapidly bioeroded (Fig. 2.24). Bioerosion on hard substrates in the shallow subtidal zone are lower. In-situ experiments with limestone slabs indicate weight losses on the order of 150 – 500 g m^{-2} y^{-1} (Vogel et al. 1996). Assuming a density of 2.5 g cm^{-3} for the limestone samples, this translates into a surface erosion rate of 60 – 200 μ/y sustained over about a year. This is several orders of magnitude less than the rate of carbonate production in shallow tropical environments (Fig. 2.19). Bioerosion increases with the nutrient content of the ambient water (e.g. Hallock, 1988). Consequently, eutrophication of reefs intensifies bioerosion of the reef rock thus decreasing the growth potential. In cool-water environments, bioerosion may exceed the rate of cementation. The result is an overall destructive diagenetic environment.

Fig. 2.24.— The cliffs of a Pacific island illustrate several principles of carbonate erosion. Erosion is most intensive in the intertidal zone where bioerosion and wave abrasion combine to undercut the island rocks, producing near-vertical sea cliffs. Intertidal erosion is also responsible for the water-covered terrace in the foreground and the overhanging beach at the foot of the cliff. Note uplifted fossil intertidal zone higher in the cliff. Karst erosion (by carbonate-dissolving rainwater and plants) creates the more gentle landforms above the cliff and modifies the upper parts of the cliff. In the shallow-marine environment, carbonate erosion is normally outpaced by precipitation. After Menard (1986).

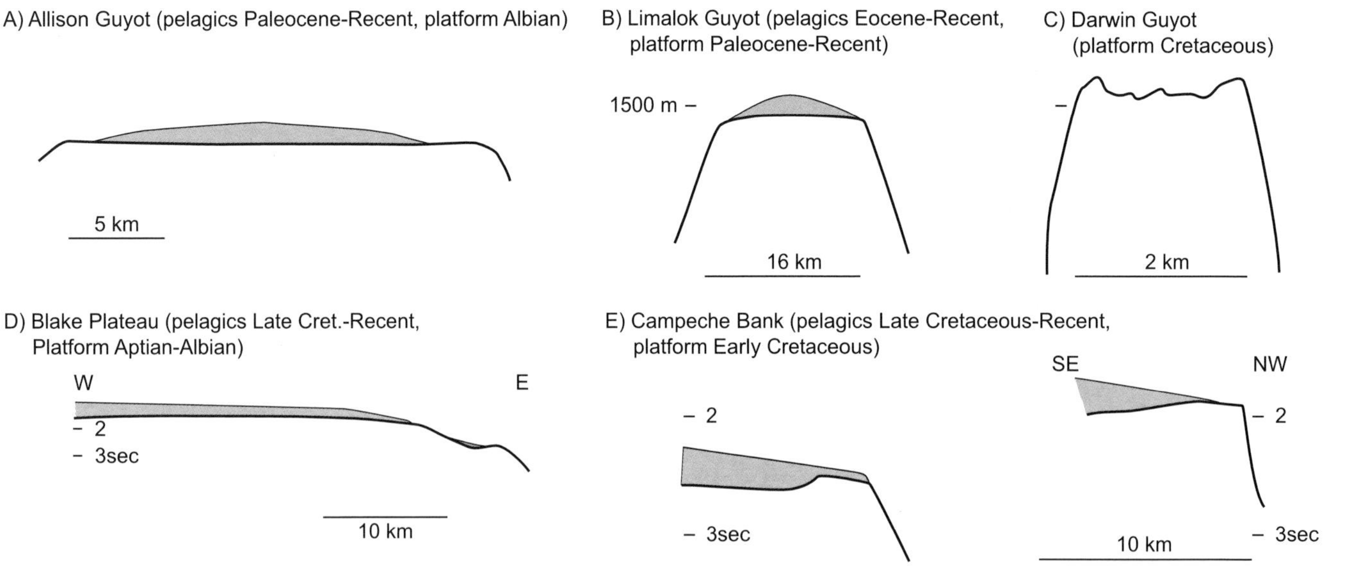

Fig. 2.25.— Drowned platforms in modern oceans commonly show thin, lens-shaped pelagic caps and significant hiatuses between platform sediments and pelagic cover. Some platforms have remained virtually bare for 100 My. After Schlager (1999b), modified.

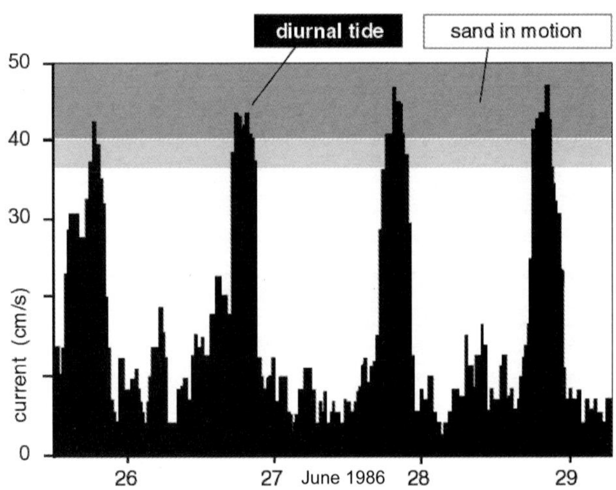

Fig. 2.26.— Current record from the top of a seamount in the east Pacific at 500 m water depth. Sluggish currents of few cm/s rise with the diurnal tide to short peaks of over 40 cm/s – enough to put sand in motion. The peaks are interpreted as resonance between the diurnal tide – generally associated with currents of few cm/s – and the sharp topography of the seamount. Current amplification by sharp topography may be one way to produce the long hiatuses on drowned carbonate platforms. After Genin et al. (1989) and Schlager (1999b), modified.

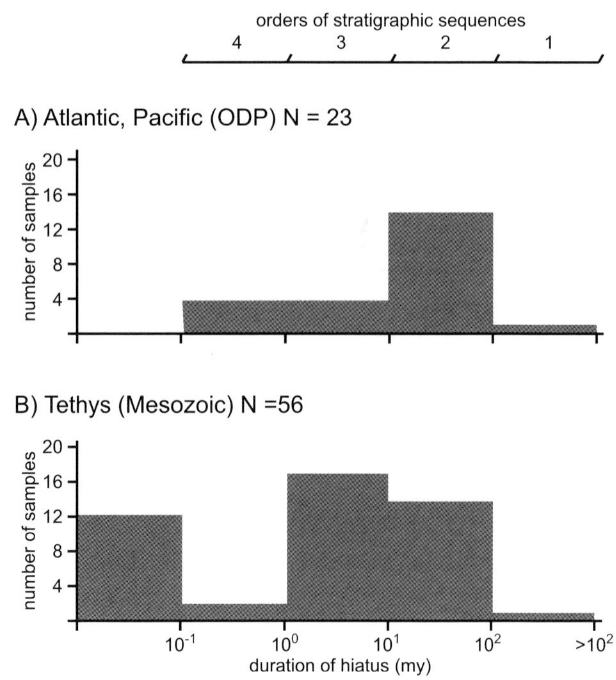

Fig. 2.27.— Hiatuses on drowned platforms in the Pacific and Mesozoic Tethys frequently extend to millions of years. Extreme examples of over 100 My are among the longest non-tectonic hiatuses in the Phanerozoic. After Schlager (1999b).

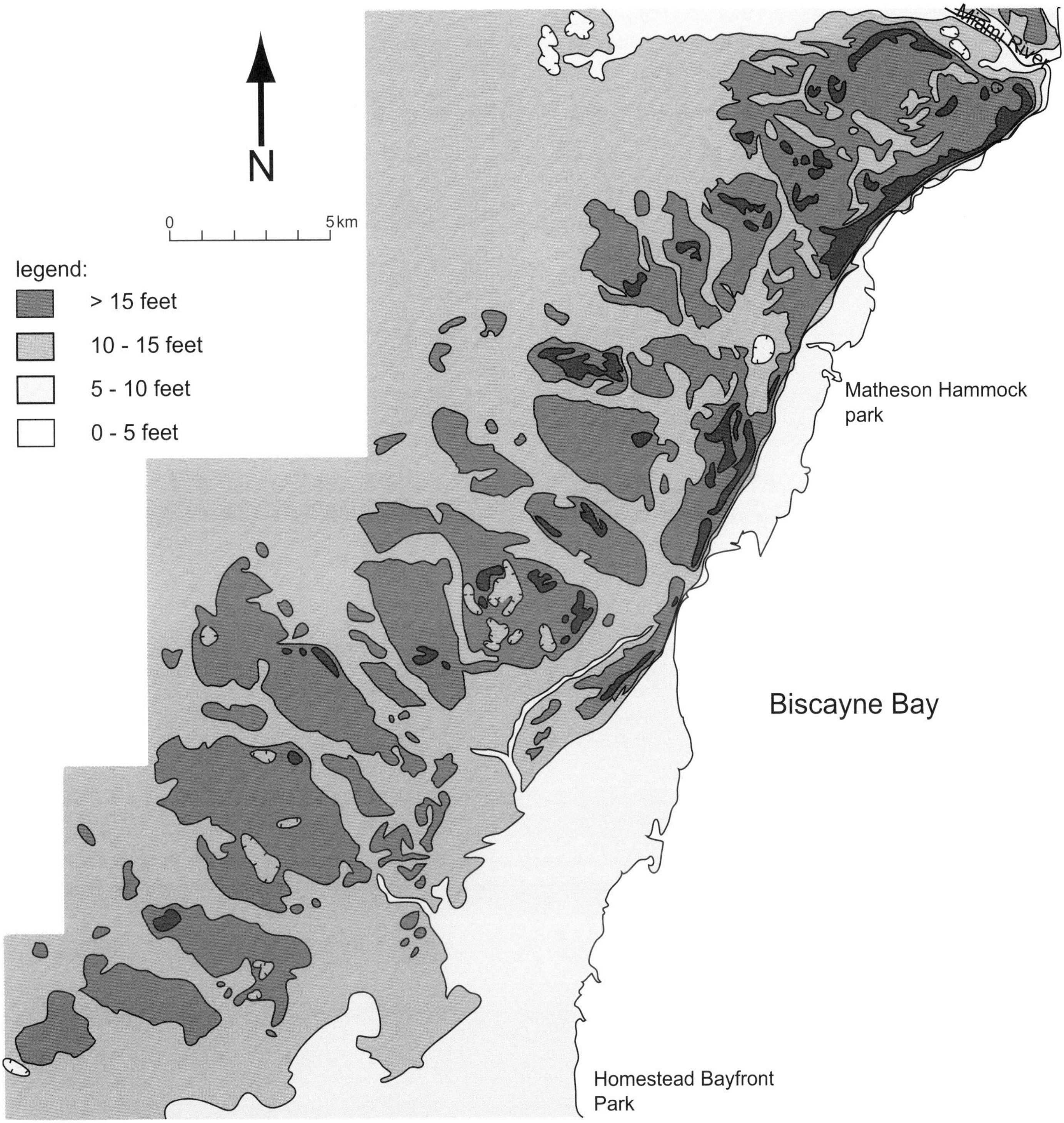

Fig. 2.28.— Topography of the Miami Oolite in SE Florida. The oolite formed during the last interglacial, about 125 ky ago and has been exposed to weathering in a humid, tropical climate (about 160 cm annual rainfall) ever since. Intensive karst weathering has produced isolated sinkholes and probably lowered the surface, but the depositional morphology of the formation still shows the pattern of sand bars and intervening channels in the northwestern part of the formation and a narrow longitudinal shoal in the southeast. The latter formed during a lower sea-level stand near the end of the interglacial. Compilation based on Halley and Evans (1983) and contour map by US Geological Survey.

Sea cliffs

Sea cliffs are not restricted to carbonate rocks but they are particularly common in limestones and dolomites. Sea cliffs form by wave erosion in the uppermost subtidal and the intertidal zones. In carbonates, intertidal bioerosion adds to the power of wave erosion by undercutting the cliff and removing the boulders on the abrasion platform. Consequently, cliffs in carbonate rocks retreat particularly fast and the adjacent abrasion platforms are nearly devoid of debris from rockfall.

The spectacular and common appearance of cliffs on the modern carbonate shores constrasts sharply with the scant evidence reported from the distant past. Fig. 2.30 summarizes personal observations from modern carbonate shores

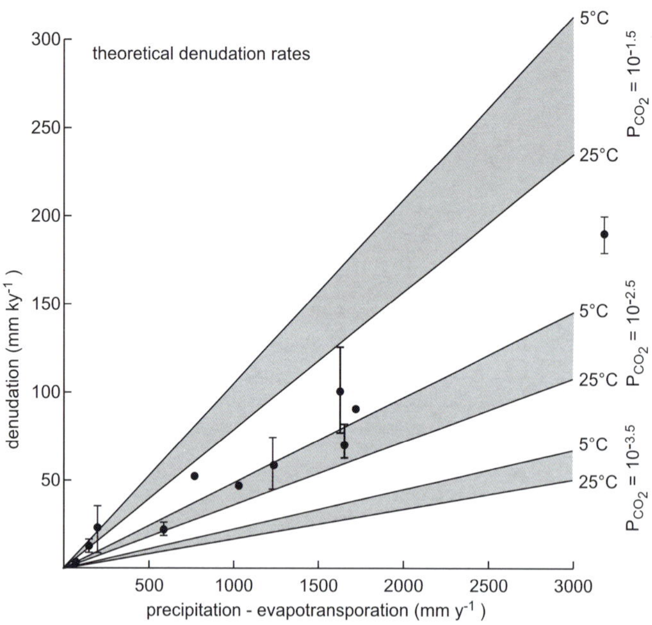

Fig. 2.29.— Denudation rates of karst surfaces. Lines represent rates calculated with the model of White (1984) that assumes net precipitation, CO_2 content of the meteoric water, and temperature to be the three most important controls. Some observed rates plotted with mean and standard deviations fit the model fairly well. After Dreybrodt (1988), modified.

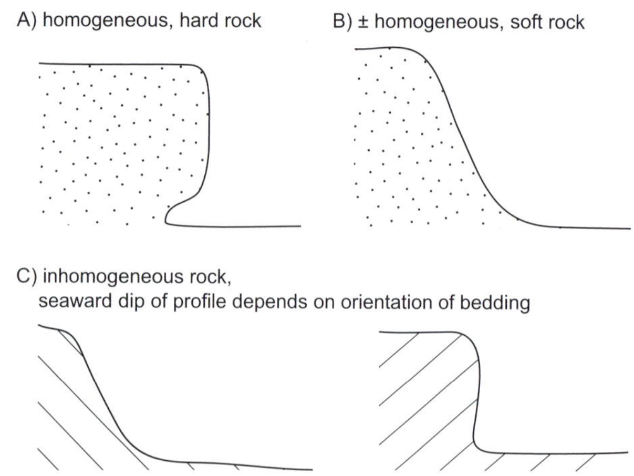

Fig. 2.30.— Shape of the cliff profile as a function of anatomy of the rock and dip of the bedding. Based on field observations of B.W. Fouke and the author.

Fig. 2.31.— Influence of bedding on cliff profile – an example from Exuma Islands, Bahamas. Rocks are Pleistocene eolianites that consist of harder and softer layers. Where bedding is flat, a near-vertical cliff develops; the varying hardness of the layers only shows in the rugged outline of the cliff. In the lower part, where bedding dips seaward, the weathering profile follows the bedding. Photo courtesy of D.F. McNeill.

that partly explain this discrepancy and Fig. 2.31 shows an outcrop example. Height and steepness of carbonate sea cliffs depend on the internal anatomy of the rock. Homogeneous hard limestones or dolomites are apt to form near-vertical cliffs, often undercut by an intertidal notch carved out by bio-erosion. Homogeneous but soft rocks cannot support a steep cliff and for inhomogenous rocks with alternations of hard and soft layers, the shore profile depends on the orientation and the angle of the stratigraphic dip. Steep cliffs can be cut into inhomogeneously bedded formations only if the dip is shoreward. Seaward dipping packages tend to develop a dipslope, possibly with a slight undercut immediately above the water line.

The overview in Fig. 2.30 shows that vertical cliffs are most likely to be found on poorly stratified, hard limestones, such as reefs and mounds. Unfortunately, such rocks are also most likely to conceal vertical stratigraphic boundaries because they abound with steep depositional boundaries and a cliff is easily masked when marine material is plastered on it during the next transgression.

Despite these problems we should intensify the search for fossil sea cliffs in the carbonate record. They are reliable indications for lowstands of the sea, providing much stronger evidence for sea-level movements than the alternation of transgressive and highstand systems tracts as will be shown in chapters 6 and 7.

CARBONATE ROCKS – THEIR DESCRIPTION AND CLASSIFICATION

The carbonate factories produce sediment in the size range of clay to boulder, similar to siliciclastics, plus sediments that are hard upon formation because of organic framebuilding or cementation within the depositional environment. These "framestones" or "bindstones" are a specialty of carbonate rocks.

Description of rock texture generally proceeds in two steps: (1) A crude characterization in terms of grain size by determining the proportion of mud (i.e. silt and clay-size material), sand, gravel and the presence of organic framework; this step, in combination with the depositional structures, is the key to the hydrodynamic environment of deposition. (2) Determination of the kinds of grains in the coarse fractions, using hand lens or microscope. In siliciclastics, grain kind is the clue to the source area of the material. In carbonates, it offers information on the depositional environment, particularly water chemistry and ecology.

Formal classification systems honor both aspects – grain size, and grain kind of the coarse fraction. Figs. 2.32, 2.33, 2.34 summarize the most common classification system by Dunham (1962) with modification by Embry and Klovan (1971). The system requires identification of fine matrix (silt, clay), particles (sand and coarser), pore space (open or filled by clear cement), and organic framework. The principles for defining the classes are, in descending order

- Presence and kind of hard organic framework.
- Presence of particles coarser than sand.
- Proportion of particles vs. fine matrix and, where both are abundant, the distinction between grains floating in the matrix (matrix support) and grains resting upon one another (grain support).

Several features in the Dunham classification of Fig. 2.32 merit discussion. The distinction between matrix support and grain support cannot be reliably answered by examining 2D sections of the rock. However, fairly good guesses are possible. The eye can be trained to recognize grain-supported fabrics by examining grainstone sections. Dunham (1962) provides a series of examples illustrating the range of textures resulting from irregular grain shapes.

A second point concerns wackestones. They consist of fine matrix, presumably originally mud, with significant admixtures of sand or coarser material. This may be a primary depositional fabric, for instance a deposit from a flow that had reached the limit of its transport capacity and was forced to deposit many grain sizes simultaneously. However, wackestones can just as easily be the result of burrow-

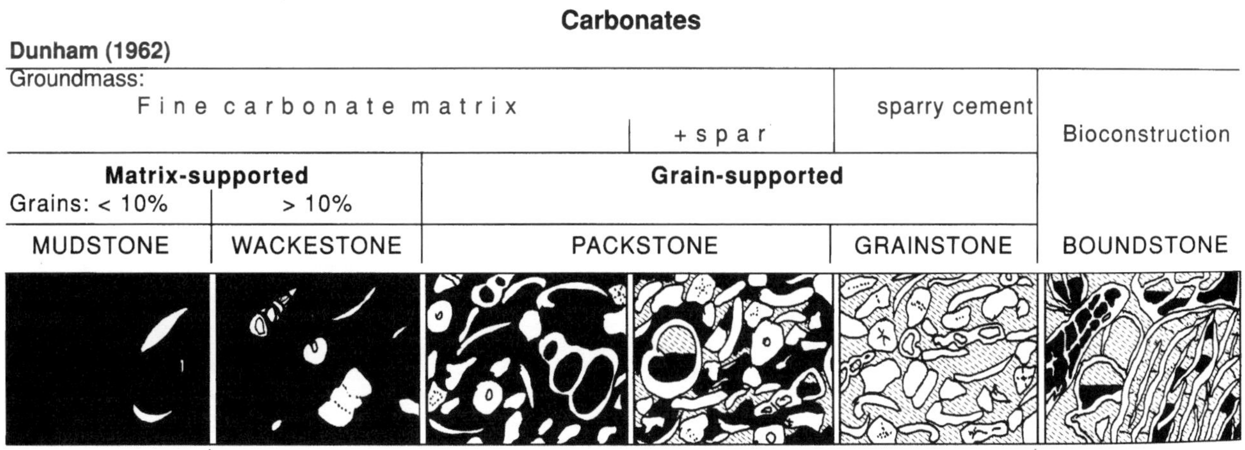

Fig. 2.32.— Summary of Dunham's (1962) textural classification of limestones consisting mixtures of sand and mud. After Flügel (2004), modified.

DESCRIPTION	ORIGIN OF GRAINS
Skeletal grains: calcareous hard parts of organisms or fragments of them.	Secreted by organisms in biotically-controlled fashion; specific shapes and textures provide important clues to the organism and the environent of formation.
Peloids: micritic, subrounded grains, usually without internal structure; soft or hard during deposition	Can originate (1) as fecal pellets of mud-digesting organisms; (2) by micritization (alteration by boring microbes) of skeletal grains or ooids; (3) by precipitation of micrite in biotically-induced fashion.
Aggregate grains: lumps of several round grains, agglutinated by sparitic or micritic cement or micrite laminae.	Delicate balance between deposition, erosion and cementation at the sea floor. Sand-size grains are first deposited, then weakly cemented in a patchy manner, and subsequently ripped apart by renewed currents such that only the best-cemented groups stay together.
Oncoids: irregularly shaped grains consisting of crudely concentric to strongly asymmetric micritic laminae wrapped around a nucleus. Oncoids may be hard or soft upon deposition.	Micrite layers are precipitated in bio-induced fashion around a nucleus and this lump is occasionally rolled around to become coated on all sides; precipitation mainly by bacteria or cyanobacteria.
Ooids: Smooth, spherical grains consisting of a nucleus and concentric laminae. Laminae may be micritic or sparitic, sometime displaying radial structure super-imposed on the concentric fabric. Most ooids are sand size, coarser ooids are often distinguished as "pisoids".	Smooth surfaces and near-spherical shape indicate formation in an environment where physical abrasion in turbulent water and growth alternate. Ooids of different mineralogies and textures have been produced by abiotic precipitation in the laboratory (Davies & Ferguson, 1978) but bio-induced precipitation cannot be excluded for natural envionments.
Cortoids: grains with the principal structure of oncoids or ooids but consisting of a large nucleus covered by very few concentric micrite layers.	Formation analogous to oncoids or ooids except that the grain reaches limiting size already after few laminae because of the large nucleus. Some cortoids form by micritization of the outermost parts of skeletal grains or ooids.
Lithoclasts: fragments of limestone or dolomite; as carbonate sediment may lithify in a few years within the depositional environment, lithoclasts may be geologically coeval and derived from the same environment. These clasts are sometimes called "intraclasts" and distinguished from externally derived, stratigraphically older "exoclasts".	Reworking of lithified layers or rocks by marine or terrestrial erosion or biotic activity.

Fig. 2.33.— Carbonate grains. Compiled from Flügel (1982, p. 124); Tucker and Wright (1990, p. 1-13); Flügel, 2004, p. 100-107.

<table>
<tr><td colspan="4">original components not bound together during deposition</td><td colspan="3">original components were bound together during deposition</td><td colspan="2">original components not bound together during</td></tr>
<tr><td colspan="4">generally smaller grains (arenite and silt size)</td><td rowspan="3">Organisms act as sediment bafflers (e.g. dendroid corals)</td><td rowspan="3">Organisms act as sediment binders (e.g. algal mats)</td><td rowspan="3">Organisms act as frame builders (e.g. intergrown reef corals)</td><td colspan="2">more than 10 % larger grains (rudite size)</td></tr>
<tr><td colspan="3">contains mud (micrite matrix)</td><td>lacks mud (sparite matrix)</td><td rowspan="2">contains mud (micrite matrix)</td><td rowspan="2">lacks mud (sparite matrix)</td></tr>
<tr><td>less than 10% grains</td><td>more than 10% grains</td><td></td><td></td></tr>
<tr><td colspan="2">mud-supported</td><td colspan="2">grain-supported</td><td colspan="3">Boundstone</td><td>mud-supported</td><td>grain-supported</td></tr>
<tr><td>Mudstone</td><td>Wackestone</td><td>Packstone</td><td>Grainstone</td><td>Bafflestone</td><td>Bindstone</td><td>Framestone</td><td>Floatstone</td><td>Rudstone</td></tr>
</table>

Fig. 2.34.— Classification of carbonate rocks based on Dunham (1962) and Embry and Klovan (1971).

INTERPARTICLE	VUGGY	
	SEPARATE	TOUCHING

Fig. 2.35.— Basic categories of pores in the classification scheme of Lucia (1995). Interparticle pores occupy space between grains or crystals. All other pores, including fractures, are classified as vugs. Since vugs are usually considerably larger than interparticle pores, it is petrophysically very important to determine if there exists a network of touching vugs in the rock.

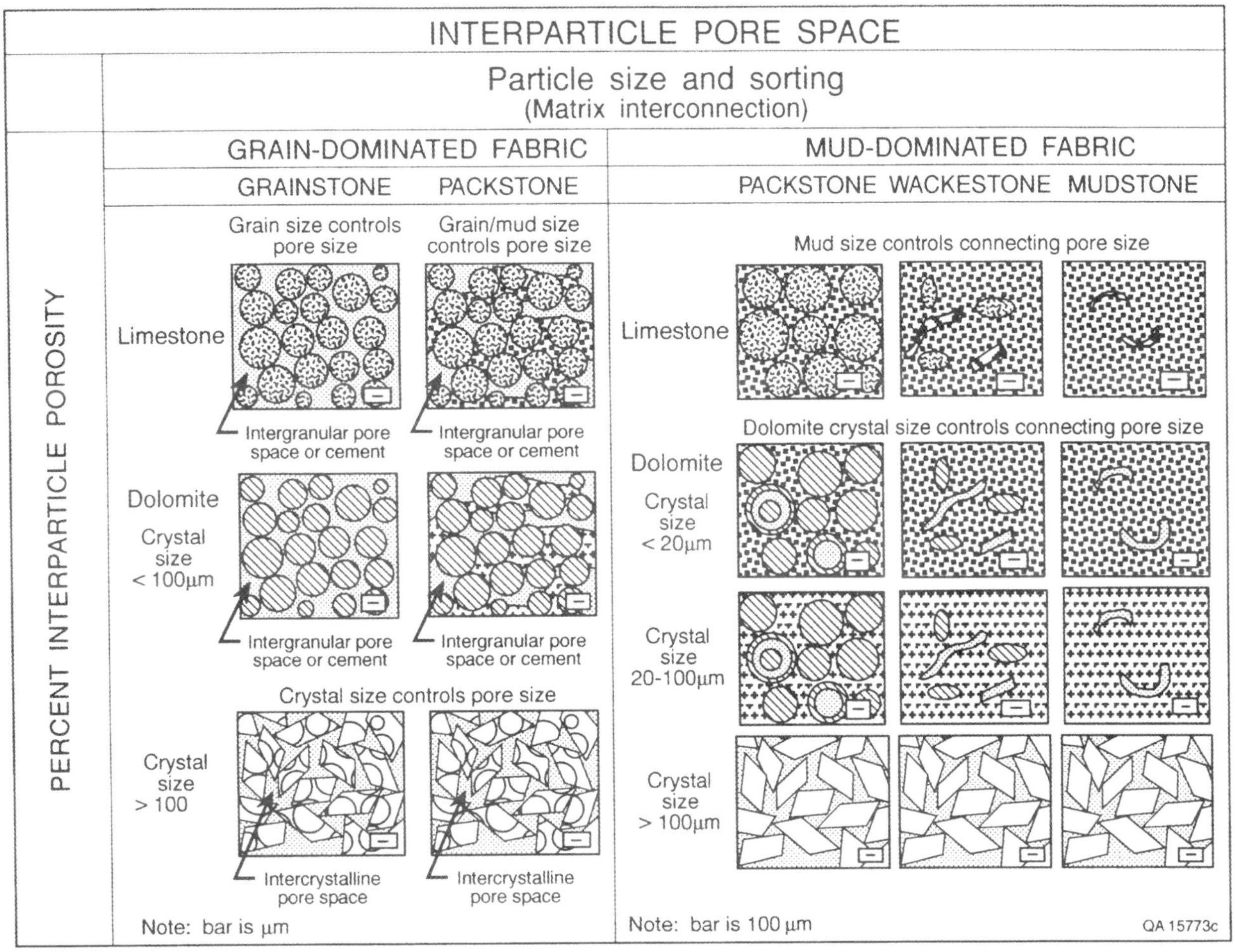

Fig. 2.36.— Classification of interparticle pore space based on size and sorting of grains and crystals.The classification honors depositional textures as well as diagenetic (crystal) textures where their effect is dominant. Vugs are classified in a separate scheme not shown here. After Lucia (1995). (Reprinted by permission of the AAPG whose permission is required for further use).

ing that mixed an original alternation of mud and sandy storm layers. Bioturbation in shoal-water carbonates is particularly intensive. Burrowing shrimps, for instance, commonly homogenize the top meter of the sediment column.

Finally, the distinction between a framework acting only as a baffle (bafflestone) and one that acts as a wave-resistant barrier (framestone) is gradational, difficult to verify in the geologic record and in most instances also of limited importance.

Careful application of the textural classification in Fig. 2.34 and attention to key grain types in Fig. 2.33 goes quite a ways towards identifying the three factories. "Index features" of the T factory are: skeletal framework, ooids, aggregate grains, green algae, abundant carbonate mud (particularly clay-size mud). The C factory is low in mud, framework and marine cement, but rich in skeletal grainstones. Typical of the M factory is a framework of micrite, supporting large cavities filled with acicular (marine) cements.

Porosity of carbonate rocks deserves special attention. It is more varied than in other sedimentary rocks but also one of the main reasons for the societal relevance of carbonate rocks. Porosity classification should include geological and petrophysical information. The system of Lucia (1995) balances these two aspects particularly well. First, porosity in carbonates is separated into to basic categories – interparticle porosity and vuggy porosity. Interparticle porosity is the space between depositional grains or crystals. Interparticle porosity is directly related to depositional texture or to crystal texture where diagenesis has obscured the depositional fabric, such as in dolomites and severely recrystallized limestones. Vuggy porosity represents all pore space that is not of the interparticle type. This includes pores within grains or crystals, pores significantly larger than grains or crystals, large irregular cavities and, finally, fractures. Lucia (1995) subdivides vuggy pores into separate vugs that are only connected via interparticle pores, and touching vugs that form a connected pore system.

Choquette and Pray (1970) proposed a genetic classification system of porosity. Pores are subdivided according to their origin, analogous to recognizing grain kind in sediment. This classification is particularly useful for recognizing the origin of porosity and reconstructing the sequence of diagenetic events.

CHAPTER 3

Geometry of carbonate accumulations

Depositional geometry of recent accumulations is an important tool for predicting the anatomy of sedimentary rocks in the subsurface. One advantage of sediment analysis by depositional geometry is that it can be performed on remote images, such as seismic or radar profiles and photos of distant or inaccessible outcrops. Geometry contains significant information on the internal structure and the depositional history of a formation. This chapter deals first with basic controls on the geometry of carbonate accumulations and the suitable terminology for their description. Subsequently, we turn to characteristic patterns associated with the three carbonate factories or specific depositional environments.

BASIC TRENDS IN GEOMETRY OF CARBONATE ACCUMULATIONS

The geometry of carbonate deposits results from the spatial patterns of production and the superposed effects of sediment redistribution by waves and currents. Four commonly occurring patterns in carbonate geometry are directly related to principles of carbonate production and the hydrodynamics of the water column.

- "The rich get richer". Carbonate factories tend to build elevated, localized accumulation because biotic and abiotic precipitation operates best where little other sediment disturbs the local environment. Once a production site has risen above the adjacent sea floor, precipitation is likely to accelerate and build up the accumulation even faster. This effect is felt in a wide range of scales, from decimeter-size stromatolite heads to Bahama-size platforms.
- "The sea is the limit". Carbonate production is highest in the uppermost part of the water column but the terrestrial environment immediately above is detrimental to carbonates (Fig. 2.3). Consequently, carbonate accumulations tend to build flat-topped platforms close to sea level. Small differences in production are leveled out through sediment redistribution by waves and tides.
- "The bucket principle". The boundary of the platform top shaped by waves and the slope shaped by gravity transport is a significant juncture in all depositional systems (Fig. 3.1). Tropical platforms tend to form a discrete rim at the platform-slope boundary. Several effects contribute to rim construction. The outer edge of the wave-swept platform top is the preferred location of framebuilders and thus of barrier reefs that form a rim (Fig. 3.2, 3.3). The organic reef structures are further strengthened by abiotic cementation that is particularly extensive there because of high primary porosity and the pumping effect of heavy seas. Furthermore, the upper slope environment is a preferred location of microbial crusts and cements that stabilize the underpinnings of the shoal-water barriers. The production of the

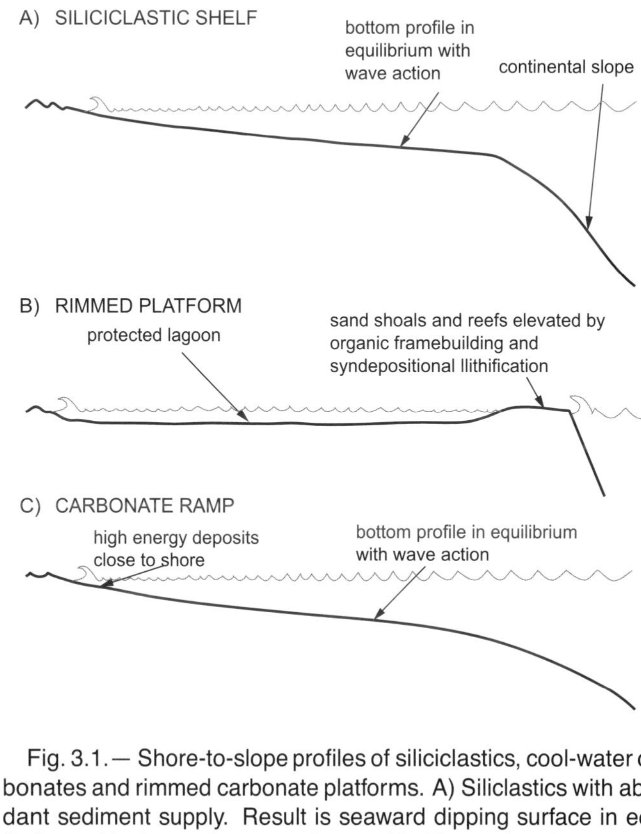

Fig. 3.1.— Shore-to-slope profiles of siliciclastics, cool-water carbonates and rimmed carbonate platforms. A) Siliclastics with abundant sediment supply. Result is seaward dipping surface in equilibrium with deepening wave base. B) Rimmed carbonate platforms. On tropical platforms, the wave-equilibrium profile is grossly distorted by the construction of wave-resistant reefs and quickly lithifying sand shoals, both occurring mainly at the platform margin. Platforms are basically dish-shaped and equilibrium profiles develop only locally in parts of the lagoon. C) Carbonate ramps are accumulations without rims that resemble the siliciclastic equilibrium profile. Cool-water carbonate follow this pattern. Tropical platforms commonly show ramps as a transient stage during rapid transgressions before a rim can develop.

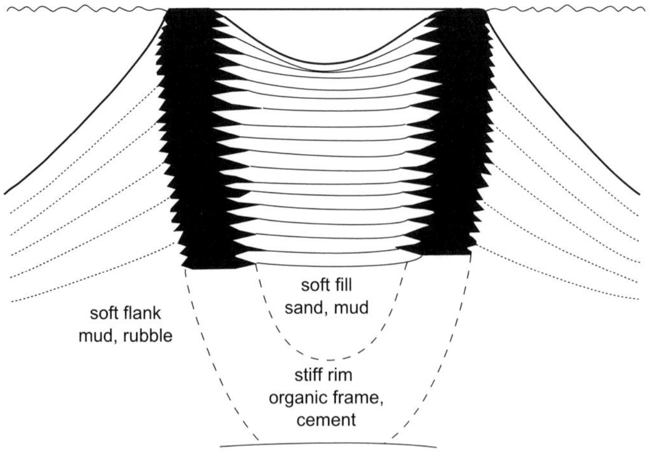

Fig. 3.2.— The bucket principle. The growth anatomy of rimmed carbonate platforms (the products of the T factory) resembles a bucket. The platforms are held together by stiff rims of reefs or rapidly cemented sand shoals, and filled with less consolidated sediments of lagoons and tidal flats. The growth potential of a platform is largely determined by the growth potential of the rim. This atoll shifts from slow progradation to retrogradation as it approaches its maximum growth rate. Raised rim and deep lagoon ("empty bucket") indicate incipient drowning. After Schlager (1981).

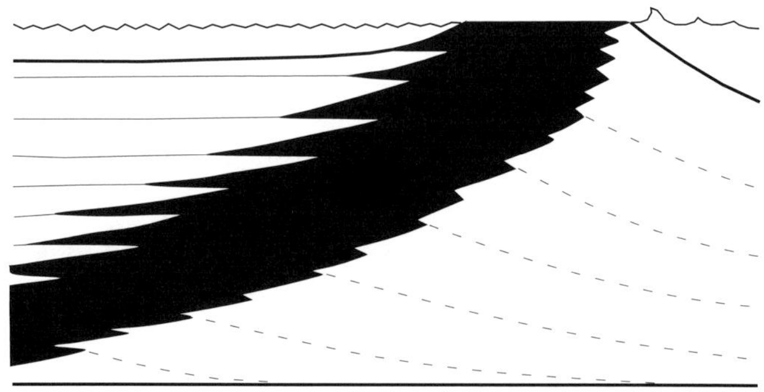

Fig. 3.3.— Bucket structure of an attached platform. Production at first comfortably exceeds accommodation creation and later slowly approaches the limit of growth. This change is demonstrated by decreasing rate of progradation and appearance of a deep lagoon.

platform rim is higher than that of the platform interior such that the rim rises above the lagoon and sheds its excess sediment both downslope and into the lagoon. Cool-water carbonates have no rims to speak of and therefore tend to form seaward sloping profiles in equilibrium with wave action. Accumulations that form below intensive wave action are convex rather than flat-topped (Fig. 3.4).

- "Steepen the slopes". Carbonate slopes steepen with height and are generally steeper than slopes of siliciclastic accumulations. Several effects contribute to this trend: Shoal-water carbonate production includes much sand and rubble; these materials cannot be carried far and have high angle of repose. Slope lithification retards slumping and stabilizes steep angles once they are formed (see further on p. 43f).

The terms used for describing the geometry of reefs and shoal-water carbonate accumulations are quite numerous and not always self-explanatory (see Wilson, 1975; Wright and Burchette, 1996). However, in this plethora of words there are some that nicely capture the principal geometric trends just discussed. They will be given preference in this book and are explained below.

Localised accumulations

Carbonate build-up. Build-up simply indicates that a carbonate body rose above the adjacent sea floor according to the principle of " the rich get richer". The term is unspecific about the origin of the carbonate material. It applies to a wide range of scales. The term has been used for features that are only a few meters in diameter and height, as well as for features measuring tens of kilometers across and over a kilometer in height. Furthermore, it may be used for flat-topped accumulations, i.e. carbonate platforms, as well as convex reefs or mounds.

Reef . In carbonate sedimentology, the term reef denotes a wave-resistant buildup formed by the interplay of organic framebuilding, erosion, sedimentation and cementation (Wright and Burchette, 1996, p. 368). Ecologic reefs are defined by texture and composition. The geometry of reefs is only a secondary attribute albeit an important one. Geometry as observed in map view or cross sections of outcrops or seismic data can aid in locating reefs and provides crucial information on the effect of the reef on its surroundings.

The basic controls on reef geometry are upward growth of the organic framework, current re-inforcement by this structure and sediment export by the reef factory. The interaction of the wave-resistant structure with the adjoining sea produces scours around the reef because the structure generates extra turbulence, much like sand or snow are blown away from rock walls. Sediment export from the reef partly compensates for the scouring effect of waves and currents. Reefs generally shed sediment of a wide range of sizes and this material forms debris aprons that become thinner and finer with increasing distance from the reef core. If major scours are geometrically manifest, the reef probably suffered an extended period of greatly reduced production or it was drowned and scoured before it became buried.

The degree of symmetry of reef core and aprons indicates to what extent a reef functioned as a barrier. Reef belts at the platform margin are the prototype of a protective barrier. They are pounded by the ocean on one side and face calm lagoon waters on the other. The result is pronounced lagoonward transport of reef debris, commonly in the form of prograding sediment tongues. Passages in the barrier may develop curved spits of reef rubble that point lagoonward. The seaward reef front, on the other hand, remains bare. Reefs with radially symmetric aprons indicate growth in a location

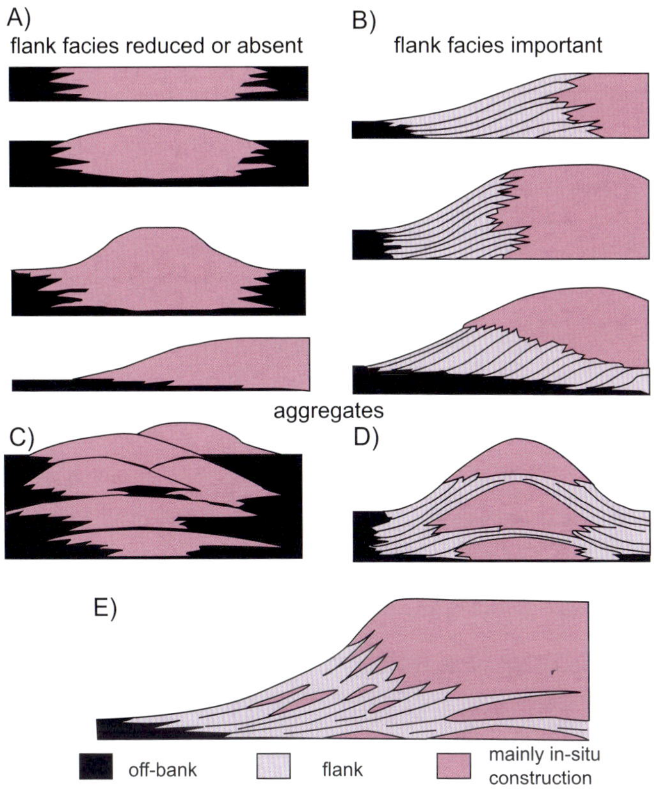

Fig. 3.4.— The M factory typically produces upward-convex accumulations, the mounds, that interfinger laterally with the surrounding facies. Flank facies are present when the factory sheds much debris. However, the km-wide debris aprons of tropical platforms are rare. Based on Lees and Miller (1995), Keim and Schlager (2001) and Schlager (2003).

where energy flux in the water column frequently changes direction dependent on weather conditions. Patch reefs in lagoons or epeiric seas commonly display this geometry.

Mound. James and Bourque (1992) divide organic carbonate buildups into reefs and mounds. In this classification, "mound" denotes a rounded, hill-like, submarine structure composed of skeletal material or automicrite of predominantly microbial origin. Mounded accumulations of skeletal material can be interpreted as hydrodynamic structures just like any other detrital accumulations. Siliciclastic and carbonate sediment drifts in the deep sea indicate that muddy sediments may still exhibit bedding structures related to bedload transport. However, the vast majority of the carbonate mud mounds of the sedimentologic literature do not fall in this category and merit special attention.

Carbonate mud mounds are isometric or elongate submarine hills. Their top is normally convex and lacks the horizontal surface of carbonate platforms or of reefs built to sea level. Like reefs, mud mounds commontly show a core flanked by debris aprons that steepen with the height of the mound and sometime dip at over 40° (Lees and Miller, 1995). In contrast to reefs, the top of the mound normally does not prograde; only the base of slope progrades to the degree required by the increase in height of the mound.

The convex top of the mounds probably reflects their formation in deeper water below the "shaving" effect of wave action. Where mounds build into the zone of intensive wave action, their top flattens and their facies changes. Conversely, the geometry of slowly drowning reefs may gradually change from flat platform to mound as they subside below wave action (see also p. 122f; Neuhaus et al., 2004; Zampetti et al., 2004).

Carbonate platform. Carbonate platform is a widely used and rather loose term. The dictionary definition precisely captures the essence of the usage of this term in carbonate sedimentology: "a horizontal flat surface usually higher than the adjoining area" (Webster's Dictionary). Shoal-water carbonates form flatter tops than other depositional systems because (1) the high-production zone is abruptly limited by sea level, (2) waves and currents efficiently redistribute sediment from high-production areas to fill up depressions, and (3) the platform margin tends to develop a wave-resistant rim that protects the sediment of the platform interior. The term carbonate platform is broadly synonymous to "carbonate shelf" (Wilson, 1975). I think platform is more appropriate because "shelf" has the connotation of being attached to land. Platform is neutral in this regard.

Shoalwater carbonate accumulations that lack a rim will strive towards a profile of equilibrium between the sediment and the wave energy in the overlying water column. In these instances, the carbonate-specific morphology is lost and the geometric discrimination of carbonates and siliciclastics may become impossible.

Size is not a rigid criterion for defining carbonate platforms. However, the term is usually applied to features that are kilometers to hundreds of kilometers across and rise many tens of meters to thousands of meters above the adjacent basin floor.

Carbonate platforms may be attached to a land mass or detached, isolated features that are surrounded by deeper water on all sides.

Platforms may prograde, aggrade vertically, or retrograde. A geometry characteristic of carbonate platforms is "backstepping" (Fig. 3.5). It implies that the platform retrogrades in discrete steps such that discrete margin positions are recognizable and separated by flat seafloor. Backstepping is common with rimmed platforms. This seems paradoxical at first because rims normally grow faster than platform interiors (p. 25). Thus, if the rim goes under, the entire platform should be lost. The main advantage of backstepping is the reduced power of destructive waves in a backstepped position. Waves that reach a backstepped margin have lost part of their energy to bottom friction as they travelled over shallow ground. Other reasons for backstepping are elevated topography or, on large passive margins, shift to areas of slower subsidence.

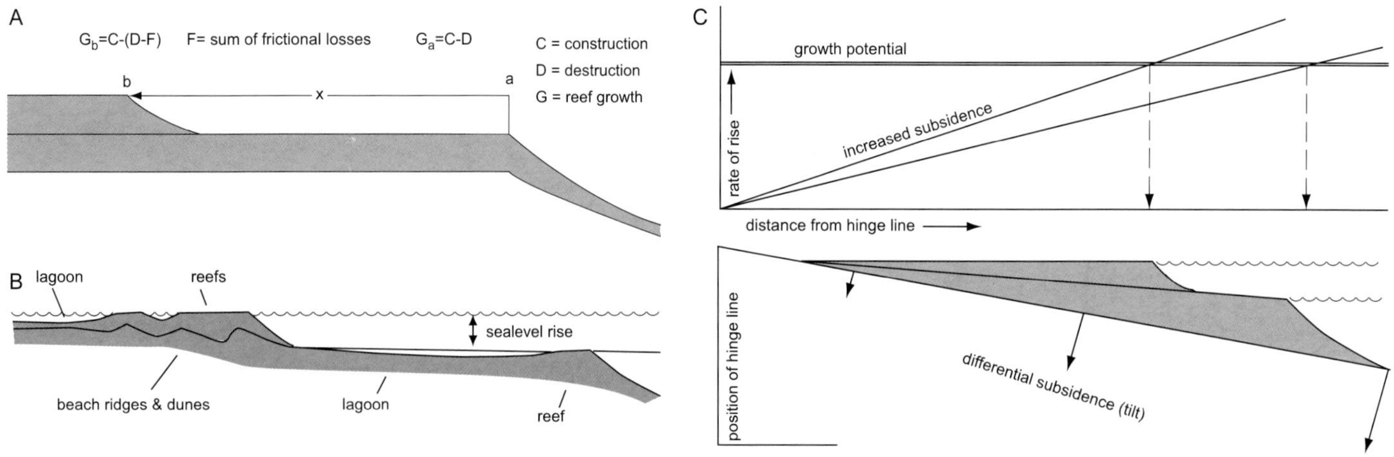

Fig. 3.5.— Carbonate platforms backstep when faced with relative sea-level rise that slightly exceeds their growth potential. This figure shows reasons why backstepping may be advantageous for a platform under stress. A) Destruction of the margin by waves is less in the backstepped position because waves have lost energy by bottom friction. B) Backstepping to higher ground. C) Backstepping to area of lower subsidence (on passive margin).

Platform rim. The geometric effect of rim-building is best illustrated by direct comparison of rimmed platforms and siliciclastic shelves that lack constructional rims (Fig. 3.1). The rim is more productive than the adjacent lagoon or upper slope. Excess sediment is shed into the lagoon and down the slope. The geometric expression of the high production and the export of vast quantities of sediment is the common pattern of bi-directional progradation away from the rim: progradation of the backreef apron into the lagoon and simultaneous seaward progradation of the rim and the slope.

The degree of wave resistance of the rim varied in time and space. During most of the Phanerozoic, shoal-water carbonate systems were able to build rims in the zone of perennial wave action. Currently, some reef communities can build into the intertidal zone even in settings that face the full power of oceanic waves in the trade-wind belt. The system can build into the supratidal zone by forming "islands" of storm ridges that may contain freshwater lenses and be capped by terrestrial (carbonate) eolianites.

The platform rim need not be a reef. Carbonate sand shoals can also form wave-resistant barriers at the platform margin. They may consist of oolites, precipitated locally in the mixing zone of normal marine and platform waters, or of skeletal debris from the outer, winnowed parts of the platform. Either type can build into the supratidal zone and is prone to early lithification in the submarine or the supratidal environment. This early lithification greatly enhances the wave resistance of the shoal and reduces the rate of lateral migration to almost zero. Consequently, the hydrodynamic effect of these shoals is similar to that of reefs and reef aprons, both are wave-resistant, stationary structures near the platform margin.

The efficiency of reefs and shoals as barriers against wave energy depends on the elevation of the crest and the continuity of the rim. A geometric criterion for elevation into the zone of wave action is the presence of a flat top. Degree of continuity of the rim may be expressed as the rim index, defined as the fraction of the platform perimeter that is occupied by reef or sand shoal (Fig. 3.6). The rim index is highly variable and should be quantitatively estimated wherever possible, for instance via seismic data, maps or large outcrops. Quantification of rim continuity provides an estimate of the fraction of oceanic wave energy that enters the lagoon. In first approximation, the fraction of wave energy that passes through a leaky rim is the inverse of the rim index.

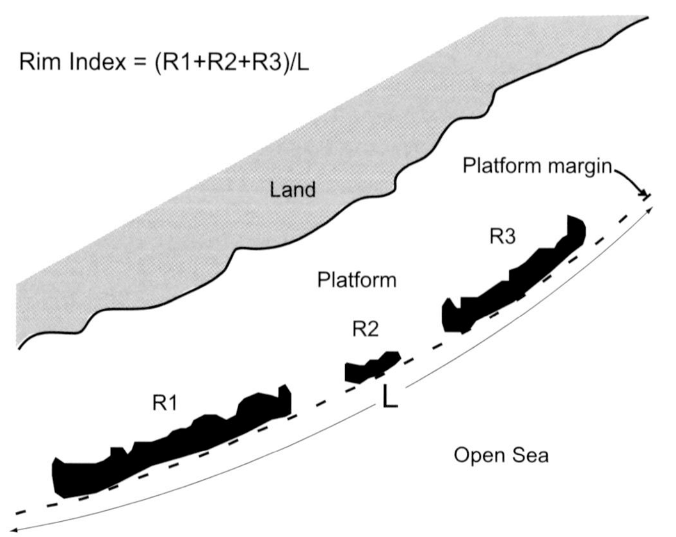

Fig. 3.6.— Rim index is the fraction of the platform margin occupied by reefs or shallow sand bars. Sketch depicts land-attached platform with leaky rim of reefs (R1 etc.) shown in black. Fraction of wave energy passing the leaky rim is the inverse of the rim index.

Ramp

Ramps are shoal-water carbonate systems that lack the steep slope seaward of the platform margin on rimmed platforms. They show a seaward-sloping surface with dips of 0.1 - 1.5° instead (Fig. 3.1). According to many authors, ramps may possess an offshore rim and a lagoon. This complicates the definition of ramp. Originally, ramps were depicted as systems devoid of an offshore rim (Ahr, 1973; Wilson, 1975. Under these conditions, the distinction of ramps and rimmed platforms is easy: rimmed platforms have a high-energy facies belt offshore, ramps lack this belt; their only belt of high-energy sediments is in the littoral zone close to shore.

Subsequent studies produced a substantial number of ramp surfaces that are not hung on the shoreline but on an offshore rim with a protected lagoon between the high-energy belt of the offshore rim and the coastal one. For those ramps, the high-energy-only-nearshore criterion does not work. These "detached ramps" are rimmed systems that do not have a slope but a ramp surface seaward of the rim. (Fig. 3.7)

Read (1985) recognized homoclinal and distally steepened ramps. This distinction is useful. Distally steepened ramps sit on top of a slope, commonly a continental slope. The slope may be the inactive slope of a rimmed platform that has backstepped. The homoclinal ramp with a uniformly dipping surface is typically found in shallow intracratonic basins or foreland basins. Even in these settings, the ramp morphology is normally a transient feature. Carbonates with rim-building capability have a strong tendency to prograde, steepen their slope and thus differentiate into a platform top, a rim and a steep slope (Fig. 3.8A). The reverse transition, from rimmed platform to ramp, occurs when a basin fills up and slope height and slope angle decrease (Fig. 3.8B; e.g. Biddle et al., 1992).

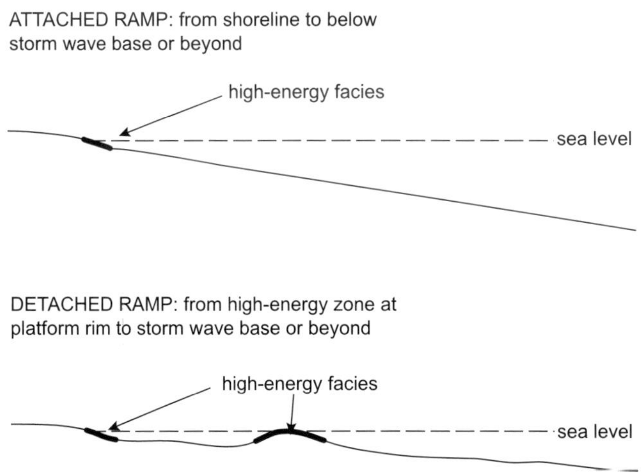

Fig. 3.7.— Number and location of high-energy sand belts allow one to distinguish between the classical attached ramp with high-energy deposits only close to shore (top panel) and the detached ramp that starts from a high-energy shoal that lies offshore (lower panel). The latter system has two high-energy belts separated by a lagoon.

Slope, rise, basin floor

Slopes and debris aprons around platforms are important elements of the edifice and largely determine the extent and shape of the top. These areas also act as sinks for much of the excess sediment produced by the platform top. In sequence stratigraphy, platform margins and slopes play a crucial role as they hold much of the information on lowstands of sea level.

Geometry and facies of slopes and rises rich in mud are governed by several rules:

- ➤ The volume of sediment required to maintain a constant slope increases as a function of platform height. The increase is proportional to the square of the height for conical slopes of isolated platforms, such as atolls, and it is proportional to the first power of the height of linear platform slopes, such as on passive margins (Fig. 3.9). These geometric laws limit platform growth, particularly on high atolls (Fig. 3.10).
- ➤ The upper parts of platform slopes steepen with the height of the slope, a trend that siliciclastics abandon at early stages of growth because they reach the angle of repose of mud. As a consequence, the slopes of most platforms, notably the high-rising ones, are steeper than siliciclastic slopes (Fig. 3.11). Changes in slope angle during platform growth change the sediment regime on the slope: the balance between erosion and deposition of turbidity currents shifts such that the slopes evolve from the accretionary to bypass and finally to erosional conditions. This change in depositional regime in turn changes sediment geometry on the slope and the rise (Fig. 3.12).
- ➤ The angle of repose of loose sediment is a function of grain size. This relationship has been quantified in the

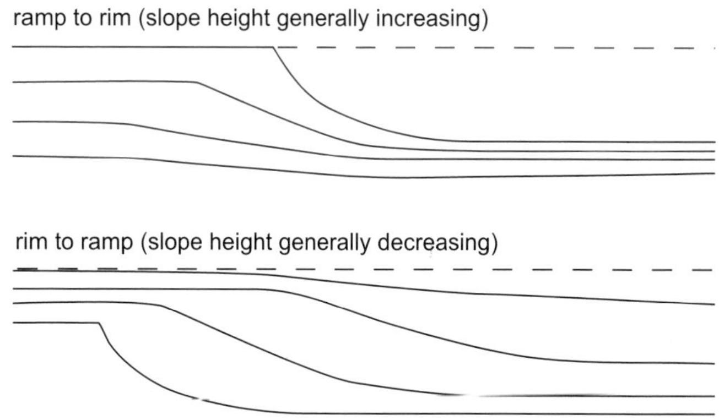

Fig. 3.8.— A) Transition from ramp to rimmed platform with sharp shelf break is commonly observed where a basin deepens and slopes become higher and steeper as they prograde. B) Transition from rimmed platform to ramp occurs where a basin fills up and the relief flattens.

Recent and the same numerical relationships have recently been shown to apply to large-scale, geological features of the past (Figs 3.13, 3.14, 3.15). The most important control on the angle of repose is the degree of cohesion of the sediment.

The change in depositional regime with changing slope angle is largely caused by the degree of sediment bypassing on the slope. The sediment source of most carbonate slopes is the platform. Thus, sediment enters the slope at the upper end and is distributed downslope by sediment gravity transport. At gentle slope angles, the competence and capacity of the transporting agents decreases steadily away from the sediment source at the platform margin. Consequently, sedimentation rates decrease and time lines converge basinward. At this stage, slope angle is a function of the rate of sediment supply and the grain size of the sediment (Fig. 3.13). As declivity increases, the vigor of sediment gravity flows also increases and significant volumes of sediment start to bypass the slope and come to rest on the adjacent basin floor. The geometric expression of a bypass slope are downslope gullies with erosional flanks, scours and gravel lags; intergully sediment is mainly mud with only few turbidites. Basinward of the gullied slope a sediment apron develops as a series of laterally coalescing turbidite fans (Schlager and Chermak, 1979; Mullins et al., 1984; Harwood and Towers, 1988). These aprons are comparable to the continental rises of deep ocean basins. They are dominated by turbidite sheets; erosional channels (fan valleys) are shallow and scarce. In a global survey, Heezen et al. (1959) suggested that the boundary between slopes and rises in the ocean typically lies at about 1.4° (tanS = 0.025). Carbonate data indicate higher variability but overall similar averages for the slope-rise boundary. (Fig. 3.16)

Steepening of slopes beyond the realm of bypassing pro-

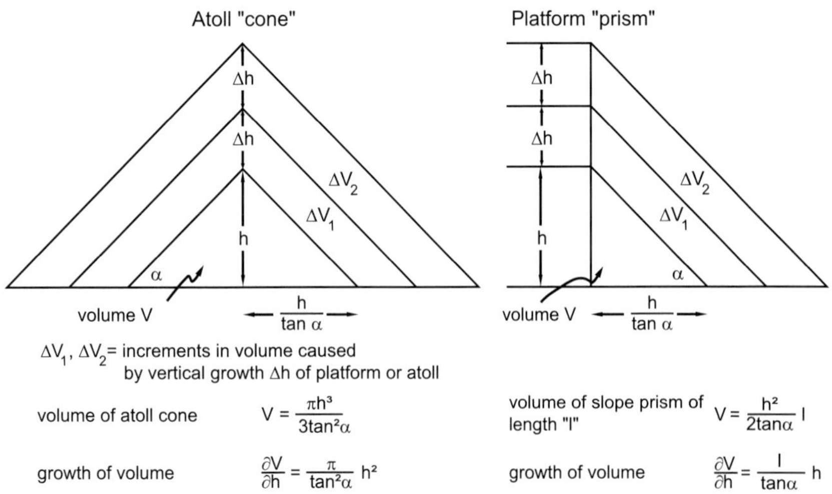

Fig. 3.9.— Upward growth of carbonate platforms with constant slope requires deposition of ever larger volumes of sediment on the flanks. In the case of a cone-shaped atoll, the growth of volume (V) is proportional to the square of the height of the cone. In case of a linear platform margin, the growth in volume of the slope prism, V, is proportional to the height of the platform. Schlager (1981), modified.

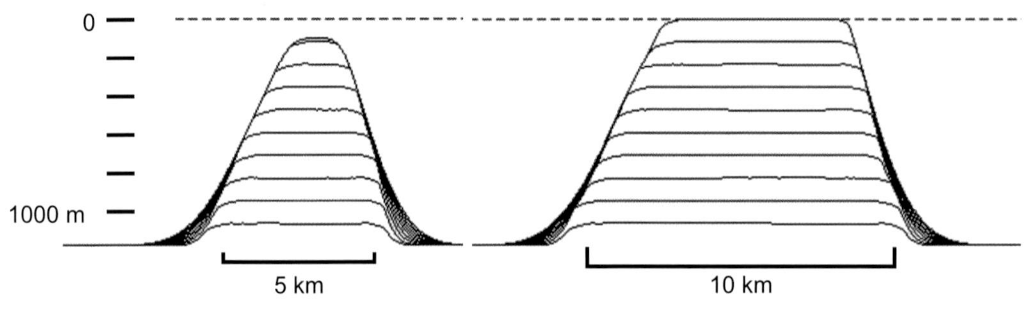

Run time = 400 ky, line spacing = 40 ky
Diffusion: $K_{nonmarine}$ = 100 m^2/y, K_{marine} = 1
Maximum production, C_1 = 7.835 m/ky at 0-5 m depth
Subsidence = 2 m/ky

Fig. 3.10.— Effect of increasing slope height on two atolls of different size, modeled by program STRATA (appendix B). Carbonate production of both atolls starts in area marked by black bars. Both atolls first aggrade, then retrograde as slopes get higher. Retrogradation reduces the production area at the top and leads to gradual drowning. Time of drowning depends on the size of the production area. After 400 ky, the large atoll still is healthy and its flat top built to sea level. The small atoll is virtually dead. Its top lies 100 m below sea level where production is so low that it no longer compensates for sediment loss by downslope transport – aggradation ceases and the flat top changes into a mound.

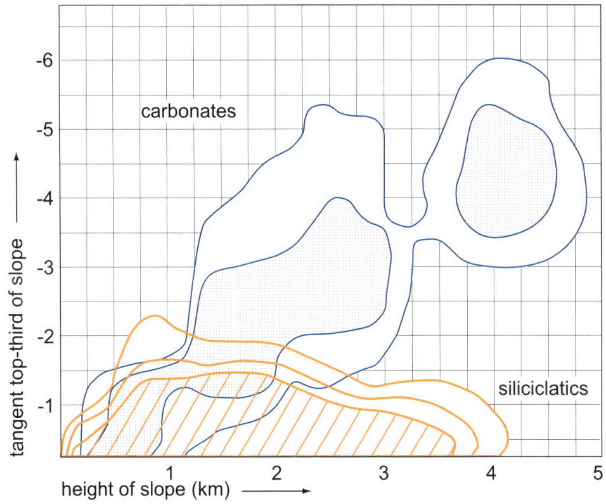

Fig. 3.11.— Modern submarine slope angles of carbonate platforms and siliciclastic systems in Atlantic and Pacific. Contours of 1, 2 and 4% of total sample (N) in unit area. N (carbonates) = 413, N (clastics) = 72. Carbonates steepen with height by building slump-resistant slopes. After Schlager (1989), modified.

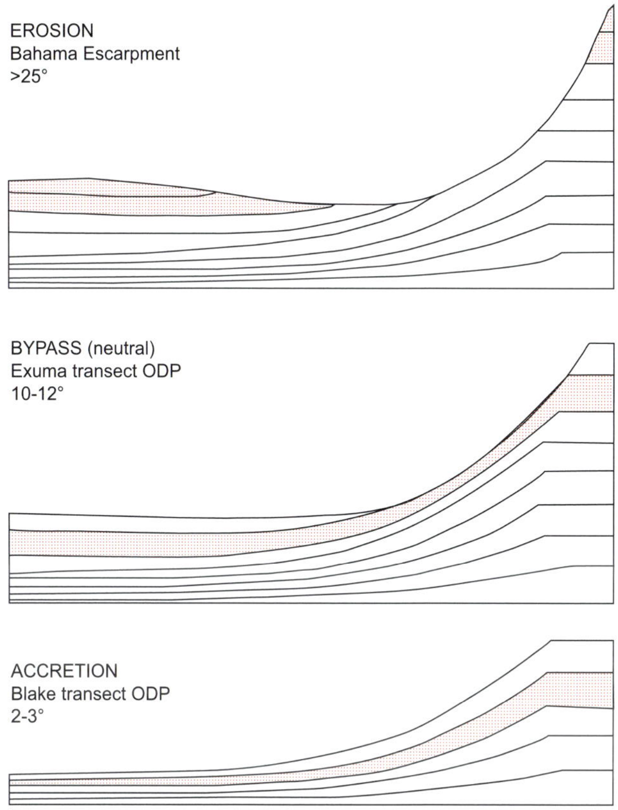

Fig. 3.12.— Slope angle and the balance of erosion and deposition on slopes. In the accretionary stage, slope angle is a function of rate of sediment supply and rate of relative sea-level rise. As slope angle increases, the vigor of turbidity currents also increases and changes the depositional regime from accretion to erosion. Bypass slopes represent an intermediate stage. They receive mud from the perennial rain of sediment, but large turbidity currents bypass them. After Schlager and Camber (1986), modified.

duces erosional slopes where the sediment budget of the slope is entirely negative: more material is exported by erosional turbidity currents and by slumping than the slope receives from the platform. The change from accretionary to bypass to erosional slopes with increasing declivity is well supported by observations on mud-rich slopes. It is unclear if slopes dominated by rubble and sand or slopes controlled by automicrite accretion change in a systematic way as slope angle increases. It is well established, however, that such slopes can maintain much steeper slope angles than the mud-rich ones (Figs 3.13, 3.14, 3.15).

Slope curvature

The curvature of many slope profiles has been shown to obey rather simple mathematical functions that provide clues to the transport regime on the slope (Kenyon and Turcotte, 1985; Adams and Schlager, 2000; Schlager and Adams, 2001). On reasonably smooth slopes, three types of curvature have been observed:

- Concave slopes normally fit a negative exponential function; this probably reflects the exponential decay with distance from source in the sediment volume moved by creep, slumping and sediment gravity flows.
- Straight-line profiles are normally steep and were found to correspond to non-cohesive sediment piled up to the angle of repose.
- Sigmoidal profiles are interpreted as a combination of an exponentially curved lower part and an upper, convex part that reflects the smoothing of the shelf break by storms and sea-level fluctuations (Fig. 3.17).

In carbonates, concave profiles that culminate in a very steep upper slope and sharp shelf break correspond to rimmed platforms with reefs or lithified sand shoals at the margin. Sigmoidal profiles, on the other hand, indicate predominantly detrital accumulations where reefs or early-lithified sands occur only as lenses. Straight-line profiles have been found to be controlled by lime sand or rubble at the angle of repose (Kenter, 1990; Adams et al., 2002). Straight slopes of the M factory may also contain significant portions of automicrite layers besides sand and rubble. These rigid layers of automicrite episodically slide and break up to form megabreccias at the toe-of-slope.

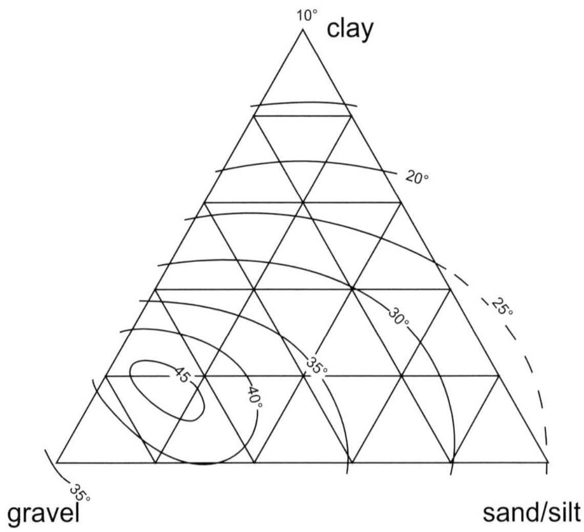

Fig. 3.13.— Grainsize vs. maximum stable slope angle of terrestrial, weathered slopes. Extrapolation to marine sediments is fairly well justified for clay-free material. Clay-rich material is highly sensitive to water content and pore water pressures; values shown here are for dry material. Stable angles for water-saturated clays are about one half of the dry values (measured by the tangent). Stable angles of clays with overpressured water are even lower. After Kirkby (1987), modified.

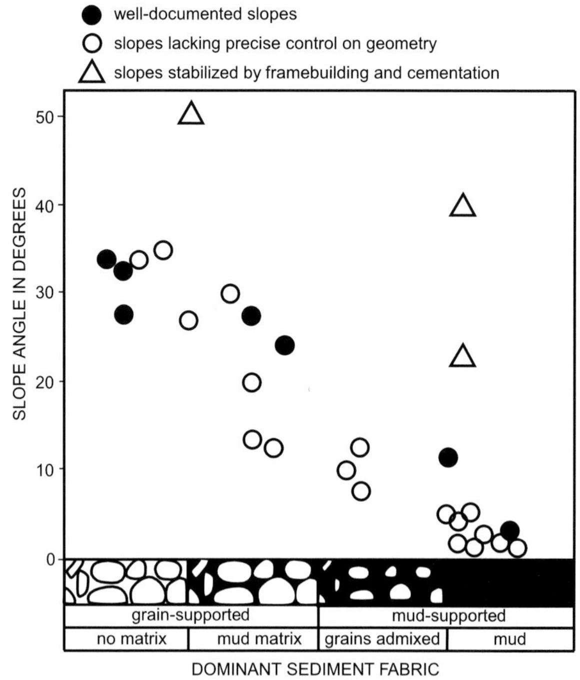

Fig. 3.14.— Sediment composition strongly influences slope angles of carbonate platform flanks. Cohesionless sediments, such as clean sand and rubble, build up to angles of over 40°. Muddy, cohesive sediments tend to develop large slumps that maintain a low slope angle. After Kenter (1990), modified.

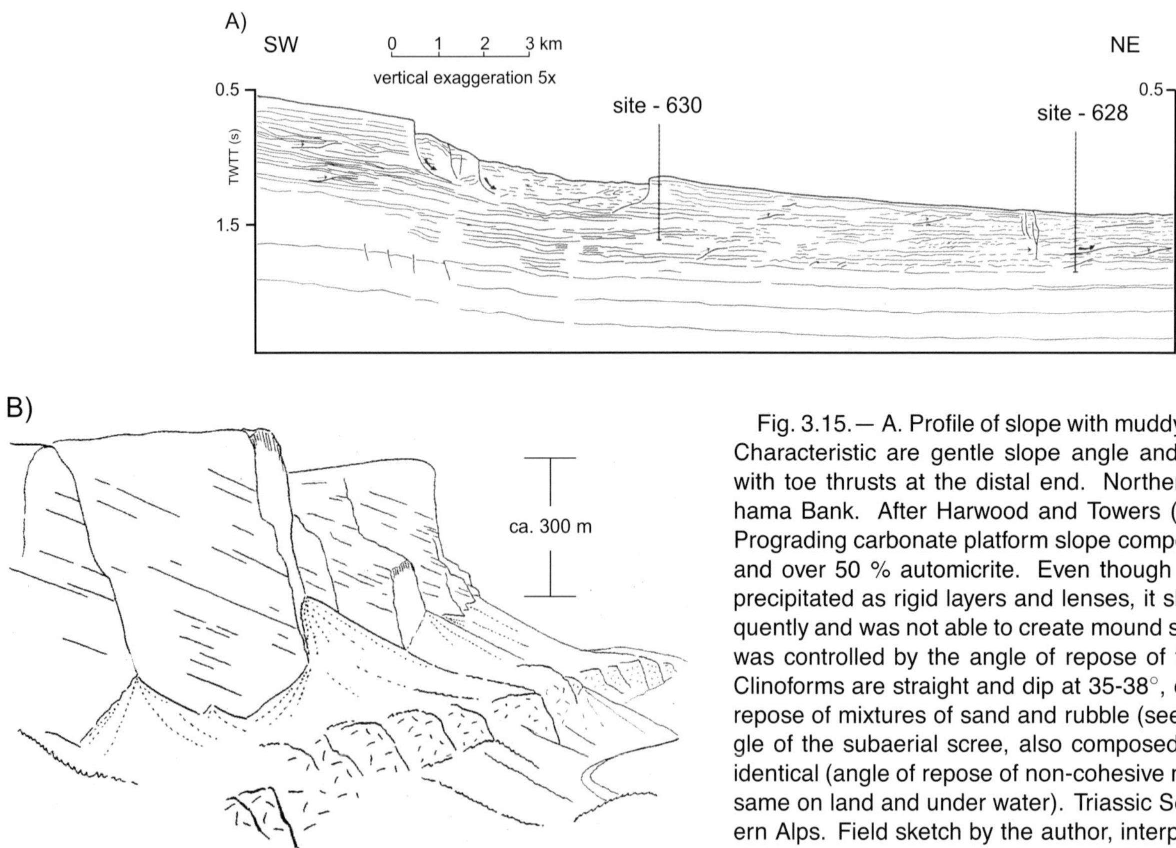

Fig. 3.15.— A. Profile of slope with muddy cohesive sediment. Characteristic are gentle slope angle and large-scale slumps with toe thrusts at the distal end. Northern flank of Little Bahama Bank. After Harwood and Towers (1988), modified. B. Prograding carbonate platform slope composed of sand, rubble and over 50 % automicrite. Even though the automicrite was precipitated as rigid layers and lenses, it slid and broke up frequently and was not able to create mound structures. The slope was controlled by the angle of repose of the detrital material. Clinoforms are straight and dip at 35-38°, close to the angle of repose of mixtures of sand and rubble (see Fig. 3.13). The angle of the subaerial scree, also composed of rubble, is nearly identical (angle of repose of non-cohesive material is nearly the same on land and under water). Triassic Sella Platform, Southern Alps. Field sketch by the author, interpretation after Kenter (1990).

Geometry of T, C and M factories

The three carbonate factories introduced in chapter 2 idffer not only by the pathways of precipitation but also by the geometry of the respective deposits. The main points are summarized in Figs 3.18, 3.19, 3.20, 3.21) and discussed in more detail below.

T factory. When left to its internal dynamics, the T factory will strive towards a platform shape, i.e. a flat top near sea level and a steep slope on the seaward side; only minuscule parts of the system will extend into the terrestrial environment. This characteristic geometry of tropical carbonate accumulations is directly related to the principles of production and destruction. A critical element in the carbonate edifice is a wave-resistant rim at the boundary of the wave-swept top and the slope shaped by gravity transport – hence the term "rimmed platforms".

The presence of a rim, often built to sea level, disrupts the seaward-sloping surface normally developed by loose sediment accumulations on a shelf. Rimmed platforms differ fundamentally from siliciclastic shelves in this respect. The growth anatomy of a rimmed platform is that of a bucket – a competent, rigid rim protects the loose sediment accumulation of the platform interior (Figs 3.2, 3.3, 3.8, 3.18, 3.19).

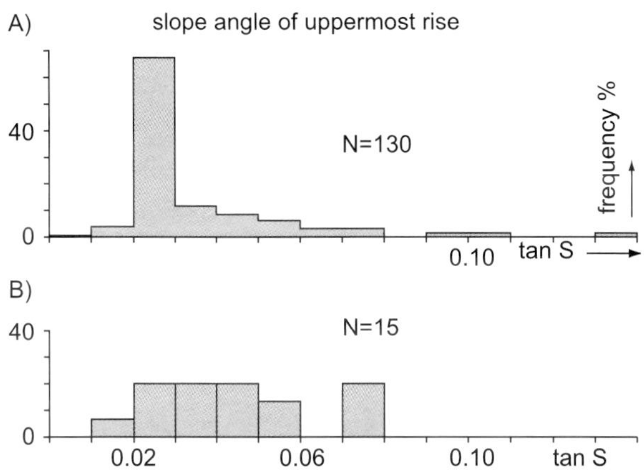

Fig. 3.16.— Boundary between slope and rise in siliciclastic and carbonate settings. A) Frequency distribution of slope angles at the upper limit of the deep-sea fans that constitute the continental rise. Note pronounced frequency maximum at tan S of 0.02 to 0.03 (1.1 – 1.7°), in agreement with the conclusions of Heezen et al. (1959). B) Analogous data from carbonate settings show much broader scatter but the lower end of the frequency distribution agrees with the siliciclastic data set. Based on seismic data from numerous sources.

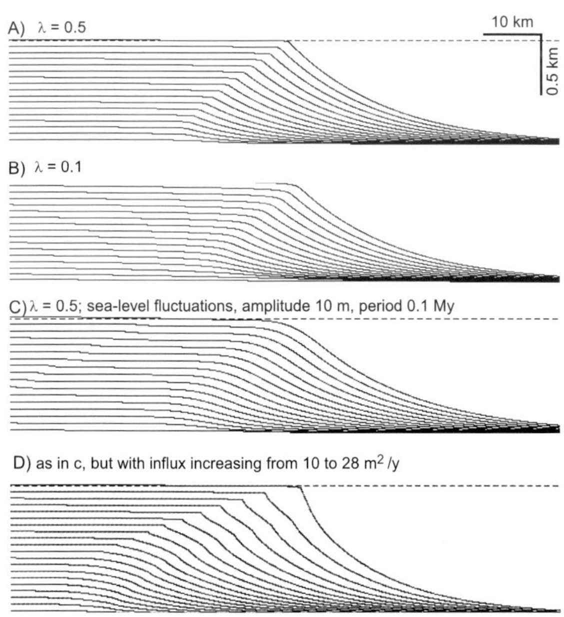

Fig. 3.17.— Exponential and sigmoidal slope profiles generated with program STRATA (appendix B). Basic setting: uniform subsidence of $1.67 \cdot 10^{-4}$ m/yr; sediment supply from left starts at 10 m^2/yr, increasing each My by 2 m^2/y; nonmarine diffusion coefficient $K_n = 1 \cdot 10^5$; marine diffusion coefficient K_m = 150. Spacing of time lines 0.3 My. A) Delta builds to sea level and progrades with exponentially curved slope and sharp shelf break because the diffusion decay constant, $\lambda = 0.5$; this means that the transition between the highly-diffusive terrestrial regime and the low-diffusive marine regime occurs in the uppermost $\sim$ 20 m of the water column. B) Setting diffusion decay constant, $\lambda = 0.1$, widens the transition zone to the upper $\sim$ 100 m. This change, tantamount to lowering storm wave base, results in rounded shelf break and a sigmoidal profile, particularly in the early stages of progradation. C) Diffusion decay constant is reset to $\lambda = 0.5$. Rounded shelf break and sigmoidal profile are now generated by small sea-level fluctuations. D) Sharp shelf break is restored by increased sediment flux. After Schlager and Adams (2001), modified.

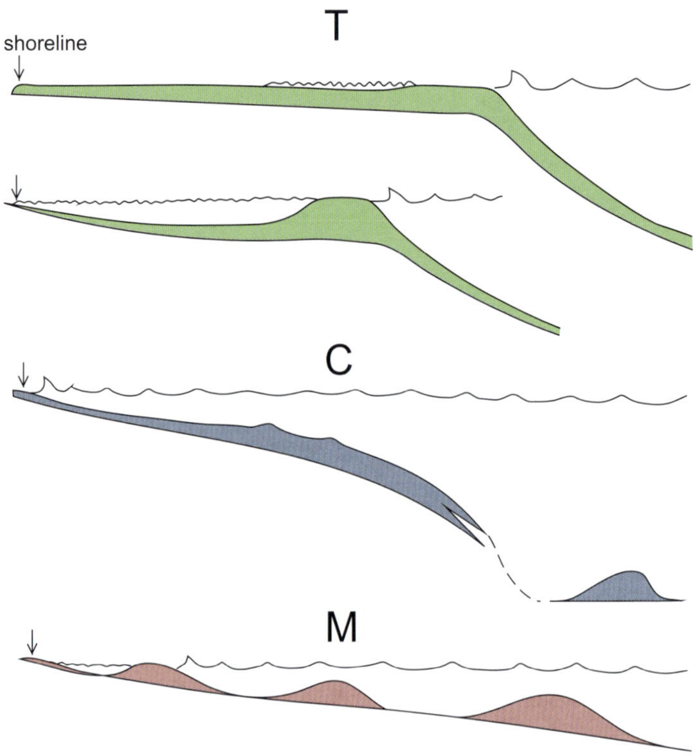

Fig. 3.18.— Cartoons of the depositional geometries of the three carbonate factories. The cross sections show sea level and wave patterns, bottom morphology and thickness variation of a typical growth increment. The arrow marks the shoreline. T factory (green) produces platforms rimmed by reefs or sand shoals; the factory exports much sediment of all grain sizes; consequently, slopes are steep and rich in shoal-water debris. C factory (blue) cannot build shallow offshore rims, only scattered deeper-water skeletal mounds. The geometry of the accumulations is that of a ramp with the highest energy conditions close to shore. M factory (red) forms convex mounds on gentle slopes below the zone of wave action. The mounds develop flat tops and caps of grainstones where they build into the zone of intense wave action. The flanks of mounds may be steeper than the maximum angle of repose of sand and rubble (about 42°) because automicrite cements and stabilizes the flanks.

Fig. 3.19.— The island of Bora Bora in the SW Pacific – an example of a reef-rimmed, tropical platform. The reef belt at the platform margin forms an effective barrier for oceanic waves, protecting a wide lagoon. High water energy on the ocean side and low energy on the lagon side of the reef causes reef debris to be shed preferentially into the lagoon. Filling of the lagoon by progradation of the reef aprons (light green) is much faster than filling with debris from the eroding volcanoe on the landward side. Consequently, the deepest parts of the lagoon are near the mountain. Photo courtesy of O. van de Plassche.

Fig. 3.20.— The Otway shelf of southern Australia, the landward end of a cool-water carbonate factory. Swell with wave lengths of 50 – 100 m breaks on the shoreface. On a tropical platform such waves (that feel bottom at 25 – 50 m) would be absorbed by the offshore rim of reefs or sand shoals. Photo by N.P. James.

Fig. 3.21.— Paleozoic mud mounds in the Algerian Sahara that formed on a ramp in an epeiric sea. The exquisitely preserved mounds in this photo are tens of meters high and over hundred meters in diameter. The rounded tops as well as biota and sediment facies suggest that even the crests lay below wave base. Photo by B. Kaufmann

Because of the higher growth potential of the rim, "empty buckets", i.e. raised rims and empty lagoons are commonly found in the geologic record. Retrogradation of rimmed platforms usually proceeds in discrete steps because once the rim is drowned much of the adjacent lagoon is also lost. Fig. 3.5 illustrates possibilities for backstepping platforms to establish a new "line of defense".

Platform morphologies are common with the T-factory but they are not diagnostic. If the T factory is weak, the M factory may occupy the niche and build platforms with flat tops (and rims) at sea level (Fig. 3.22).

C factory. Modern cool-water carbonates essentially lack the ability to build rims. Consequently, their characteristic depositional geometry is that of a ramp (Fig. 3.1). Many cool-water accumulations lie in zones of strong winds and high wave energy (such as the "hailing fourties" of the southern hemisphere, Fig. 3.20). Consequently, accumulation in the shallow part of the ramp may be low and hardgrounds abundant (see chapter 4).

M factory. The favorite setting for accumulations of the mud-mound factory are deeper-water environments in the thermocline with low lateral influx of sediment. When left to its own dynamics in this setting, the M factory will produce mounds, i.e. circular or elliptical, upward-convex accumulations that rise above the adjacent sea floor (Figs 3.4, 3.21). What sets these mounds apart from purely mechanical accumulations of muddy sediment is the abundant evidence for syndepositional lithification of the structure. This evidence consists primarily of microborings, cracks and large cavities filled by coeval sediment and clasts of automicrite. Additional arguments for pervasive lithification of the mound edifice are the steep dip (commonly 40-50°) of the micritic flank deposits, meter-size, angular boulders of automicrite facies at the toe-of-slope (Lees and Miller, 1995) and fractures filled by coeval sediment in automicrite deposits. The debris aprons of mud mounds are small. In most instances, the calcareous flank deposits of automicrite mounds extend only meters to few hundreds of meters into the flat-lying basin deposits surrounding the mound (e.g. Lees and Miller, 1995).

In settings with high external sediment input, the mud-mound factory may be unable to create mounds. One such situation are steep platform slopes where automicrite layers have been found to alternate with layers of platform debris. The mud-mound factory betrays its presence by large boulders of automicrite boundstone that accumulate at the toe-of-slope. (e.g. Russo et al., 1997).

The M factory has no monopoly on mound geometries even though they are very common in this factory. Algal mounds of the T factory (e.g. Roberts et al., 1988) and bryozoan mounds of the C factory (e.g. James et al., 2000) illustrate that the mound geometry is not diagnostic of a factory.

Fig. 3.22.— Sella Mountain in the Southern Alps, Italy, a carbonate platform created by the M factory. Top was built to sea level as documented by abundant supratidal deposits. Slope consists of a 1:1 mix of automicrite and cement *in situ*, and reworked sand and rubble. Note planar slope clinoforms built to over 35°, the approximate angle of repose of this material. Interpretation based on Keim and Schlager (2001), photo by J.A.M. Kenter.

THE EMPTY BUCKET

At the beginning of this chapter it was pointed out that rimmed platform have the anatomy of a bucket, composed of a stiff rim of stacked reefs or lithified sand shoals, and an interior of loose sediment (Ladd, 1973; Schlager, 1981). The rim is not only the stiffer, more competent element of the system, it is also more productive than the platform interior. This can be shown by sediment accumulation rates (Fig. 2.22) but also by sediment anatomy: Sand shoals and reefs at the platform margin commonly shed excess sediment into the lagoon in the form of a debris apron that progrades away from the rim (Fig. 3.19). One also observes rims that keep up with rising sea level while the lagoon becomes submerged to near the limits of the photic zone. This selective upbuilding leaves the platform bucket partly empty.

The origin of raised rims and deep lagoons by selective upbuilding of the reef rims goes back to Charles Darwin, who included this principle in his atoll model (Darwin, 1842). An alternative explanation of empty-bucket morphology is not quite as old. It postulates that the morphology of modern atolls is the result of differential karst weathering during lowstands of sea level (Ladd and Hoffmeister, 1945; MacNeil, 1954; Purdy, 1974; Winterer, 1998; Purdy and Winterer, 2001). Both hypotheses rely on theoretical concepts as well as supporting observations. The discussion in the literature is in full swing and, as the empty bucket is a characteristic carbonate growth form, the state of the argument is reviewed below.

The conceptual basis of the construction hypothesis is the measurably higher growth potential of the rim and the disposition of reefs to grow upward rather than migrate laterally. The observational support is briefly summarized in Figs 2.22, 3.23 and 3.24. Sea-level rises during the deglaciations of the Late Quaternary were near the limit of the growth potential of the modern T-factory; several well-documented successions of drowned reefs from the deglaciation periods attest to that (see Montaggioni, 2000, for review). Thus, it is not surprising that raised rims and deep lagoons are common among extant carbonate platforms. The rim-to-lagoon relief often exceeds hundred meters and normally represents the cumulative effect of several glacial-interglacial cycles. Composite records of this type are only shown here if the contribution of one phase, e.g. the Holocene, can be clearly identified. Numerous examples in the Holocene and the distant past show how reefs and, less frequently, sand shoals, rose above the platform interior by differential upbuilding (e.g. Belopolsky and Droxler, 2004). Commonly, high rims represent the last growth increment before platforms drowned. A variation on this theme is the record of the Holocene transgression on seaward-dipping shelves: barrier reefs were repeatedly established and drowned, creating a record of raised rims without lagoon fill (Fig. 3.24). This pattern clearly illustrates the tendency of reefs to stand tall (and die) rather than gradually retreating upslope.

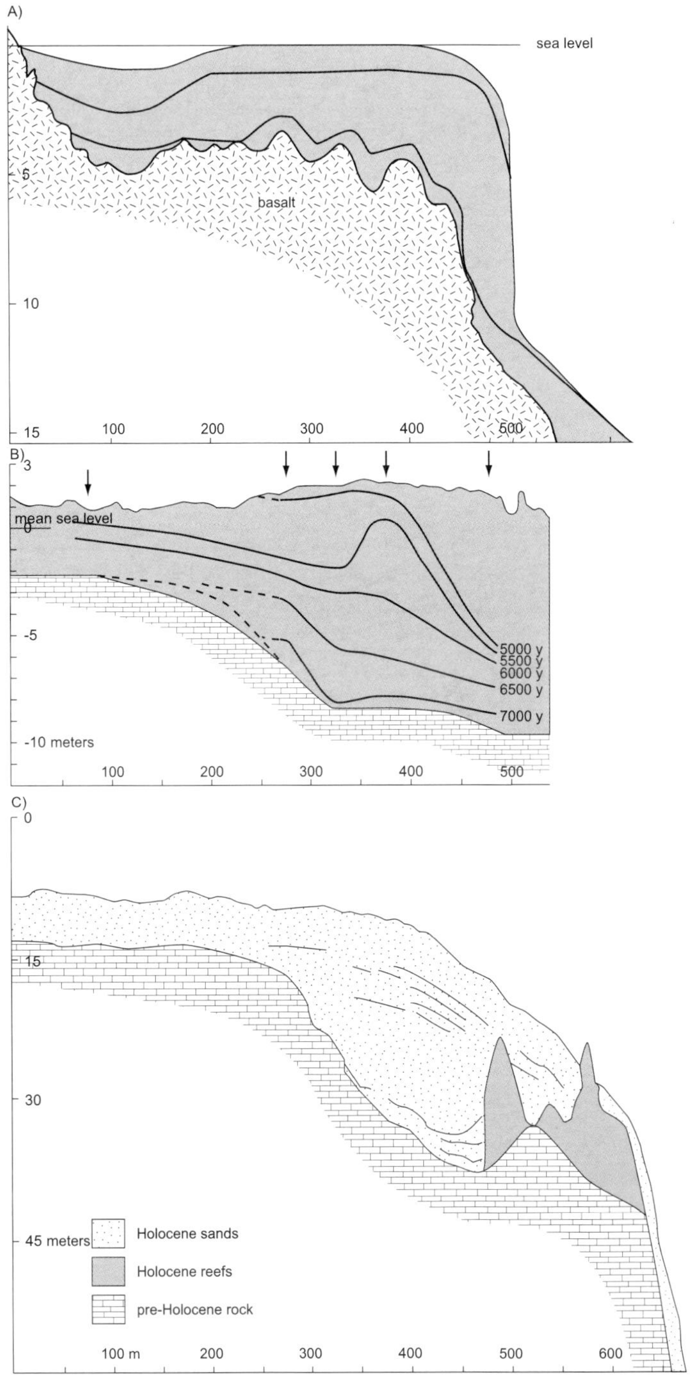

Fig. 3.23.— Bucket morphology of Holocene reefs created by differential growth, compiled from Cortes et al.(1994), Takahashi et al. (1988), Hine et al. (1983). A) Pacific fringing reef Punta Islotes, Costa Rica. Growth history constrained by 13 bore holes and radiocarbon dates. B) Kume Island, Ryukyu archipelago, western Pacific. Reef growth constrained by 5 boreholes and numerous radiocarbon dates. Empty bucket developed 5500 y ago and was subsequently filled by reef debris. Modern karst erosion (top surface) is also accentuating the reef rim. C) Raised reef that was drowned and subsequently buried by carbonate sands, Great Bahama Bank. Line drawing of seismic profile.

The karst hypothesis has a conceptual basis in what one may call the "edge effect of dissolution": a fluid runs off the edge of a horizontal surface faster than from the center of this surface. In his dissolution experiments on limestone blocks, Purdy (1974) observed that – as the rate of acid rain was gradually reduced – , the acid film on the limestone was pierced first at the edge. This reduced the etching time at the edge and produced elevated rims on orginally flat surfaces of homogeneous limestone. A second principle that has been invoked but has not been demonstrated in this particular application, is case hardening: extra cementation of the steep outer cliffs of limestone islands by evaporation of mineral-bearing water.

Observational support for the karst hypothesis is the discovery of Purdy and Winterer (2001) that on Pacific atolls lagoon depth positively correlates with rainfall.

Geologic field observations provided a number of examples of karst islands with raised rims and low centers (see Purdy and Winterer, 2001 for summary). The problem with most of them is the lack of detailed maps of the limestone facies. Thus, it is generally impossible to determine whether the ridges were carved from homogeneous limestone as in the laboratory experiments of Purdy (1974), or whether karst dissolution merely accentuated concentric patterns of depositional facies. An example of the latter is the well-mapped island of Makatea (Fig. 3.25; Montaggioni et al., 1985). Maps and sections leave little doubt that karst weathering has significantly enhanced the relief between rim and lagoon but the data also show that what now constitutes the morphologic rim was originally a barrier reef with its debris apron.

Purdy and Winterer (2001) offer the raised Miocene limestones of Tuvutha (Fiji) as a particularly convincing example of concentric ridges carved from homogeneous limestone. Indeed the entire limestone cap was mapped as one formation, mainly composed of algae and foraminifers (Ladd and Hoffmeister, 1945). However, the mapping was done in the 1920's, before modern reef research had demonstrated that the geologic record of most reefs is a pile of rubble. Consequently, Ladd and Hoffmeister (1945) emphasized the lack of outcrops or samples with *in-situ* framework of corals. They did report coral fragments to occur regularly (Ladd and Hoffmeister, 1945, p. 216-217). Furthermore, the authors interpret the vertical walls on the seaward side of the ridges as sea-cliffs formed during the gradual uplift of the island. Thus, the present data do not exclude the possibility that the ridges are remnants of platform margin facies just like on Makatea.

The dissolution experiments of Purdy (1974) were performed on tight limestones (erroneously referred to as "marble" by Schlager, 2003, p. 451). Thus, the experiments demonstrate that a saucer shape is created if the acid runs off the edges rather quickly but collects in the center of the sample surface and continues to lower the surface there. In contrast, much of the dissolution caused by rainwater on young carbonate rocks takes place below the surface because the rocks are highly porous and permeable. Examples of well-preserved depositional morphology after nearly 100 ky of karst weathering (Fig. 2.34) suggest that extrapolation of the laboratory experiments to natural settings needs to proceed with caution.

The karst interpretation of raised rims on Cretaceous platforms in the Pacific has created a lively debate. The hypothesis was advanced by Winterer et al.(1995), Winterer (1998), Van Waasbergen and Winterer (1993) and Purdy and Winterer (2001). It faces the problem that there is practically no isotopic evidence of meteoric diagenesis. This would imply that during millions of years of exposure, carbonate rocks were extensively dissolved but no meteoric cements precipitated in these formations – an assumption that runs counter to extensive observations on Quaternary rocks (Jenkyns and Wilson, 1999). Erosional morphology is apparent in seismic lines and sea-floor maps. However, similar patterns can be observed in the overlying pelagics and the stratigraphy and facies of the pelagic caps indicate extremely low sedimentation rates and numerous hiatuses – in agreement with the widespread evidence for current-enhancement by sharp topography (see Figs 2.26, 2.25).

In summary, I conclude that the formation of raised rims by differential growth can be demonstrated by a number of examples from the Holocene and the distant past. The theoretical underpinning for this phenomenon is the measurably higher growth potential of the rim and the disposition of reefs to grow upward rather than shift laterally to more favorable locations. Regarding the karst hypothesis, I think differential lowering of karst surfaces during lowstands has been shown to at least significantly enhance the relief between rim and lagoon. Purdy (1974) has demonstrated that in laboratory experiments raised rims can be created on horizontal surfaces solely by dissolution without lateral heterogeneities in the limestone. Field observations in support of this extreme form of the karst hypothesis are scant and not compelling. However, the discussion should be continued. It forms an important part of the broader issue of the effects and vestiges of exposure in carbonate rocks.

Constructional "empty buckets" observed in outcrop or seismic data indicate that the platform was stressed by a relative sea-level rise that closely approached the growth potential of the factory. It remains open, of course, if the sea-level rise was exceptionally fast or if the platform's growth potential was reduced by environmental factors.

The bottomline of the discussion is that both hypotheses, selective construction and karst weathering, can be counted on to produce raised rims and deep lagoons. In fact, at times of large and rapid sea-level fluctuations both processes are likely to alternate. The record of the Late Quaternary indicates that the rates of sea-level rise during deglaciations approached the growth potential of healthy modern reefs. Constructional empty buckets are a likely result. On the other hand, differential karst weathering almost certainly occurred during glacial lowstands. Both conditions must have alternated repeatedly in the past 700 ky, creating a composite of constructional and karst effects. One task before us is to discriminate between the constructional and the karst hypothesis in the distant geologic past and determine the

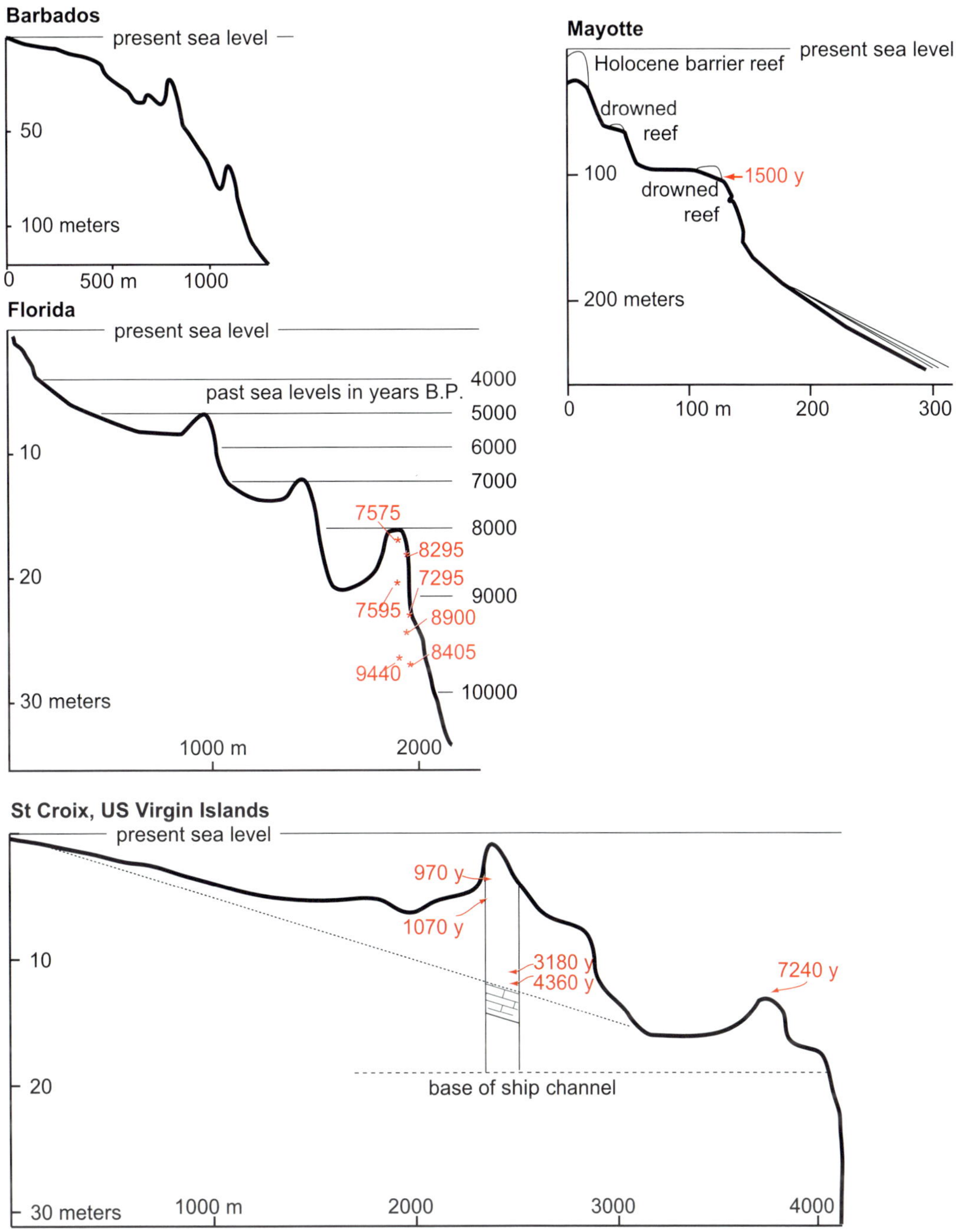

Fig. 3.24.— Reefs that were outpaced and drowned by the Holocene transgression show the raised-rim morphology in pristine form, illustrating the strong inclination of reefs to grow upward rather than gradually shift sideways upslope. Compiled from Macintyre et al. (1991), Barbados; Lighty et al. (1978), Florida; Adey et al. (1977), St. Croix; Dullo et al. (1998), Mayotte.

balance between the effects of differential upbuilding and selective dissolution. This balance almost certainly varied in space and time.

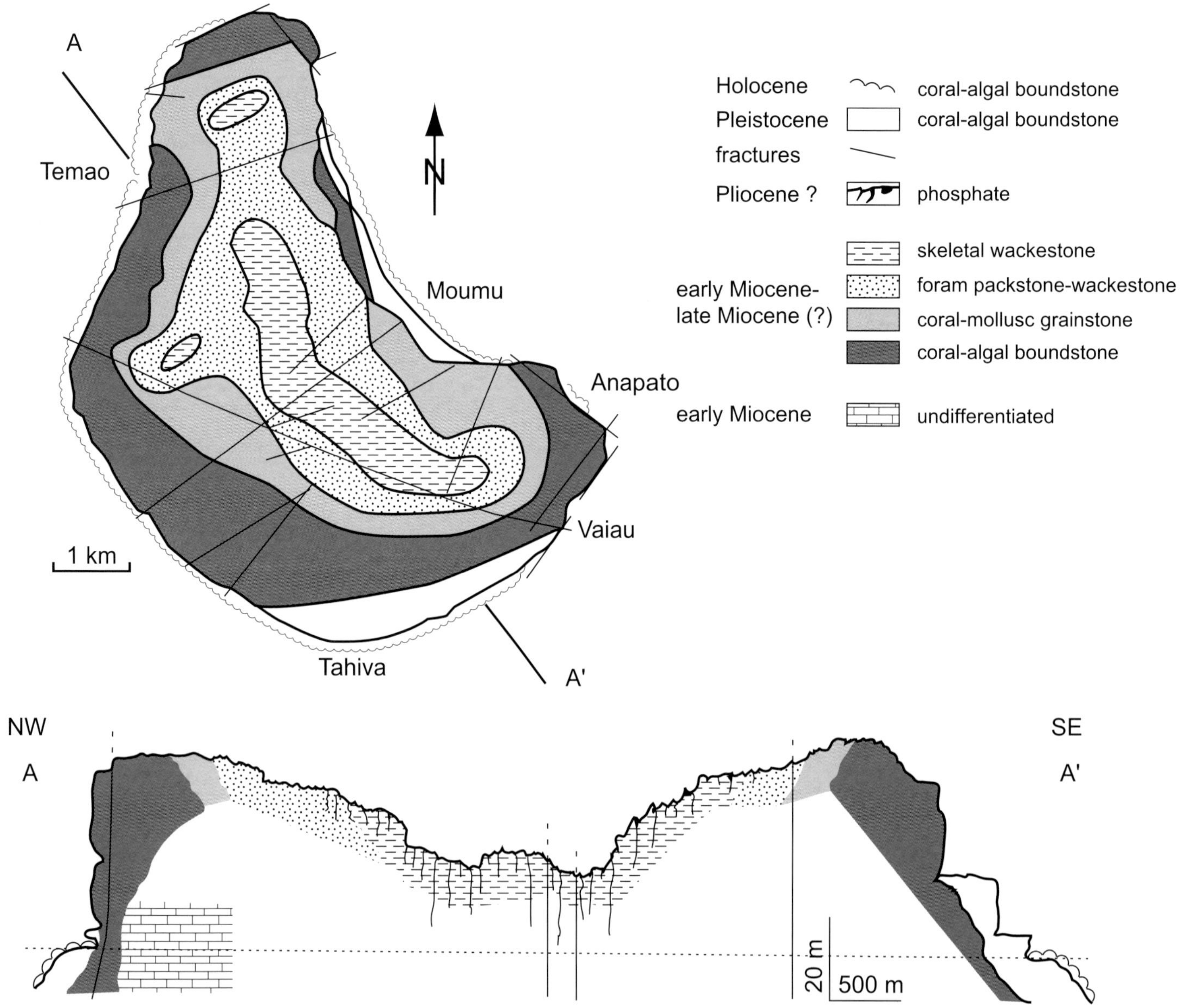

Fig. 3.25.— Island of Makatea in the SW Pacific is an uplifted Neogene atoll with some Holocene-Pleistocene reefs attached on the sides. The saucer-shaped topography of the island is largely due to karst processes. However, terrestrial dissolution is selective and closely follows the facies pattern of the original atoll – the present topography was not cut out from a homogenous limestone formation. After Montaggioni et al. (1985), modified.

CHAPTER 4

Carbonate facies models

Carbonate rocks often overwhelm the untrained eye by a bewildering variety of textures, structures and grain types. Patchy diagenesis adds to the impression of almost chaotic diversity and irregularity. Upon closer inspection, the situation is not nearly as bad. If carbonate sediments are characterized by sedimentary structure, texture and grain kind, a recurring succession of facies belts can be recognized in shore-to-basin transects. These facies appear throughout the Phanerozoic and with only slight modification also in the late Precambrian. This surprising persistence indicates that the evolution of organisms in this time interval had only a modifying effect on the basic carbonate facies. The standard carbonate facies seem to capture trends dictated by other parameters such as the carbonate growth function, i.e. the distribution of growth rates as a function of depth and distance from shore, the degree of protection from waves and tidal currents, and the degree of restriction in the water exchange with the open sea. On the slopes, the declivity and the the balance between sedimentation and erosion are crucial controls. These principles are discussed in the next section, followed by a presentation of facies on ramps and rimmed platforms.

UNDA, CLINO AND FONDO ENVIRONMENTS

One of the most fundamental classifications of depositional environments and facies is the subdivision by Rich (1951) into unda (shallow, wave-swept), clino (slopes shaped by gravity transport), and fondo (basin-floor) environments. Rich (1951) proposed the suffix "-form" for the morphologies associated with these environments and the suffix "-them" for the respective deposits. In practice, the distinction between morphology and sediment body has rarely been made. I will use the terms undaform, clinoform and fondoform for the deposits and the morphology inferred from these deposits. Rich's (1951) classification is broader (but analogous) to the subdivision of deltas in topset, foreset and bottomset beds (Barrell, 1912). The definitions of Rich (1951) apply to all depositional systems where transport of particulate sediment is important. In carbonate rocks, the expression of the unda, clino and fondo domains varies somewhat among the three factories.

The T factory produces nearly all its sediment in a narrow depth range that normally extends only tens of meters down from sea level. The seaward perimeter of this highly productive zone is often protected by an elevated, wave-resistant rim. This production system generates a platform geometry with a particularly flat unda environment swept by waves and tidal currents, and a rapid transition, across the rim, into the clino environment. The clinoforms can be much steeper than in siliciclastics (Fig. 3.11). The clino environment passes basinward into the fondo environment of the flat basin floors. Both clinoforms and fondoforms may contain abundant slumps and debrites, often with meter-size clasts. Where the M factory replaces the T factory, for instance after major extinctions, it builds platforms with the same unda, clino and fondo differentiation as the T factory. In its typical development, the M factory lacks the unda facies. The mud-mounds are upward-convex constructions that are not planed by waves and form in the clino environment, even though the slope declivity may be very low (ramp setting).

In the C factory, the ability to build rims is weak and the facies pattern resembles that of siliciclastic sediments. The unda environment is a seaward sloping shelf that gradually bends down into the clino domain. Sediments reflect these gradual changes. Slumps and debrites are scarce, as are large clasts. Deep-sea accumulations of the C factory lie in the path of contour currents and are streamlined by these currents.

In the unda environment of carbonate platforms, two parameters control the further subdivision of depositional environment and facies – the degree of *protection* from waves and currents, and the degree of *restriction* in the exchange of water with the open sea. Both parameters are related to a third characteristic of the system – the elevation and continuity of the platform rim. The rim index (p.42) provides a measure of the fraction of wave energy that passes through the rim. Sediments deposited behind the rim obviously "feel" this energy flux.

In the clinoforms, the most important control on both geometry and facies is the balance between sediment input from above and sediment output onto the basin floor. Variations in the material balance lead to the subdivision into accretionary, bypass and erosional slopes decribed in chapter 3.

FACIES PATTERNS – FROM RAMP TO RIMMED PLATFORM

Chapter 3 treated the shore-to-slope profile of shoal-water carbonate accumulations as a spectrum with two end members – the seaward-dipping profile in equilibrium with wave action and the rimmed platform with horizontal top, wave-resistant rim and steep slope. The seaward- sloping equilibrium profile is strictly analogous to that of siliciclastic systems whereas the morphology of rimmed platforms is peculiar to tropical carbonates the T factory and to the M factory where it builds to sealevel. The same situation exists with regard to depositional facies.

Tucker and Wright (1990, p. 49) pointed out that the succession of textures and sedimentary structures on carbonate ramps can be directly matched with siliciclastic shelf models. Fig. 4.1 depicts the most simple siliciclastic shore-to-shelf model. It extends from the beach to the shelf break and assumes onshore wind and associated wave action but no longshore currents by tides or oceanic circulation. Wind and waves induce a strong onshore currents at the surface and a weaker (but thicker) return flow near the bottom. Major facies boundaries are at sea level, at fair-weather wave base and at storm wave base. The first-order facies belts are nearshore sand and offshore mud but the episodic activity of storms creates a broad zone of interfingering of shelf mud and nearshore sand.

Depositional environments on simple carbonate ramps are analogous to Fig. 4.1. Fig. 4.2 shows the principal environmental subdivisions on a carbonate ramp as conceived by Burchette and Wright (1992). The analogy with the silciclastic systems in Figs 4.1 is obvious. Individual cases may differ from this most simple model by the effects of tidal currents flowing parallel or oblique to shore, and by the presence of biotic constructions such a mud mounds, shell beds and the likes. We shall return to the details of ramp facies after examination of the better known facies belts of rimmed platforms.

There is no siliciclastic analogue for rimmed platforms. However, the facies pattern of rimmed platforms can be de-

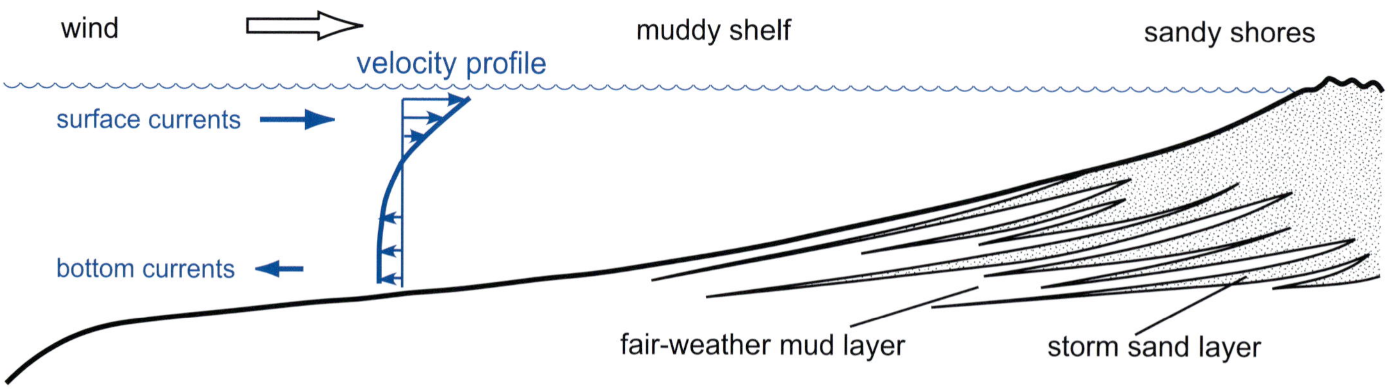

Fig. 4.1.— Model of a "tide-less" sea, supplied with sand and mud and dominated by onshore winds that set up a strong, shoreward current at the surface and a weaker return flow below. Wave action creates a nearshore zone of sand that interfingers with offshore muds. In the zone of interfingering mud accumulates during calm weather, sand during storms. After Allen (1982), modified.

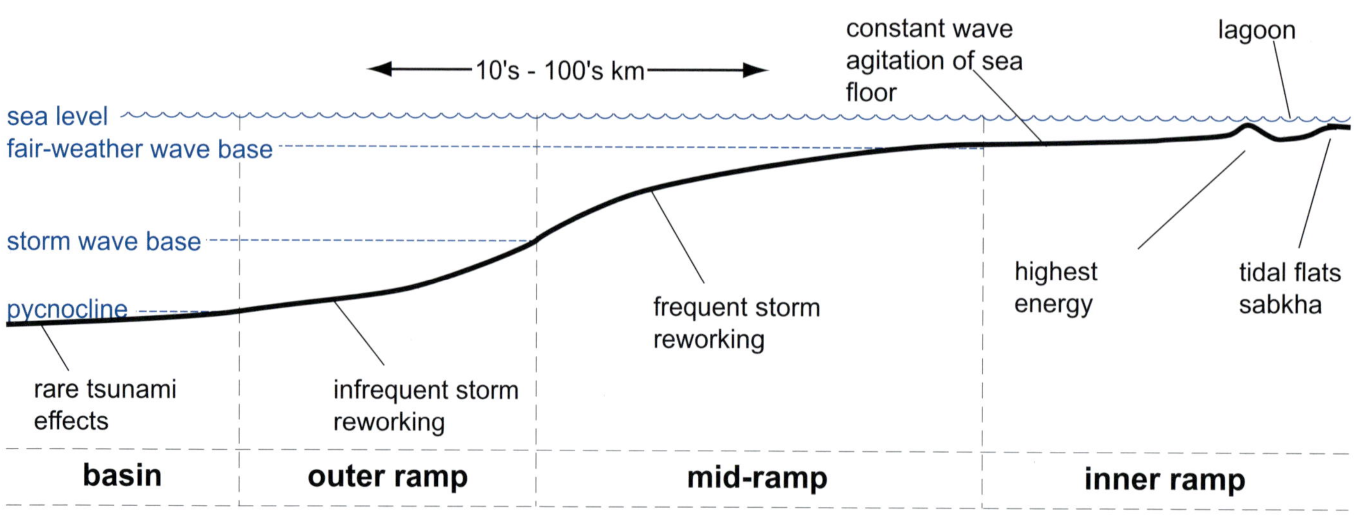

Fig. 4.2.— Carbonate ramp setting according to Burchette and Wright (1992). Note similarity with the beach-shelf model of Fig. 4.1.

rived from the siliciclastic beach-shelf model. All one needs to do is insert a wave-resistant rim (composed of reefs or partly lithified sand shoals) and adjust the facies for the effects of this rim.

The morphologic transition from ramp to rimmed platform is characterized by a gradual increase in slope and an accentuation of the platform margin. It has been observed in seismic data (e.g. Sheriff, 1988, Fig. 1; Harris and Saller, 1999,p. 40), in correlated boreholes (e.g. Burchette and Wright, 1992, Fig. 10) and in outcrops (e.g. Stanton and Flügel, 1989; Kerans and Tinker, 1999). The observational data do not constrain the transition in all details but some general trends emerge.

It seems that the formation of a rim normally starts at the inner ramp in one of two ways: (1) growth of patch reefs that form in the inner ramp seaward of the nearshore sand and subsequently coalesce to a shore-parallel belt; (2) the high-energy shoal itself becomes an quasi-stationary barrier by high sediment production and syndepositional lithification. Both reefs and sand shoals are likely to produce more sediment than they can store, they therefore fill the lagoon and then prograde seaward. As the crest remains a sea level but the ramp surface dips basinward, a slope gradually forms and its declivity increases with increasing height (Fig. 3.8). When should we call this prograding system a rimmed platform? If the geometry is well preserved and slope angles well measurable, such as in many seismic data, I recommend drawing the boundary at approximately 1.5°. Systems with a barrier belt and a foreslope distinctly greater than 1.5° should be called rimmed platforms (see chapter 3). If slope angles are not available, facies may serve as a proxy for morphology. On rimmed platforms, the rim belt should be reasonably continuous (rim index of ≥ 0.25) and its seaward slope should show evidence of slumping or of bypassing by sediment gravity flows, as indicated by gullies or canyons, increase in thickness and abundance of debrites and turbidites at the toe of slope etc.

In the climax stage of platform evolution, the wave-resistant rim sits at the platform edge, directly atop the slope. This position is rather stable because of the dynamics of platform growth: a rim that originally forms in a more bankward position will rapidly prograde if it produces more sediment than is needed to match relative sea-level rise; such excess production is likely considering the high productivity of rims. At the platform edge, the rate of progradation slows down because the high slope requires much larger sediment volumes for the same amount of progradation (Fig. 3.9). Thus, further progradation will be slower but the slope will tend to steepen to the angle of repose because of the high sediment supply from the productive rim. On the platform side, the rim sheds excess sediment ranging in size from clay to boulder. Beyond the reach of rim debris, an open lagoon may develop with a bottom profile in equilibrium with wave action. However, at the climax stage of platform evolution, this lagoon is largely filled and replaced by tidal flats that expand seaward almost to the platform rim.

FACIES BELTS OF THE T FACTORY

The succession of facies on rimmed tropical platforms has been cast into a standard facies model by Wilson (1975). The model, based on two decades of case studies by numerous researchers, has passed the test of time. It has become a widely accepted framework for presentation of carbonate facies (e.g. Tucker and Wrigth, 1990; Wright and Burchette, 1996; Flügel, 2004) and is reproduced in Fig.4.3. Despite the remarkable success of the model, some modifications are called for. They are shown in Fig. 4.4 and discussed below.

The standard model is "overcomplete". It contains more facies belts than one normally finds on any one platform. Platforms with a reduced number of facies belts may be perfectly normal and healthy. In the clinoform and fondoform settings, facies 2 (deep shelf) will only be present in addition to facies 1 if the platform has recently backstepped or has been structurally deformed. If a deep shelf exists, the structure of the lithospere virtually dictates that it be connected with the deep basin floor of facies 1 by a slope. In epeiric seas, the facies succession may terminate with facies 2. In the undaform setting, the rim need not consist of facies 5 plus 6. Many healthy platforms have either facies 5 (reefs) or facies 6 (sand shoals) as a rim.

In one instance, the Wilson model lacks a facies belt. Facies 9 of Wilson (1975) was defined for arid settings only. However, the equivalent deposits in humid climates have been described and are added here as facies "9-humid". It should be noted that facies 9-arid and 9-humid are alternatives that will not occur side by side in one shore-to-basin succession.

Another characteristic of the Wilson model is that it uses a discrete horizontal scale with sharp boundaries between facies. In nature, these facies boundaries may be gradational and irregular. For instance, the subdivision of slopes in facies 3 and 4 is often impossible and a combined belt 3/ 4 may be more appropriate. The boundary between facies 7 and 8 is often very gradual. In these instances it is preferable to designate a combined facies belt 7/ 8 and express increasing restriction by biotic indices or a larger number of subfacies.

The standard model says nothing about windward-leeward differentiation. Most platforms develop asymmetries in response to dominant wind directions (Fig. 4.9 and chapter 7). Seismic surveys reveal these asymmetries better than most other techniques.

Discription of the standard facies belts

1A) Deep Sea. Setting: Below wave base and below euphotic zone; part of deep sea, i.e. reaching through the thermocline into the realm of oceanic deep water. Sediments: Entire suite of deep-sea sediments such as pelagic clay, siliceous and carbonate ooze, hemipelagic muds including turbidites; adjacent to platforms we find mixtures of pelagic and platform-derived materials in the form of peri-platform oozes and muds. Biota: Predominantly

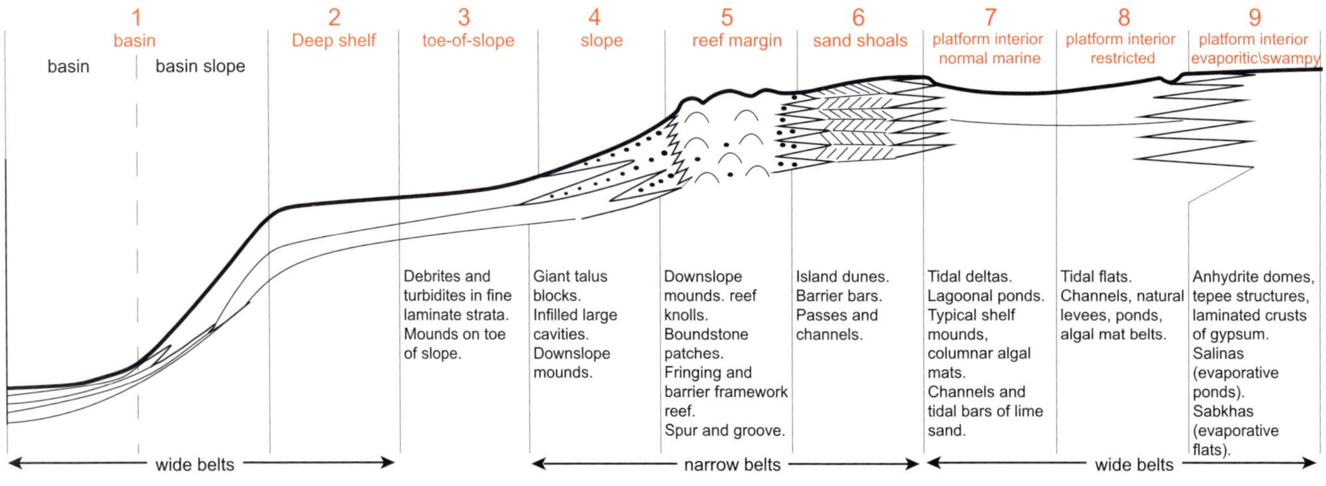

Fig. 4.3.— Synopsis of standard facies belts displaying name and number of facies belt (red), cross-section with large-scale sediment geometry and list of fine-scale features in each belt. After Wilson (1975), modified.

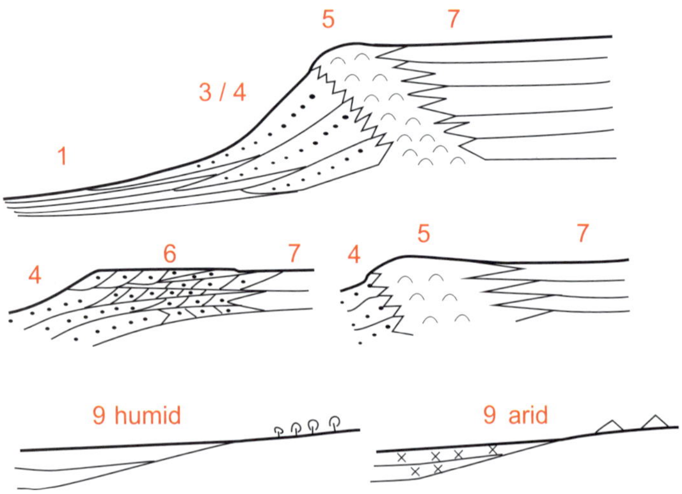

Fig. 4.4.— Commonly observed modifications of Wilson's (1975) standard facies in shore-to-basin transects. After Schlager (2002), modified.

plankton, typical oceanic associations. In peri-platform sediments up to 75% shallow-water benthos.

1B) Cratonic deepwater basins. Setting: Below wave base and below euphotic zone but normally not connected with the oceanic deepwater body. Sediments: Similar to 1A but in Mesozoic-Cenozoic rarely ever pelagic clay; hemipelagic muds very common; occasionally anhydritic; some chert; anoxic conditions fairly common (lack of bioturbation, high organic content). Biota: Predominantly nekton and plankton, coquinas of thin-shelled bivalves, occasionally sponge spicules.

2) Deep shelf. Setting: Below fair-weather wave base but within reach of storm waves, within or just below euphotic zone; forming plateaus between active platform and deeper basin (these plateaus are commonly established on top of drowned platforms). Sediments: Mostly carbonate (skeletal wackestone, some grainstone) and marl, some silica; well bioturbated, well bedded. Biota: Diverse shelly fauna indicating normal marine conditions. Minor plankton.

3) Toe-of-slope apron. Setting: Moderately inclined sea floors (>1.5°) basinward of a steeper slope. Sediments: Mostly pure carbonates, rare intercalations of terrigenous mud. Grain size highly variable; typical are well-defined graded beds or breccia layers (turbidites or debris-flow deposits) intercalated in muddy background sediment. Biota: Mostly redeposited shallow-water benthos, some deep-water benthos and plankton.

4) Slope. Setting: distinctly inclined sea floors (commonly 5° to near-vertical) seaward of platform margin. Sediment: Predominantly reworked platform material with pelagic admixtures. Highly variable grain size; end members are gentle muddy slope with much slumping and sandy or rubbly slope with steep, planar foresets. Biota: Mostly redeposited shallow-water benthos, some deep-water benthos and plankton.

5) Reefs of platform margin (Figs 4.6, 4.7). Setting: (a) wave-resistant barrier reefs rimming the platform, or (b) belts of knoll reefs and skeletal sands. Sediments: Almost pure carbonate of very variable grain size. Most diagnostic are masses or patches of boundstone or framestone, internal cavities with fillings of cement or sediment, multiple generations of construction, encrustation and boring and destruction. Biota: Almost exclusively benthos. Colonies of framebuilders, encrusters, borers along with large volumes of loose skeletal sand and rubble, including fragments of boundstone/framestone.

6) Sand shoals of platform margins (Fig. 4.7). Setting: Elongate shoals and tidal bars, sometimes with eolianite islands; above fair-weather wave base and within euphotic zone,

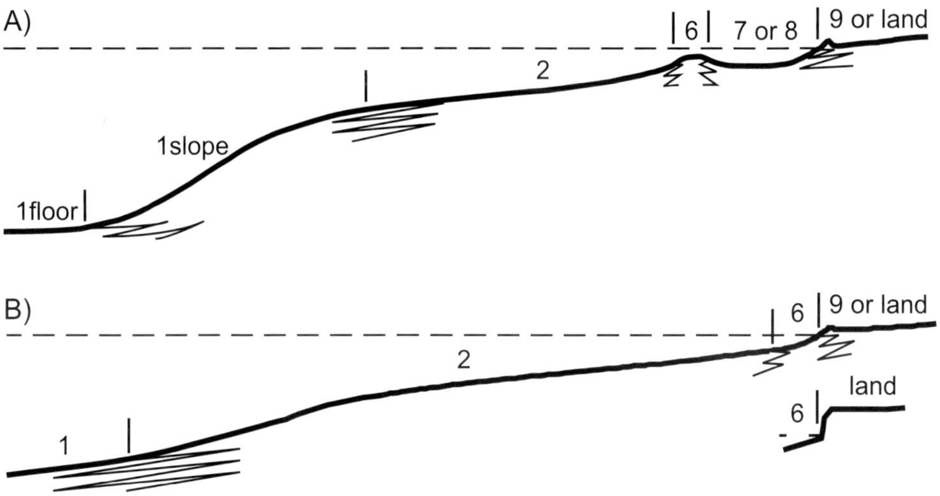

Fig. 4.5.— Wilson's standard facies belts applied to a ramp. A) Complete facies succession. Facies 2, deeper-water shelf, is the dominant element. Landward of belt 2 follows a high-energy sand shoal of facies belt 6 that protects a lagoon; depending on the degree of restriction caused by the sand shoal, the lagoon falls in facies belt 7 or 8. The spectrum ends with supratidal flats of belt 9. On the seaward side, belt 2 grades into facies belt 1. If the sea floor becomes steep enough for significant sediment-gravity transport, facies 1 may be differentiable into slope facies, 1-slope, and basin floor facies, 1-floor. B) Common variations of the ramp theme. Facies 7 through 9 need not be present. The succession may terminate with a high-energy sand belt that ends in a beach or a sea cliff. On the seaward side, belt 2 may imperceptibly grade into the basin floor.

strongly influenced by tidal currents. Sediments: Clean lime sands, occasionally with quartz; partly with well-preserved cross bedding, partly bioturbated. Biota: Worn and abraded biota from reefs and associated environments, low-diversity in-fauna adjusted to very mobile substrate.

7) Platform interior – normal marine. Setting: Flat platform top within euphotic zone and normally above fair-weather wave base; called lagoon when protected by sand shoals, islands or reefs of platform margin; sufficiently connected with open sea to maintain salinities and temperatures close to those of adjacent ocean. Sediments: Lime mud, muddy sand or sand, depending on grain size of local sediment production and the efficiency of winnowing by waves and tidal currents; patches of bioherms and biostromes. Terrigenous sand and mud may be common in platforms attached to land, absent in detached platforms such as oceanic atolls. Biota: Shallow-water benthos with bivalves, gastropods, sponges, arthropods, foraminifers and algae particularly common.

8) Platform interior – restricted. Setting: As for facies 7, but less well connected with open ocean so that large variations of temperature and salinity are common. Sediments: Mostly lime mud and muddy sand, some clean sand; often tidal flats (Fig. 4.8); early diagenetic cementation common; terrigenous admixtures common. Biota: Shallow-water biota of reduced diversitym e,g, cerithid gastropods, miliolid foraminifers.

9A) Platform interior – evaporitic. Setting: largely arid supratidal flats with only episodic influx of normal marine waters such that evaporites commonly alternate with the carbonates. Sabkhas, salt marshes and salt ponds are typical features. Sediments: Calcareous or dolomitic mud or sand, with nodular or coarse-crystalline gypsum or anhydrite; intercalations of red beds and terrigenous eolianites in land-attached platforms. Biota: Little indigenous biota except mats of cyanobacteria, brine shrimp, abnormal-salinity ostracodes, reworked marine biota.

9B) Platform interior – brackish. Setting: poor connection with the open sea just like 9A but with a humid climate such that fresh water runoff dilutes the small bodies of ponded seawater and marsh vegetation spreads in the supratidal flats. Sediments: Calcareous marine mud or sand with occasional freshwater lime mud and peat layers. Biota: shoal-water marine organisms washed in with storms plus abnormal-salinity ostracodes, freshwater snails and charophytic algae.

The internal variability of facies belts may be considerable. Gischler and Lomando (1999) contributed an instructive example on variability of the platform interior facies (belt 7) of the isolated platforms off Belize and its connection to antecedent karst relief and variations in subsidence.

Wilson's (1975) standard facies can also be applied to carbonate ramps (Fig. 4.5). On attached ramps, the succession includes facies 7 (platform interior) passing into facies 2 (deep shelf). On detached ramps, the succession starts on the landward side with facies 8 or 7, depending on the degree of restriction, includes the barrier as facies 6, followed on the seaward side by facies 2. The critical difference to rimmed platforms remains even if the same facies categories are used to describe the situation. The diagnostic criteria are the slope facies 3 and 4. On rimmed platforms, facies 2 is separated from the rim facies 6 by a slope with facies 3 or 4. Homoclinal ramps lack slope facies altogether, whereas distally steepened ramps show slope facies seaward of facies 2. Using Wilson's (1975) standard facies for description of ramps is advantageous because ramps often grade upward into rimmed platforms. Rimmed profiles and ramp profiles may also alternate in space where the rim index is low.

In most instances, the analogy with siliciclastic shelf models is not perfect because carbonate systems maintain some rim building capability. Offshore highs may be capped by reefs, such as in the Persian Gulf, and mud mounds or pinnacle reefs may develop on the homocline of facies belt 2. If these buildups are isolated features in facies belt 2, their presence is compatible with classifying the system as a ramp. If the buildups coalesce to a continuous belt, the system becomes a rimmed platform. Whether a belt of reefs or mounds acts as an effective rim can be determined from the facies pattern in plan view. If the facies patterns within each buildup are nearly concentric and the surrounding facies shows no systematic difference between the landward and the seaward sides of the buildup belt, the barrier function problably is negligible. In effective barriers, the facies patterns of buildups show landward-seaward asymmetry and the surrounding sediments are different on the landward and seaward side.

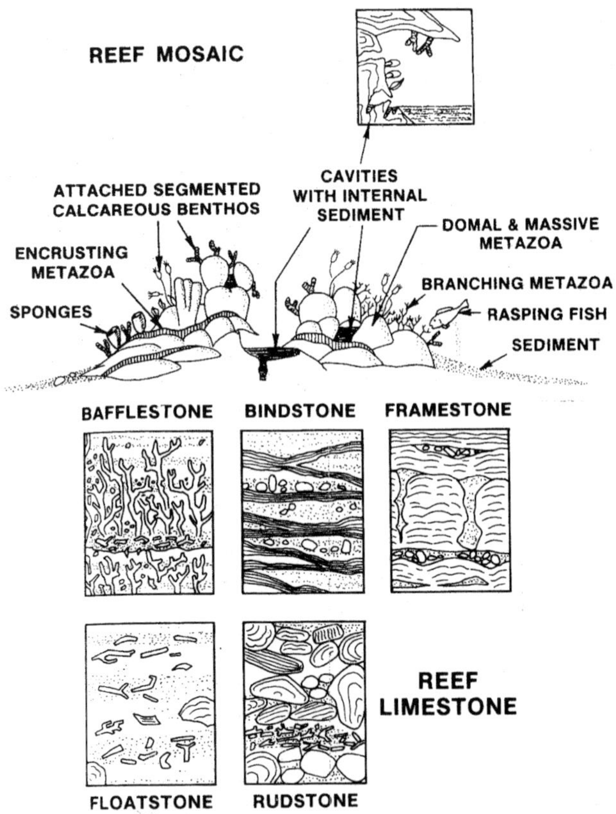

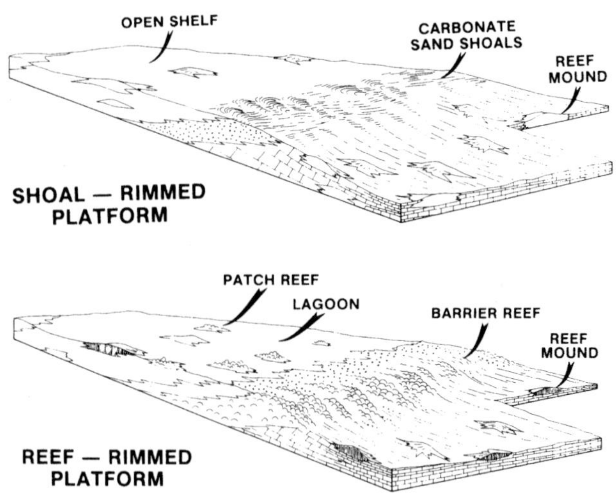

Fig. 4.6.— Recognition of ecologic reefs in the geologic record depends on a variety of features; most of them require outcrops or cores as an observational basis. After James and Macintyre (1985). A) Characteristic aspects of reefs. Note particularly the multiple successions of different encrusters and frame builders as well as cavities with internal sediment and cement. B) Characteristic sediment textures in reef deposits.

Fig. 4.7.— Rimmed margins are common on carbonate platforms. Reefs (facies 5) are important rim builders, but early-lithified, stacked sand shoals (facies 6) can be equally effective in defending platforms. A rim's effectiveness in protecting the interior depends largely on its continuity, measured by the rim index (see p. 42). Reef and sand shoals may also occur together on a platform margin. After James and Macintyre (1985).

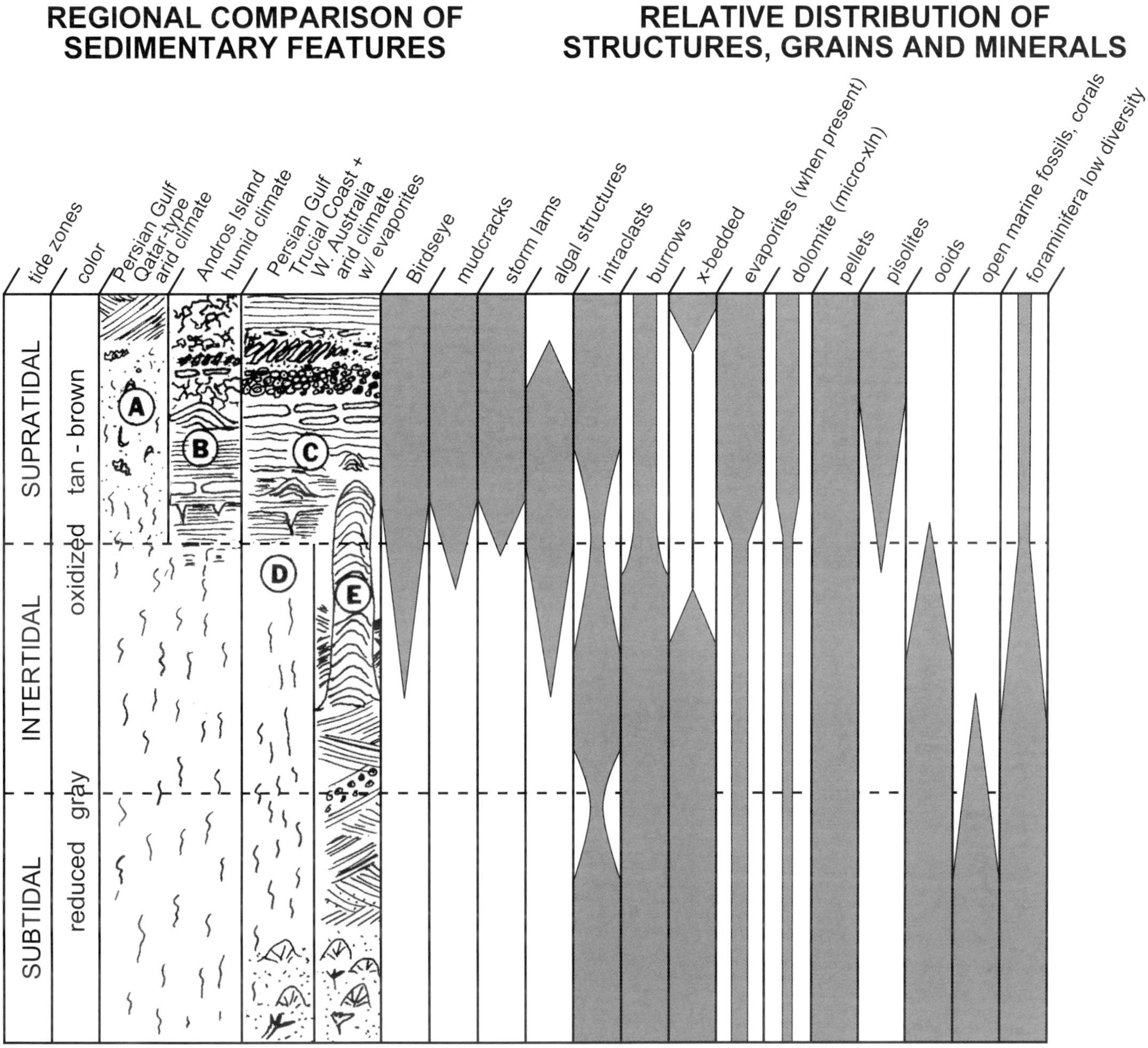

Fig. 4.8.— Left part: Overview of facies successions from subtidal to supratidal in several modern carbonate provinces. Encircled letters denote regional variations: A - Quatar, Persian Gulf (moderately arid); B - Andros, Bahamas (humid); C - Trucial Coast, Persian Gulf (very arid); D - intertidal Persian Gulf; E - intertidal Shark Bay, Australia. Right part: distribution of sedimentary structures, sediment grains and early-diagenetic minerals. Note that the subtidal/intertidal boundary is rather subtle, the only significant feature being the disappearance of certain sessile open-marine biota. The intertidal/supratidal boundary is much more pronounced. Sediments classified as subtidal (or shallow-marine) may often include intertidal deposits. After Shinn (1983), modified.

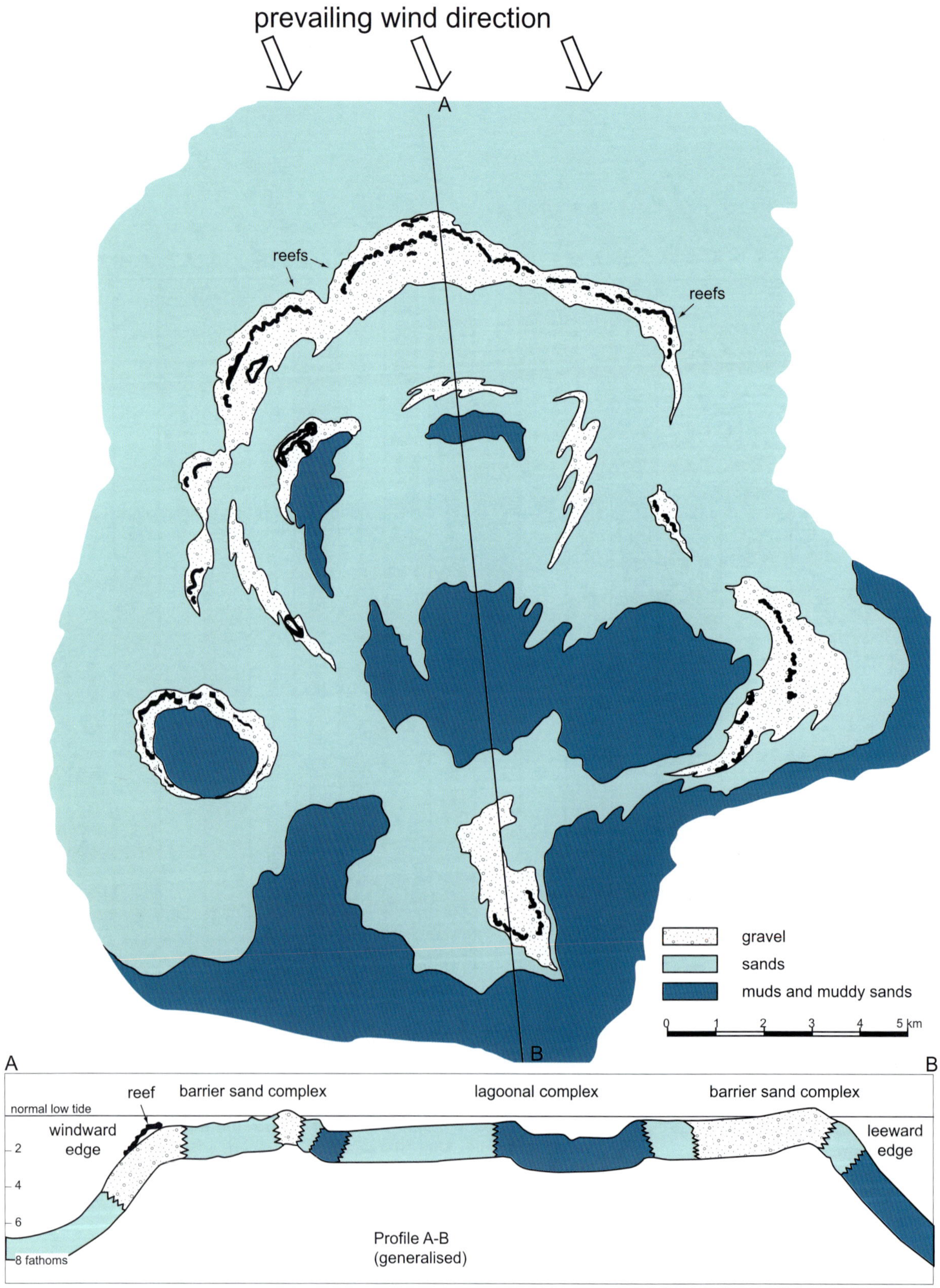

Fig. 4.9.— Sediment types and patterns on an offshore high in the Persian Gulf. Windward-leeward asymmetries such as this one are probably common in the geologic record but are difficult to recognize in outcrops because of insufficient coverage, seismic data are often more effective. After Purser (1973), modified.

FACIES BELTS OF THE C FACTORY

The facies patterns of the C factory essentially follow the ramp model as shown in Fig. 4.10. However, ramps of cool-water carbonates differ from their tropical counterparts in several ways:

- Ramps of tropical carbonates have a strong tendency to evolve to rimmed platforms; in cool-water carbonates, the ramp is a stable configuration also during climax conditions. Differences in rim building and in the depth window of production are the reason for the difference. In the T factory, maximum production is in the uppermost water column and the factory always has some rim-building capability. Together, these attributes quickly drive the system to form a rimmed platform whose slope steepens as it progrades. In the C factory, high production is maintained over a wide depth range and the ability to build wave-resistant rims is weak or absent. Thus, the system's equilibrium profile resembles that of a siliciclastic shelf.
- Mud content on cool-water ramps is significantly lower than in tropical ramps. There is less primary production of mud in the C factory and mud produced by abrasion is easily moved to greater depth because of the high turbulence. The coastal zone of cool-water ramps often consists of sandy beaches, flanked by calcareous eolian dunes.
- Wave abrasion on Neogene cool-water shelves of the southern hemisphere may become so intensive that parts of the middle part of the shelf produce sediment but cannnot accumulate any of it (Fig. 4.10b; James et al., 2001). It is not clear if this pattern was common in the past.

FACIES BELTS OF THE M FACTORY

In chapter 3 we have seen that the most common morphology of the M factory is a ramp studded with mounds. However, rimmed platforms may develop if competition by the T factory is subdued. In both instances, one needs to adjust the respective facies models developed for the T factory.

On ramps (Fig. 4.11A), sediment production in the form of peloids, micritic oncoids and layers of automicrite is widespread. Locally favorable conditions lead to growth of mud-mounds. The mounds may occur isolated or in clusters and belts parallel to the strike of the slope. Clusters may coalesce to a network of ridges (see chapter 8). The sunlit and wave-agitated part of the ramp usually is dominated by skeletal carbonate. The mounded belt typically lies in the thermocline below the oceanic mixed layer, in the thermocline where strong density gradients impede vertical mixing, oxygen is scarce and nutrient concentration high (Fig. 1.1 and Fig. 2.18; Stanton et al., 2000; Neuweiler et al., 2003). There is circumstantial evidence that upward growth of mounds has been governed by the limits of the oxygen-minimum zone (Stanton et al., 2000).

Rimmed M platforms (Fig. 4.11B) develop where the T factory is hampered by environmental or evolutionary factors; examples are the Famennian of the Canning Basin and some Triassic platforms in the Alps (Playford, 1980; Playford et al., 2001; Russo et al., 1997; Keim and Schlager, 2001). Where studied in detail, the rim consists of automicrite (thrombolite, stromatolitic crusts) and fibrous marine cements. Skeletal framebuilders, such as stromatoporoids, sponges or serpulid worms occur in highly variable amounts and indicate the gradual boundary to the T factory. The platform interior of the M factory shows facies belts 7, 8 and 9 based on sedimentary structures and textures. Determining restriction is difficult because of the low content of skeletal grains. However, extreme salinities can be recognized by the presence of evaporites. Slopes of the M platforms still are part of the factory such that layers of automicrite alternate with layers of mud and deposits of sediment-gravity flows (Blendinger et al., 1997; Keim and Schlager, 1999). Megabreccias are common. Their clasts, often over 10 m in diameter, stem from the platform rim or the automicrite layers of the slope. In contrast to the the spectacular breccias, turbidite aprons in the basin usually are inconspicuous. It seems that the M platforms export little fine material because most micrite is already hard upon formation and skeletal detritus is scarce.

CARBONATE FACIES OF EPEIRIC SEAS

Epeiric seas are shallow seas on continents. I consider the term epeiric sea synonymous to epicontinental sea. These epicontinental seas differ clearly from the pericontinental shelf seas overlying continental margins (Leeder, 1999). Process-oriented carbonate sedimentology emerged mainly from the study of two recent systems: pericontinental shelves and detached carbonate platforms in the middle of the ocean, usually on volcanic basement. Carbonate deposits of epeiric seas are common in the geologic record but scarce at present. Moreover, recent epeiric seas are small compared to some of their ancient counterparts. Irwin (1965) summarized the situation as follows: "The present is the key to the past' may be a misleading statement when considering epeiric sedimentation. We simply have no existing models of epeiric seas to guide our investigations ... ". Carbonate researchers have echoed these concerns (e.g. Aigner, 1985; Wright and Burchette, 1996). The question is both legitimate and important. Epeiric sediments are important because they are more likely to be preserved than deposits of continental margins or ocean interiors. Thus, in most instances epeiric deposits are our principal resource for reconstructing the conditions of the distant past. The strong regional differentiation of epeiric seas (e.g. Simo et al., 2003) puts serious limitations on attempts to reconstruct the conditions in the open oceanic domains.

Personally, I believe that actualism should be used with discretion but remains a powerful concept, also for interpreting the record of ancient epeiric seas. To start with, carbonate epeiric seas do exist at present and can be used for tying process to product. The most important example is the Timor-Arafura Sea with the Gulf of Carpentaria, entirely located on the Australian continent. The Persian Gulf is a foreland basin but its Arabian flank shows many characteristics of an epeiric sea: it lies on a mature, gently warped craton, has very low topographic gradient and water depths in the axial zone are less than 100 m. Finally, there is the "oceanic continent" that includes the Indonesian Archipelago, the Philippines and the South China Sea. Tectonically, this area is not a continent and anything but stable. However, it includes large shelf seas, many of them dominated by carbonates. In addition, the recent epeiric seas with siliciclastic sedimentation can be used to pursue questions of hydrodynamics and oceanography: examples are the Arctic shelves and the Hudson Bay of North America, the Yellow Sea of SE Asia, and the Barents Sea, the North Sea and the Baltic Sea of Europe. From the recent examples of epeiric seas we can derive the following guidelines for interpretation of ancient epeiric deposits.

- *Tides* in epeiric seas are highly variable, depending on size, shape and average depth of the sea (see Leeder, 1999, for overview). The basic trends related to the distance from the open ocean may be drastically altered by resonance between the oceanic tide and the water body on the shelf. Epeiric tides may also rotate around fixed points ("amphidromic points") determined by the interplay of Coriolis force and friction along the coastlines. The testimony of recent epeiric seas certainly indicates that the model of "tide-less" seas entirely dominated by storms is not generally valid (Fig. 4.12). Growing evidence of tidalites in ancient epeiric seas support this view (e.g. Pratt and James, 1986; Willis and Gabel, 2003).
- *Restriction* is also variable but probably better predictable by the distance from the open ocean than the tides. For instance, salinity in the Persian Gulf - in

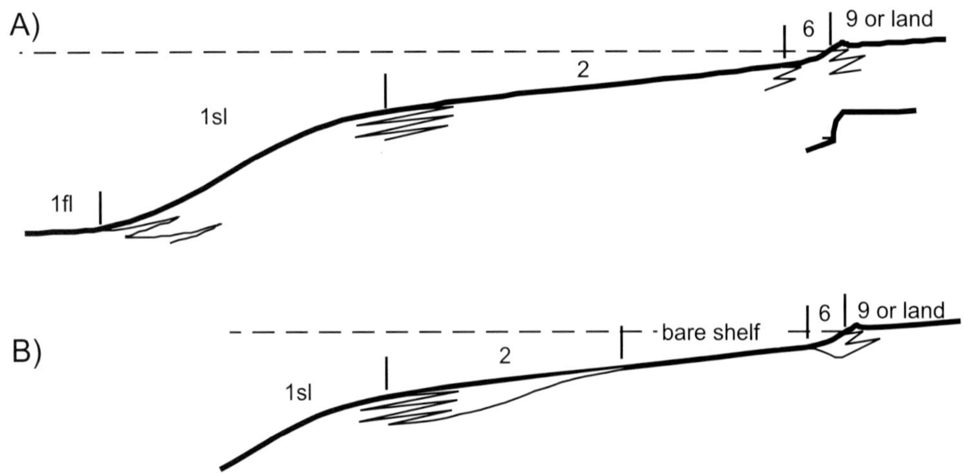

Fig. 4.10.— Facies of the C factory are well described by the ramp model. (A) Standard distally steepened ramp. On the landward side, the C factory ends in high-energy skeletal sands bordered by cliffs or by eolian dunes and sand flats. On the seaward side, the factory may extend into what is morphologically the continental or island slope. (B) Extremely intensive wave action may preclude sediment accumulation on the middle part of the shelf. Sediment produced in this area is deposited in the nearshore zone and on the outer shelf and the slope.

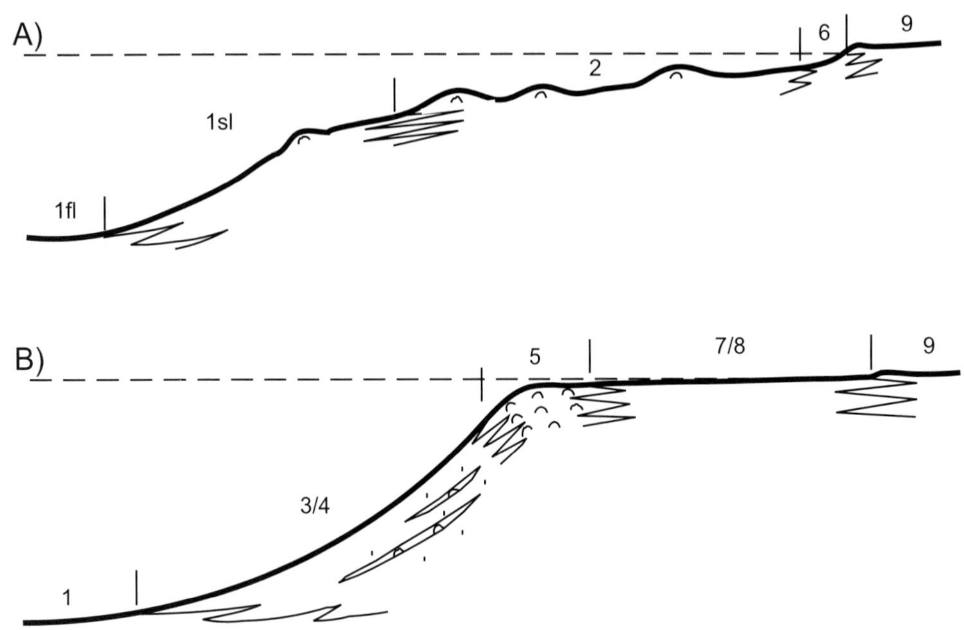

Fig. 4.11.— Facies models of the M factory. A) Ramp model. The patterns differ from a T-factory ramp by the occurrence of mud-mounds on the shelf. In plan view, the mounds form clusters or belts perpendicular to the dip of the sea floor, or they may coalesce to a network of ridges. The landward end of belt 2 is normally dominated by skeletal carbonate, the high-energy sand belt in the littoral zone may be skeletal or oolitic. B) Platform model. Facies belts are analogous to the T factory platforms. However, the rim facies may extend far down the slope and re-appear in layers and lenses on the slope because the depth window of production is much wider than in the T factory. Debris usually does not extend far into the basin but megabreccias on the slope and toe-of-slope are common.

this case a reasonable proxy for restriction - is rather straightforwardly related to water depth and distance from the open ocean (Purser (1973). The same holds for the Gulf of Carpentaria (Wolanski, 1993), albeit in the opposite sense – salinity decreases shoreward because of freshwater run-off.

- *Topographic gradients* are very low and more irregular than on shelves of continental margins. An important implication of the low slopes is that facies belts shift far and rapidly as sea level changes. The Gulf of Carpentaria in northern Australia is a case in point. The gradient from shore to the 60 m isobath is about 0.01°; thus, a 50 m drop in sea level will shift the shoreline and related facies by 230 km. Under these circumstances, the migration of facies belts with time is difficult to track and sequences and systems tracts are difficult to assemble. The low topographic gradients also imply that rates of progradation during stable sea level are higher than on continental margins because less accommodation space needs to be filled.
- *Sedimentation rates* are typically low in epeiric formations because the continental interiors are stable and only subject to slow epeirogenic movements (see Harrison, 1990, for overview). Epeiric sediment formations are, therefore, on average thinner than coeval deposits of continental margins and their hiatusses are longer. This characteristic hampers physical stratigraphic correlation, particularly in combination with far shifts of facies belts during changes of sea level.

The pecularities of tides, restriction and slope gradients in epeiric seas should be kept in mind when interpreting the facies belts of epeiric carbonates. Particularly critical is the correct interpretation of facies belts 5 and 6 – reefs and sand shoals at the platform margin. On ocean-facing platforms, these belts lie between a high, steep slope on the seaward side and a flat lagoon on the landward side. The lack of high slopes in epeiric seas complicates matters as water depth on the seaward side may differ only insignificantly from that in the lagoon. In this situation, it becomes difficult or impossible to distinguish between sand shoals rimming a lagoon and sands of a tidal-bar belt that formed on the epeiric shelf in zones of amplified tides (Fig. 4.12). Similar ambiguities may develop with regard to reef belts (facies 5). As the depth

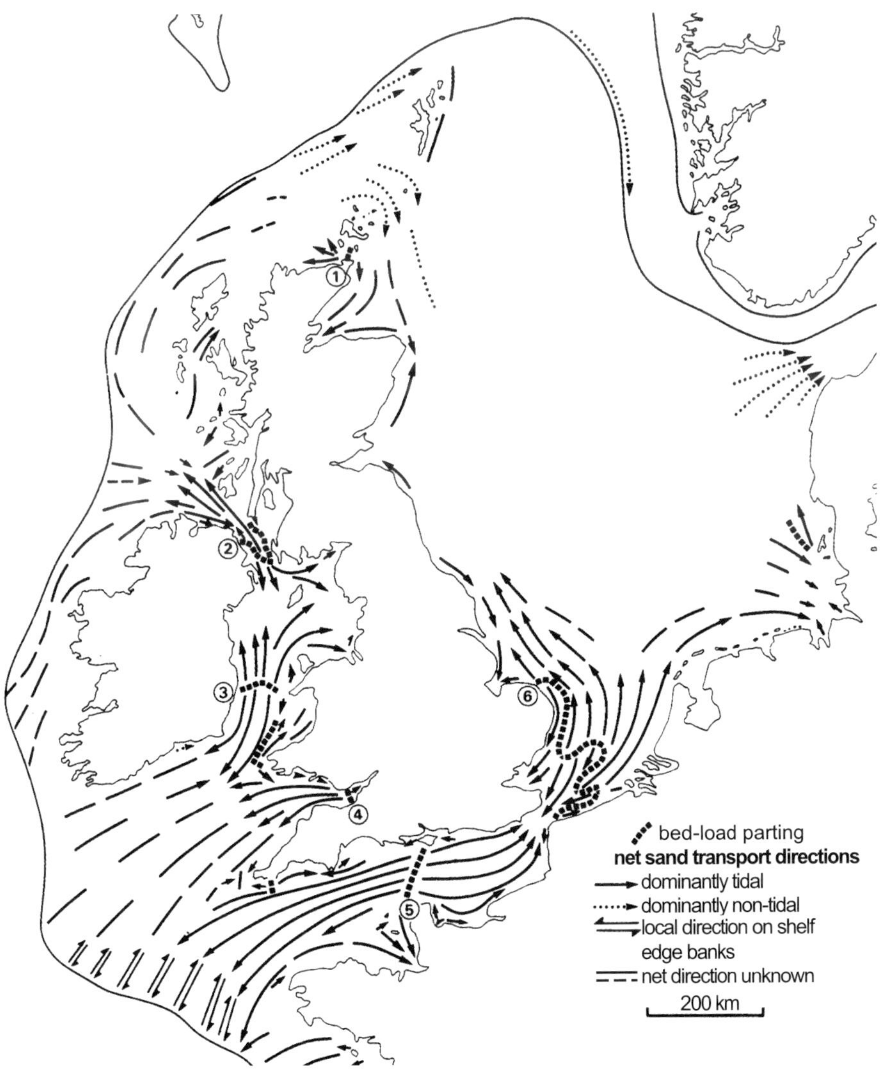

Fig. 4.12.— Paths of sand transport on the NW European shelf - a recent epeiric sea with dominantly siliciclastic sedimentation. Transport is almost entirely controlled by tides, even in the southern North Sea where the shelf break is over 1000 km away. Distribution of sand bodies is largely controlled by tidal currents and poorly correlated with distance from shore or from the shelf break. Note frequent occurrence of bed-load partings - sediment-starved areas that represent "watersheds" in the transport system where sand is carried off in opposite directions. After Johnson (1982).

difference seaward and landward of the reef belt decreases, wave energy levels become similar, too. Under these circumstances, the barrier reef may be replaced by a zone of isolated, radially symmetric shelf atolls, on which standard facies belts still can be recognised (Fig. 4.13). In extreme situations, reefs and shelf atolls may spread over large areas of the epeiric sea and the facies belt 2 may gradually pass into facies 7, i.e. a lagoon-like inland sea with patch reefs. This configuration, not uncommon in epeiric seas, lacks facies 3/4, the slope, and facies 5/6, the rim. All these belts are characteristic of rimmed platforms in open-ocean settings.

In summary, actualism and the standard facies belts remain key concepts for the interpretation of ancient epeiric carbonates. The actualistic interpretation should rely on three sources of insight: recent open-ocean carbonate systems, recent carbonate epeiric seas and sediment dynamics in recent siliciclastic epeiric seas.

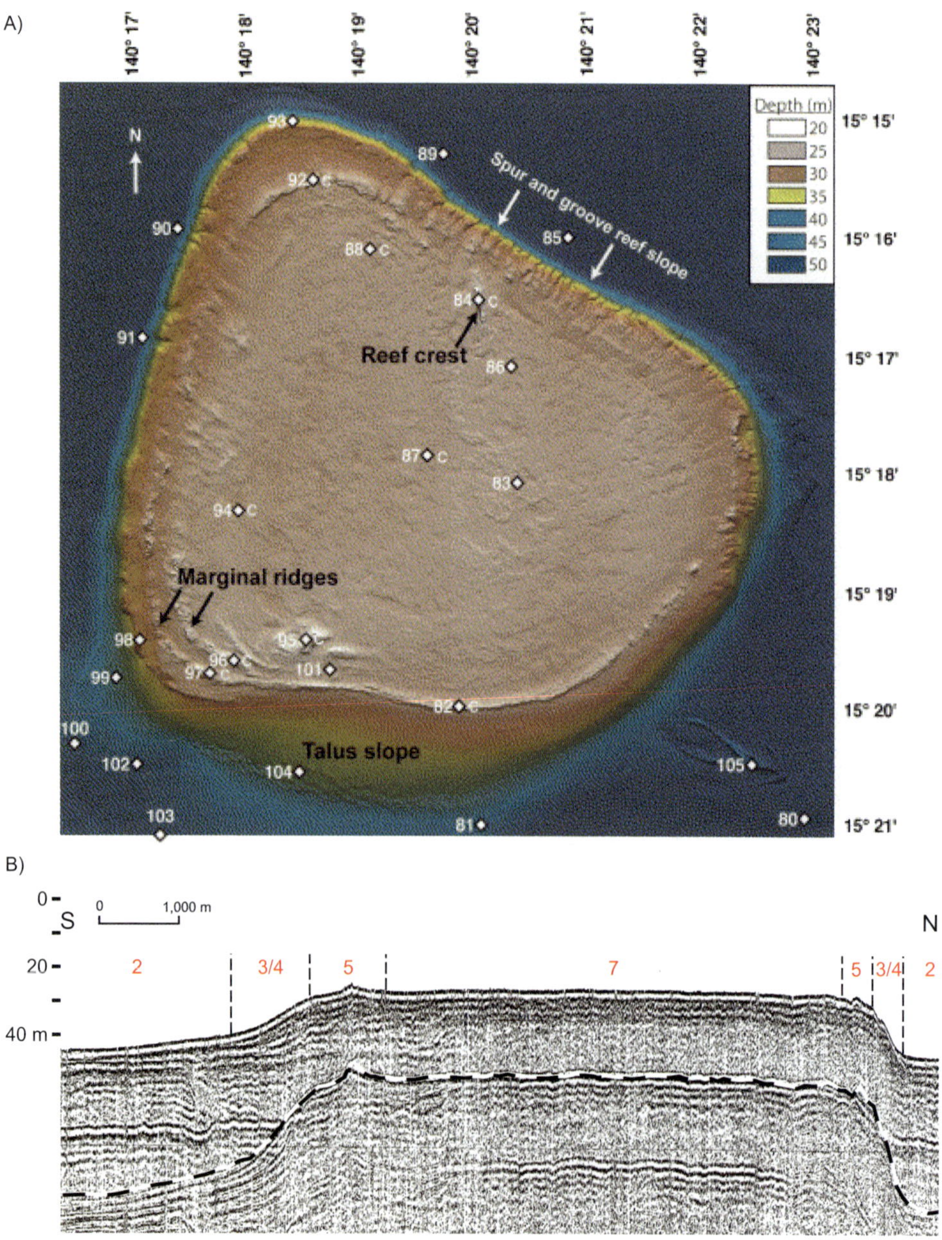

Fig. 4.13.— Holocene-Pleistocene shelf atoll in the Gulf of Carpentaria, a recent epeiric sea. A) Bathymetric map based on multibeam sonar survey. Flat top at 27 m depth and lack of reefs built to sea level indicate incipient drowning of the reefs. The atoll was thriving in the early Holocene (and intermittently in the Pleistocene) when sea level was lower. B) Seismic profile (Sparker) of shelf atoll in A. Tentative interpretation in terms of standard facies belts (in red, added by the author) refers to early-Holocene climax of reef growth and is based on rock samples and microtopography shown in A. After Harris et al. (2004), modified. Copyright 2004, reprinted with permission from Elsevier.

STABILITY OF FACIES PATTERNS

The standard facies belts of Wilson (1975) were conceived as static patterns. This is a gross simplification and stratigraphers certainly are aware of this. Wilson (1975) himself dedicated most of his book to documenting the change with time of carbonate facies and chapter 5 of this report deals with the same topic. This section, by contrast, addresses a very specific aspect of facies change with time: The question of which patterns are transient and which are more stable considering the internal dynamics of facies belts and the feed back between sedimentation, erosion and morphology.

The rimmed platform with flat top and steep flanks is a very stable pattern of the T factory. It is maintained (or re-established after external disturbance) by the following positive feedbacks:

- Production is high but constrained to a narrow part of the water column.
- High and rather uniform, wave energy leads to fairly even sediment distribution.
- Excess sediment can be dumped into the adjacent, nearly infinite accommodation of the open ocean; in this way the flat top is expanded by lateral progradation.

M factory platforms are stabilized in a similar way but their development in the Phanerozoic requires (temporary) shutdown of the T factory as a starting condition.

The empty bucket is an unstable pattern. It can develop only if the growth rate of the rim exceeds the rate of accommodation creation whereas the growth rate of the platform interior trails it. As the rim rises above the lagoon, the latter turns into a natural sediment trap that tends to fill up and counteract the effects of differential growth.

Ramps are stable or transient, depending on the factory. T-factory ramps tend to be transient. High but narrowly constrained production and formation of wave-resistant structures allow the system to quickly deviate from the hydrodynamic equilibrium profile. The T factory tends to fill the shallow part of the ramp to sea level, then prograde seaward and steepen the slope. The result is an expanding shallow-water platform. Ramps of the C factory are rather stable. The production window is wide, and frame building and cementation are minor. The factory produces in a wide depth range. The sediment is easily reworked and molded into a sigmoidal profile in equilibrium with waves and currents.

BIAS IN THE FACIES RECORD

It is widely accepted that sedimentary rocks chronicle earth history in a rather imperfect way. One distortion of the record is what I termed "depositional bias" - the tendency of different sediment families to record the same events differently, just like newspapers report on human history with a certain editorial bias. Depositional bias in sequence stratigraphy will be discussed in chapter 5. Besides depositional bias there exists bias of preservation and this is the topic of this section.

Bioturbation of carbonate sediments may lead to "wholesale destruction of evidence" as Bathurst (1971, p.128) called it. The sediments of the T factory in particular suffer intensive bioturbation by crabs in the upper 1 -1.5 m (Shinn, 1968; Bathurst, 1971). One consequence of bioturbation is the loss of hydrodynamic structures in all but the most energetic environments on carbonate platforms. Two implications of this intensive turnover have already been mentioned:

- In most instances we cannot tell if the wackestones and packstones of platform limestones were indeed deposited by simultaneously dumping a wide range of grain sizes or whether a primary alternation of mud and coarse storm layers was later mixed by burrowers.
- Hydrodynamic structures of the intertidal zone are almost entirely lost. These structures were already less well developed in carbonates than in siliciclastics because the carbonate seabed is partly covered by algal mats and seagrass; the structures that did form were subsequently erased by bioturbation.

Selective dissolution is another source of preservational bias (Flügel, 2004, p.106). Shoal-water carbonate sediments are deposited as mixtures of aragonite, magnesian calcite and calcite (see chapter 1). Aragonite and magnesian calcite normally are the most abundant minerals and both are metastable. Dissolution-precipitation reactions during diagenesis yield stable end members in the form of dolomites and calcitic limestones. The vestiges of selective dissolution during diagenesis can be observed in the rocks, but they are only preserved if the sediment was stiff or hard during the process and the voids did not collapse. Dissolution of metastable grains during early diagenesis, when soft sediment was being bioturbated and mechanically compacted, may go unnoticed. Recently, several authors proposed that this is exactly what has happened (Cherns and Wright, 2000; Peterhänsel and Pratt, 2001; chemical considerations in Morse, 2004,p.80). Work has not gone far enough to estimate how much has been lost but there can be no doubt that the problem is real and needs attention.

TERRESTRIAL EXPOSURE

In chapter 2, it was concluded that the window of marine carbonate production ends in the supratidal zone. In the terrestrial environment, carbonate dissolution prevails and carbonate precipitation is limited to small fresh-water bodies and other special environments. A relative sea level fall that exposes a marine carbonate environment to terrestrial conditions shuts down the carbonate factory immediately and creates a break in carbonate sedimentation. As sea-level fall and exposure play a pivotal role in sequence stratigraphy, the facies record of terrestrial exposure needs to be discussed in some detail.

First, some definitions need to be clarified. The term "subaerial exposure" is somewhat ambiguous. It is preferable to use the following terms:

- intertidal - between normal high and low tide, alternately flooded and awash;

- ➤ supratidal - above normal high tide, flooded only by storms;
- ➤ terrestrial - beyond the reach of the sea.

Transitions between these environments are gradual. The supratidal/terrestrial boundary in particular has been blurred in inhabited coastal lowlands as humans have settled in the supratidal marshlands and turned them into terrestrial environments. For carbonate sedimentologists, however, the distinction between supratidal and terrestrial is of prime importance. Carbonate depositional systems build into the supratidal zone by themselves, without any outside forcing. A change to terrestrial conditions, on the other hand, requires external forcing in the form of a relative sea-level fall. Theoretically, it should also be possible to transform a supratidal flat into a terrestrial environment by long-distance progradation of the shoreline without relative fall of sea level. However, I am not aware of any hard data demonstrating this effect and its significance in the geologic record.

The formation of soil is a diagnostic feature of the terrestrial environment. It requires alteration of rock or marine sediment by a complex array of abiotic and biotic processes. These alterations characteristically take on the order of 10^3 to 10^4 years to produce geologically observable results (e.g. Birkeland, 1999, p. 215).

Soils develop a variety of characteristic textures as well as chemical signals (e.g. Figs 4.14, 4.15, 4.16 ; overviews in Esteban and Klappa, 1983; Immenhauser et al., 2000).

Carbon isotopes of carbonate cement are useful chemical fingerprints of soil. The rationale is as follows. In natural substances, carbon consists of a mixture of two stable isotopes, ^{12}C and ^{13}C. Organisms take up proportionally less ^{13}C. Soils obtain most of their carbon from organic processes and are, therefore, depleted in ^{13}C. Cement precipitating from water that percolated through soils inherits this fingerprint and shows a low δ^{13}C. (Fig. 4.14; Allan and Matthews, 1977)

Cements related to exposure may contain fluid inclusions of very low salinity, again indicating fresh-water lenses or percolation of rainwater through the rocks (e.g. Goldstein et al., 1990; Fouke et al., 1995; Immenhauser et al., 1999).

Besides soils, the morphology of the exposure surface may provide evidence of exposure. In all but the most arid climates, carbonates develop a dissolution morphology, known as "karst", with sinkholes, caves and irregular corrosion surfaces (chapter 3). However, similar morphologies develop on patchily lithified sea floors swept by currents.

Finally, biota indicating fresh-water or dry conditions is an excellent indicator of terrestrial exposure. The seeds of Characean algae are particularly useful because they are easily preserved, have characteristic shapes and do not require detailed taxonomic identification because the entire group is restricted to brackish or fresh-water conditions. Microcodium is a very characteristic microfeature in Cretaceous and Cenozoic soils, probably formed by intracellular calcification of root cells (Kozir, 2004).

MEGABRECCIAS

Breccia beds are common on the slopes and in the basins adjacent to the T and M factories, i.e. in facies belts 1, 3 and 4. In the field, these breccias often are very conspicuous and rival reefs as eye catchers. The beds are usually much thicker than the background sediment and may contain outsize clasts, exceeding 10 m in diameter or, in elongate slabs, 100 m in length. The term "megabreccia" has been widely used for these beds (see Wrigth and Burchette, 1996, p.365, and Spence and Tucker, 1997 for overviews). There is general agreement that the material was transported by variable combinations of sliding, slumping, debris flows and turbidity currents, with prominent roles of sliding and slumping at the beginning and the end of the event, debris flows governing the main phase of transport, and turbidity currents developing late and emplacing graded beds on top of the coarse breccia beds and downstream of them.

Sedimentary structures and postulated transport mechanisms are not significantly different from those of coarse siliciclastic deposits in tectonically active settings. The peculiar twist with carbonate megabreccias is that they frequently occur in tectonically quiet settings and lack evidence of deep-cutting, tectonically driven submarine erosion. Yet the size of the clasts borders on that of tectonic klippen. I think that two peculiarities of the T and M factories largely explain the megabreccias. First, both factories tend to build steep slopes. Second, lithification does not require long time and deep burial; rather, lithification often takes place at the sea floor or under shallow burial, controlled as it were by the content of metastable carbonate minerals, pore-fluid composition and microbial activity. This variability of diagenetic pathways frequently produces alternations of hard and soft intervals in a wide range of spatial scales. In this set-up, small masses starting to move on the upper, steeper slopes may destabilize larger masses on the lower, flatter slopes; cushions of fluidized sediment may facilitate gravity transport on the rise and basin floor with only few degrees of dip (see Hine et al. ,1992, for a Pleistocene example and p. 119f for sequence-stratigraphic consequences).

ENVIRONMENTAL MESSAGES FROM ORGANISMS

Fossil organisms are an important source of environmental information and the literature on this subject is extensive. Most of it is beyond the scope of this book with its focus on physical sedimentology and large-scale anatomy of carbonate rocks. However, even with this bias organisms are a very important source of information.

In examining distribution patterns of organisms in Phanerozoic carbonate rocks we must remember that what we see is the result of two controls: changes of environmental conditions, and changes imposed by organic evolution. As sedimentologists we strive to subtract purely evolutionary effects and isolate the environmental information in order to formulate models that are as widely applicable as possible. However, environmental change is an important control on evolution; thus, complete separation of environmental and evolutionary controls on the sediment record is impossible. Analysis of the functional morphology of organisms is one of the most successful attempts to extract purely environmental messages from fossils (Dodd and Stanton, 1981, p. 222–261).

Fig. 4.17 show examples of information on water depth and that can be extracted from fossils in carbonate rocks. Note that the presentation remains at the level of high taxonomic categories. These categories (for instance, gastropods, echinoderms etc.) existed through most or all of the Phanerozoic. For more detail see the excellent exposes in Flügel (2004) and Dodd and Stanton (1981, p. 17–115).

SILICICLASTICS AND EVAPORITES IN CARBONATE FACIES

Siliciclastics and evaporites (e.g. gypsum, halite) may appear in carbonate deposits as in-bed admixtures, as beds that alternate with carbonate beds and as formations that interfinger laterally with carbonate formations. At the scale of entire basins, phases of carbonate deposition may alternate with phases of siliciclastic or evaporite deposition. The details of these contacts and the effects of siliciclastics and evaporites on carbonate systems differ and need to be examined separately.

Siliciclastics are transported into carbonate environments from external sources. They may occur in any one of the carbonate facies belts but the most occmon occurrence is at the landward and seaward ends of the carbonate facies spectrum - in tidal flats and lagoons, and in the basin centers. Siliciclastics are least common in platform rim facies, probably because the environment is highly turbulent and deviates furthest from the hydrodynamic equilibrium profile that siliciclastic accumulations strive to.

The direct influence of *siliciclastics* on carbonate production depends on clastic grain size and on the carbonate factory. The most negative effect is caused by influx of clay in the T factory. Clay stays in suspension for long time, dampening the sunlight and reducing photosynthetic carbonate production. Moreover, organic matter, often associated with clay, increases the nutrient level of the environment and further damages the T production system. C and M factories are rather insensitive to the light reduction caused by terrigenous fines. It seems, however, that the C and M factories also function better in clay-poor settings because most benthic carbonate production depends on firm, clean substrate.

Coarse siliciclastic material does not seem to negatively affect carbonate production in the T and C factories. With regard to the M factory, the situation is not entirely clear. An obvious effect for all three factories is that coarse siliciclastics take away accommodation space from the carbonate system.

Evaporites precipitate from sea water, like marine carbonates. However, at the saturations required for evaporite precipitation, biotic carbonate production is already greatly reduced (e.g. Fig. 1.16). Thus, the sites of carbonate production and evaporite precipitation are adjacent to each other but do not broadly overlap. Consequently, one observes facies interfingering, particularly between restricted carbonate lagoons and sabkha evaporites (e.g. Sarg, 2001). Also, diagenetic admixtures of gypsum are common in arid carbonate tidal flats.

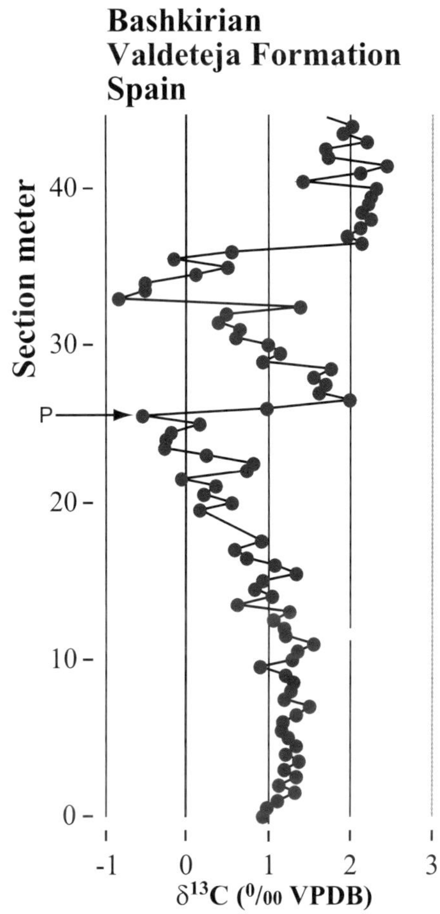

Fig. 4.14.— Variation of $\delta^{13}C$ in Carboniferous shoal-water limestones. Pronounced decrease of $\delta^{13}C$ strongly suggests influx of isotopically light soil carbon and therefore exposure. Asymmetric shape of the excursion is an additional indicator. P = location of pendent cement of Fig. 4.15. After Immenhauser et al. (2002), modified.

Evaporites in the deep basin centers usually represent distinct phases of deposition that sometimes alternate with phases of carbonate deposition. In these instances, the entire basin oscillates between an evaporite mode and a carbonate mode such that shelf, slope and basin floor alternatingly receive either carbonate or evaporite material. An example of this relationship is the Permian Zechstein Basin of NW Europe (Fig. 4.18).

Fig. 4.15.— Pendent cement (P) under a crinoid fragment, subsequently encased by turbid marine cement. Convincing evidence of exposure is created by the combination of pendent cement geometry, brown stain from humic substances and the decrease in the carbon-isotope ratio at this level (see Fig. 4.14). Pendent-cement geometry by itself indicates supratidal or terrestrial conditions. Carboniferous, NE Spain. field of view is 5 mm across. After Immenhauser et al.(2002).

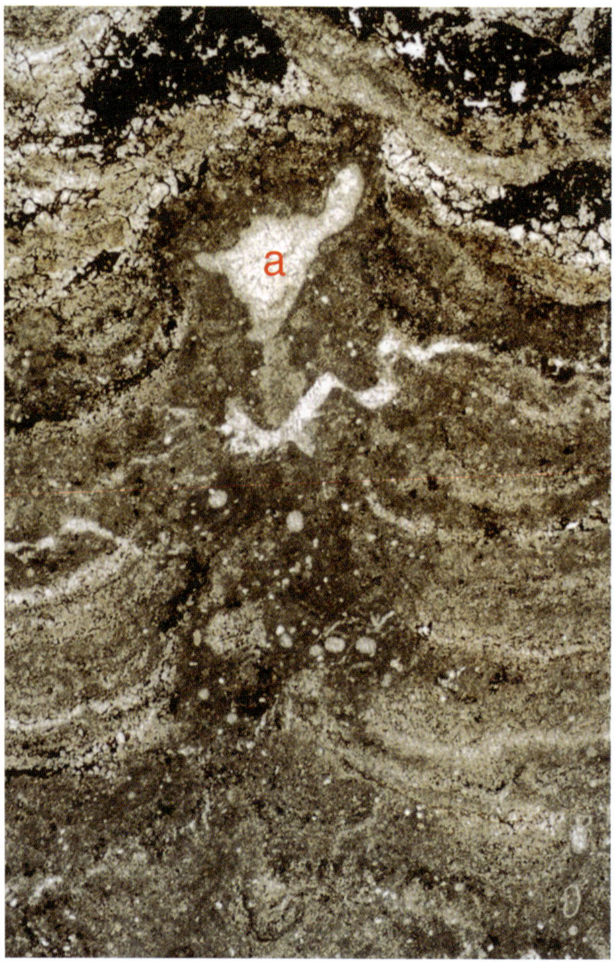

Fig. 4.16.— Thin section of layered calcrete (or caliche) - a typical carbonate soil. Upward bending of layers, wiggly cracks and alveole, a, filled with whitish calcite spar indicate early cementation. Dark iron oxides/hydroxides at the top. Exposure surface in mid-Cretaceous Nahr Umr Formation, Oman. After Immenhauser et al. (2000).

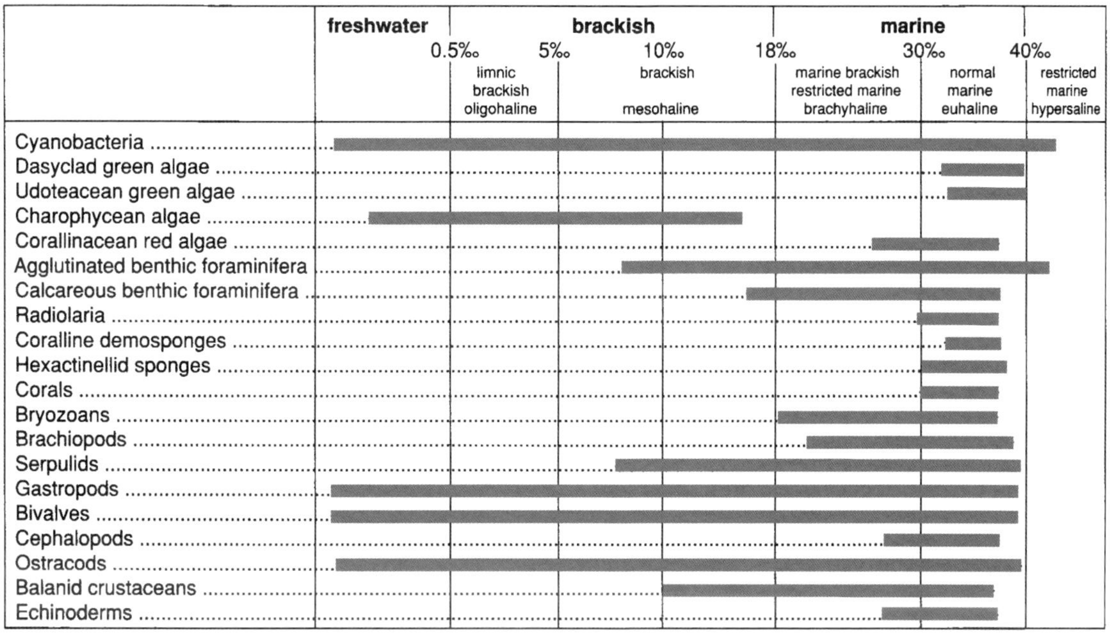

Fig. 4.17.— Distribution of major taxonomic groups of invertebrates with respect to salinity. After Flügel (2004), modified.

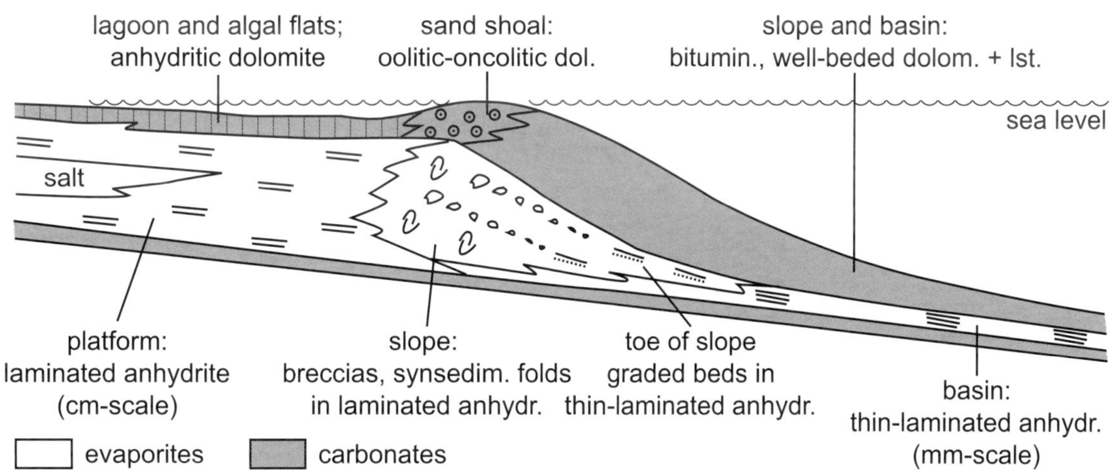

Fig. 4.18.— Model of successive phases of carbonate, anhydrite and halite deposition in the Permian Zechstein basin of NW Europe. Carbonates are differentiated into platform, slope and basin floor. Anhydrite bodies show the same differentiation as indicated by depositional structures and variations in thickness. After Schlager and Bolz (1977).

CHAPTER 5

Rhythms and events in carbonate stratigraphy

The standard facies model depicts depositional systems where sedimentation is in approximate equilibrium with intrinsic feedbacks and extrinsic controls. The assumption of equilibrium conditions is often unjustified and this chapter examines important causes of change in carbonate sedimentation with time.

Chapter 2 presented evidence that sedimentation is inherently episodic or pulsating and that the record is riddled with hiatuses in a wide range of scales. The scaling of sedimentation rates and the intense lamination of unbioturbated sediments are two major arguments in favor of a non-steady model of sedimentation. This "Cantor model" of sedimentation (Plotnick, 1986) does not invalidate facies models but limits their use. Facies models should be viewed as idealized equilibrium states that depositional systems strive to but do not always reach before being disturbed by extrinsic factors. The episodic nature of sedimentation implies that stratigraphic documentations solely in terms of standard facies belts, is inadequate and cannot do justice to the complexity of vertical successions. For carbonates, in particular, we need to consider changes through time imposed by system-internal feed back ("autocycles"), orbital cycles of the ocean-atmosphere system, organic evolution and long oscillations induced by plate tectonics and cosmic processes.

AUTOCYCLES

Most depositional systems are complex and endowed with many internal feedbacks. Consequently, stable steady-state operation is rarely ever reached. In place of that, the systems oscillate within certain limits, comparable to a heater coupled to a thermostat. The heater kicks on when the room temperature reaches the lower limit and turns off at the upper limit whence the room slowly cools to the lower limit and the cycle is repeated. Among carbonates, the T factory is particularly prone to act as a limit-driven oscillator. The depth window of production is narrow, production potential is very high and the production curve is sigmoidal with a lag phase at the beginning (see chapter 2).

Autocycles are easily produced in computer models by assuming that carbonate production starts very slowly when a supratidal area is re-flooded and that tidal flats rapidly prograde (Drummond and Wilkinson, 1993; Demicco, 1998). Fig. 5.1 illustrates shoaling cycles generated by the computer program STRATA (appendix B). The cycles form by the interplay of linear subsidence and a T factory growth function with a time lag.

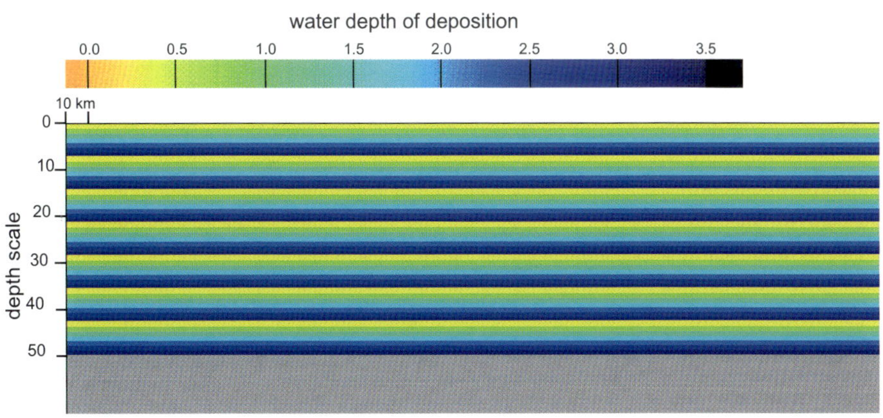

Fig. 5.1.— Shoaling autocycles on a carbonate platform generated with program STRATA (appendix B). Thickness scale on the left. Colors refer to depth of deposition. Depositional environment becomes shallower by sediment accumulation. When the sediment pile reaches sea level, production ceases and is resumed with a time lag of 7 ky. Thus, the sharp boundaries from yellow to blue represent a 7 ky hiatus. Autocycles arise from the interplay of sediment overproduction, production shut down at sea level, and resumed production after a time lag during which continued subsidence creates accommodation for the next cycle. Lag phase corresponds to the slow initial growth phase in the logistic equation (Fig. 1.12).

An observation-based model of shoaling autocycles in shoal-water carbonates was proposed by Ginsburg (1971) and summarized by Hardie and Shinn (1986) (Fig. 5.2). The model is compatible with observations in the Holocene but the Holocene record provides incomplete proof of the model because it covers only part of the postulated cycle. The most critical parts, - the halting of progradation because of lack of supply and the subsequent turnaround into a transgressive phase that floods the supratidal flats - have not been observed. Whether this can happen fast enough and on a sufficiently large scale to match the observations in the geologic record is open to debate. Holocene observations do show that prograding carbonate tidal flats tend to develop high beach ridges on the seaward side. When this defense system is overwhelmed, large parts of the landward flats must be flooded almost simultaneously - one of postulates of the Ginsburg model. The resulting autocycles oscillate between shallow marine and supratidal; based on extensive evidence from the Hoolocen, the system is assumed to not be able to build into the terrestrial environment.

Besides this non-linear response of an entire system in the time domain, we also find space rhythms within a system - mud banks, tidal passes, delta lobes etc. typically form rhythmic patterns in space. Migration of these spatial patterns can also produce a rhythmic sediment record (e.g. Reading and Levell, 1996 , 15). Fig. 5.3 shows a siliciclastic example where shore-parallel migration of subtidal mud banks leads to stepwise accretion of the beach, separated by phases of erosion.

ORBITAL RHYTHMS

The Earth's orbit around the Sun is perturbed by the motions of the Moon and the other planets; these perturbations induce subtle changes in the solar radiation received by the Earth that modulate the climate. There are three important orbital perturbations (Fig. 5.4):

- Changes in eccentricity. The Earth's orbit is an ellipse whose elongation changes with important cycles around 100 and 400 ky and weaker modulations around 1.2 and 2 My.
- Changes in tilt (or obliquity) of the Earth's rotational axis. At present, the axis of rotation is not perpendicular to the plane of orbit but tilted by 23°. Important cycles of tilt are 40 and 54 ky.

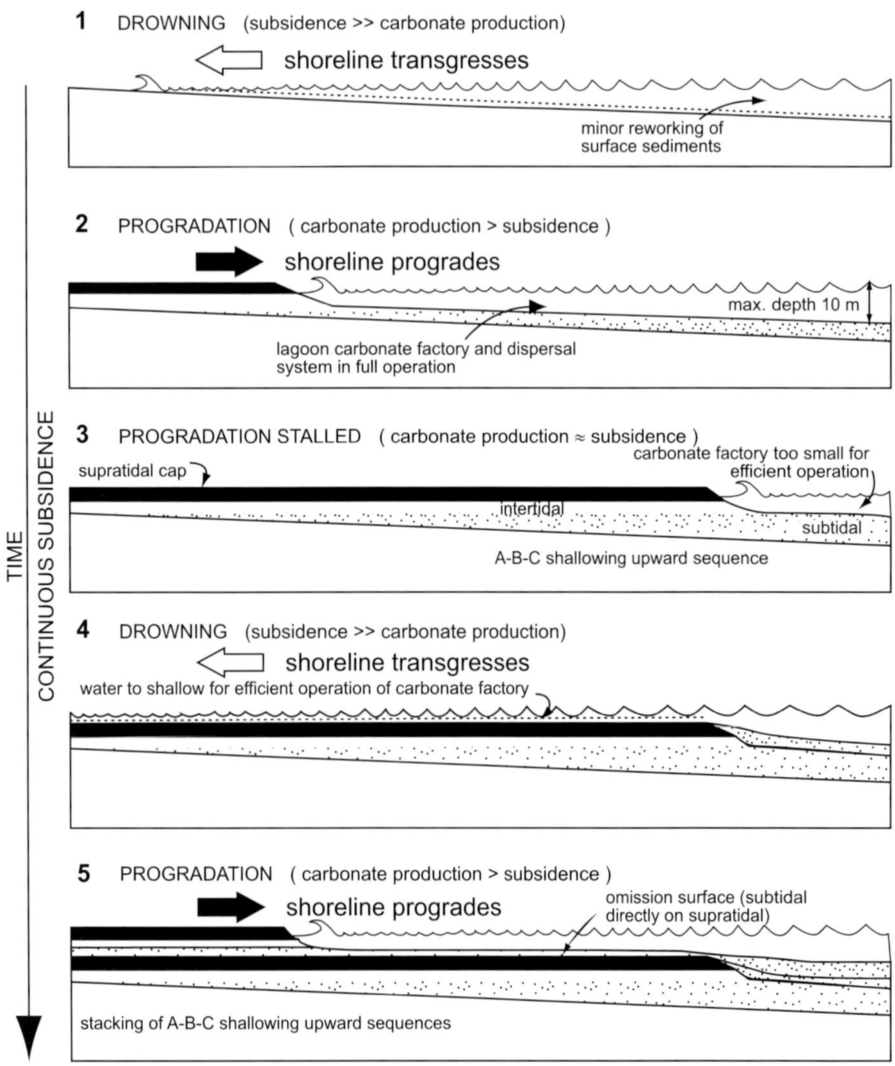

Fig. 5.2.— Ginsburg's autocyclic model of sedimentation on carbonate platforms assumes steady subsidence on a seaward dipping platform and a depth-dependent carbonate production; production is near-zero in supratidal flats (shown in black) and high in the shallow-marine lagoon (dotted). Cycle starts with rapid transgression and formation of a lagoon; sediment production in the lagoon starts slowly (start-up phase in Fig. 1.12); after several thousand years sediment production in the lagoon increases rapidly, sediment migrates shoreward, the inner part of the lagoon fills up to tidal flats and these flats begin to prograde (catch-up phase in Fig. 1.12).The progradation of tidal flats continually reduces carbonate production as productive lagoon floor is replaced by non-productive tidal flats. Progradation ceases when the lagoon has become too small to support further growth of the tidal flats. At this point, the system will stagnate until subsidence lowers the beach ridges on the seaward side of the flats and a new cycle starts with rapid transgression of the supratidal flats. After Hardie and Shinn (1986), modified.

➤ Changes in precession. The Earth's axis wobbles like that of a spinning top, i.e. its tilted axis describes a circular motion and points in different directions at different times. Important precession cycles fall in the range of 19 - 23 ky.

The climatic effect of obliquity is the seasonality of climate. If the axis of rotation were perpendicular to the orbit, there would be practically no seasons on Earth. Precession and eccentricity combine to determine the warmth of the seasons. Warmest summers occur if the eccentricity is high and the Earth is in perihelion, i.e. the position closest to the Sun during one orbit.

The orbital perturbations are quasi-periodic and the most important periods fall in the range of 20 – 400 ky. Orbital rhythms are highly interesting for stratigraphers because they are global signals and their periodic nature offers a means of measuring time beyond the normal resolution of biostratigraphy or radiometric techniques. Thus, "cyclostratigraphy" (House, 1985) has become a widely used technique. Orbital cyclicity in the sediment record is frequently inferred but the claim is less frequently backed by good data.

Shoal-water carbonates offer special challenges, mainly related to the transformation from thickness to time. Stratigraphic successions represent rhythms in space while the orbital cycles are rhythms in time. Testing a stratigraphic record for orbital rhythms requires a transformation of the data from stratigraphic thickness to geologic time. If the time control is 10^4 y or better, the transformation is rather straightforward. In most other instances, one has to assume that thickness is approximately proportional to time. For certain deposits, e.g. fine-grained pelagics, this condition is approximately satisfied. Shoal-water carbonates, particularly sediments of the T factory, are notoriously problematic is this respect. The juxtaposition of extremely high production in the uppermost water column and zero production above sea level (Fig. 2.3) induces a stop-and-go rhythm in sedimentation and makes the record very sensitive to relative changes of sea level. Even minor sea-level falls may lead to large gaps and "missed beats", i.e. orbital oscillations that remain unrecorded because sea-level remained below the platform top (Fig. 5.5). Tests for orbital rhythms, like any time-series analysis, are very sensitive to such gaps (e.g. Hinnov, 2000) and sedimentology and standard stratigraphy need to be used to the fullest to identify such gaps beforehand.

Problems can be minimized by selecting records from areas with high subsidence. Fig. 5.6 shows an example from the Triassic of the Alps where subsidence was on the order of 100 m/My. The bundling of platform cycles into groups of 4-5 was one of the first observations suggesting possible orbital control of these cycles (Schwarzacher 1954). A hierarchy of bedding with bundles and superbundles, i.e. bundles of bundles, with ratios of 4 or 5, remains one of the best field indicators of possible orbital control. To build a strong case for orbital control, time-series analysis is essential. Recent studies indicate that the display of series of data in spectrograms, i.e. continuous series of spectra, is a particularly powerful tool as it reveals the variation of rhythms with time (Fig. 5.7). Fig. 5.8 shows the spectrogram-technique applied to the analysis of bedding in a geologic example (Preto and Hinnov, 2003). There is a strong suggestion of orbital rhythms but the result is less than definitive because of the limited length of the section.

Problems with recognizing orbital rhythms are compounded in situations of low subsidence and long-term fall of eustic sea level. The Bahama platforms, for instance, have not yielded anything even remotely resembling orbital rhythms because the record abounds with hiatuses and missed beats (e.g. McNeill et al., 1998). The slopes and basins, on the other hand, have yielded excellent orbital signals both by direct correlation with the pelagic standard (Droxler et al., 1988) as well as by time-series analysis (Williams et al., 2002).

Orbital cyclostratigraphy has great potential in stratigraphy and is rapidly expanding into the realm of sequence stratigraphy. Recently, many sequence stratigraphers reported much higher numbers of sequence boundaries than

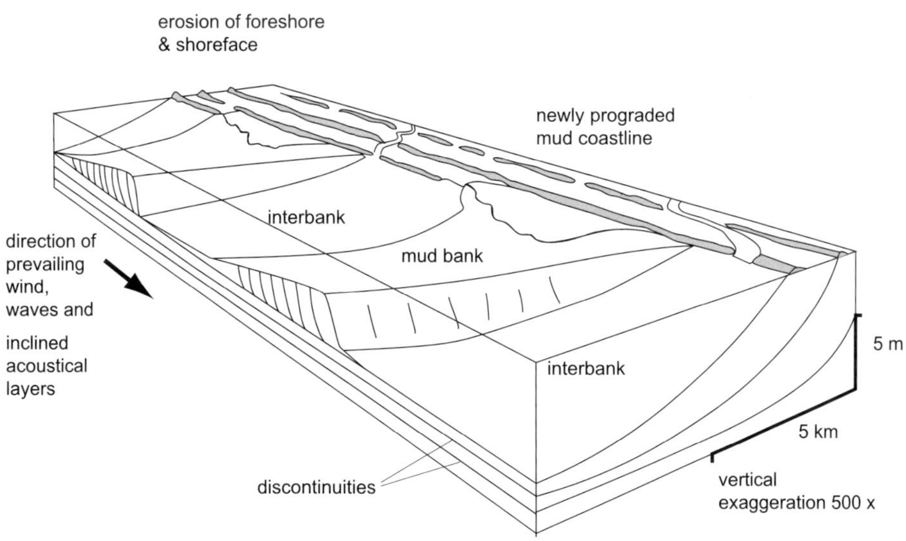

Fig. 5.3.— Siliciclastic autocycles generated by migrating mud banks on the coast of Suriname. Each time a mud bank passes, some of its sediment stays behind and becomes attached to the shore; between mudbanks, the shore is being gently eroded. In this way, the shore progrades in steps and the record consists of marine muds punctuated by erosional surfaces. A space rhythm (mud banks and interbank troughs) has been turned into a stratigraphic time rhythm. After Rine and Ginsburg (1985), modified.

previously observed. For instance, Van Wagoner et al. (1990, p. 52) estimate that type-1 unconformities are spaced at intervals of 100,000 - 150,000 years and that global curves such as the one by Haq et al. (1987) refer to sets of sequences. The carbonate record confirms this notion in many instances (e.g. Fischer, 1964; Read and Goldhammer, 1988; Goldhammer et al., 1990). These observations imply that the frequency bands of orbital cycles and stratigraphic sequences broadly overlap and the two approaches complement each other. In the Neogene and Quaternary, where a continuous orbital time scale exist, many standard sequences have been correlated to and dated by the orbital clock (e.g. Lourens and Hilgen, 1997).

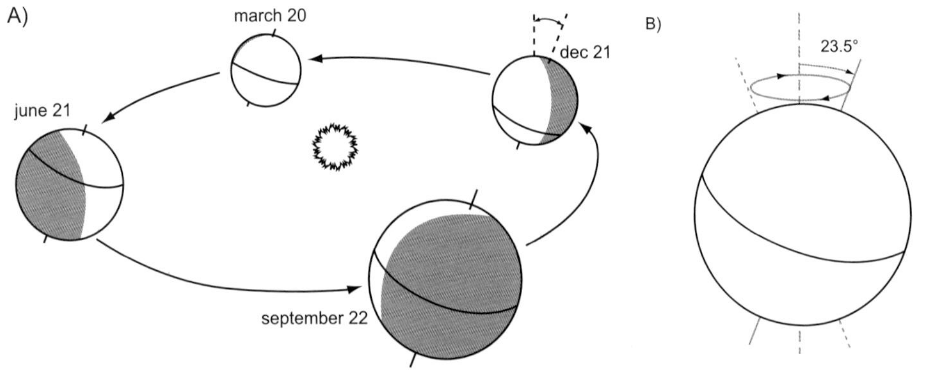

Fig. 5.4.— Principles of orbital perturbations. A) The Earth in its elliptical orbit around the Sun. Tilt of the Earth's rotational axis causes the seasons. The angle of tilt varies with cycles of 40 and 54 ky. B) The direction of tilt of the Earth's axis slowly changes, completing a circular loop every 26 ky. The climatically observable result of this cycle of axial precession is a bundle of "precession cycles" of 19 to 23 ky. After Imbrie and Imbrie (1979), modified.

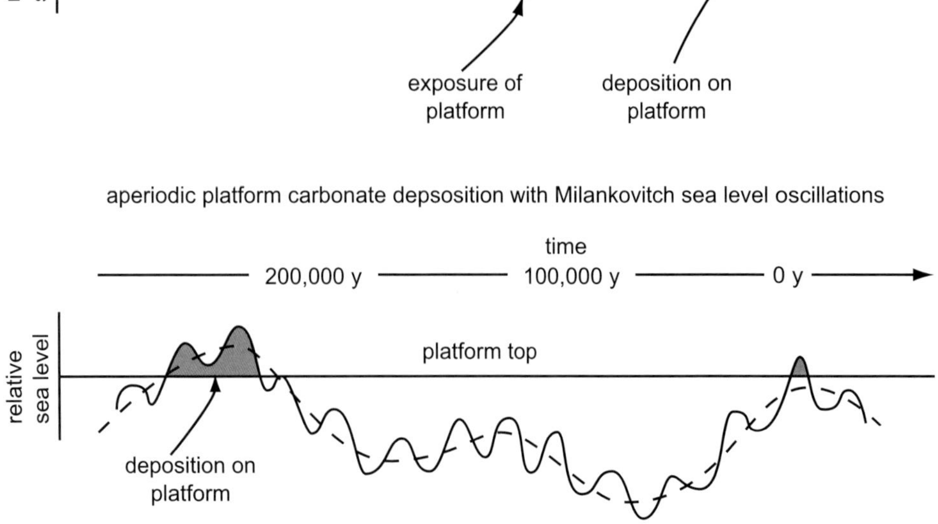

Fig. 5.5.— Scenarios for deposition of carbonate cycles dictated by the Earth's orbital perturbations (Milankovitch cycles). Top panel shows the record of flooding and exposure of a platform top by cycles that consist of a basic rhythm of 20,000-y (orbital precession) plus a 100,000-y eccentricity cycle. The 100,000-y cycle has low amplitude and consequently all sea-level fluctuations in the 20,000-y rhythm are being recorded by deposition on the platform top (shaded peaks of cycles). Bottom panel shows the same superposition of cycles but with high amplitude oscillation in the 100,000-y cycle. This leads to a very erratic record whereby only few 20,000-y cycles are recorded on the platform top and most of the time is represented by exposure of the platform top. After Hardie and Shinn (1986).

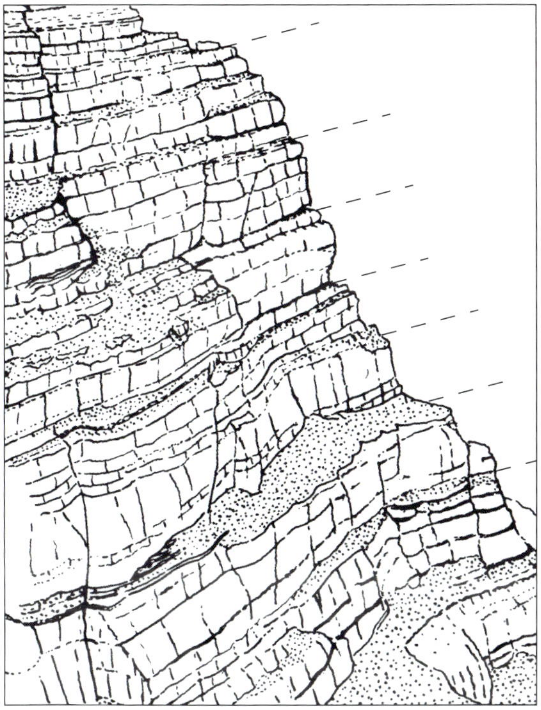

Fig. 5.6.— Bundling of beds into sets of 4 or 5 is often a first indication of their orbital origin if their estimate duration is in the range of 10^4 to 10^5 y. Example is from the Dachstein Fm., Late Triassic, Northern Calcareous Alps. After Fischer (1964).

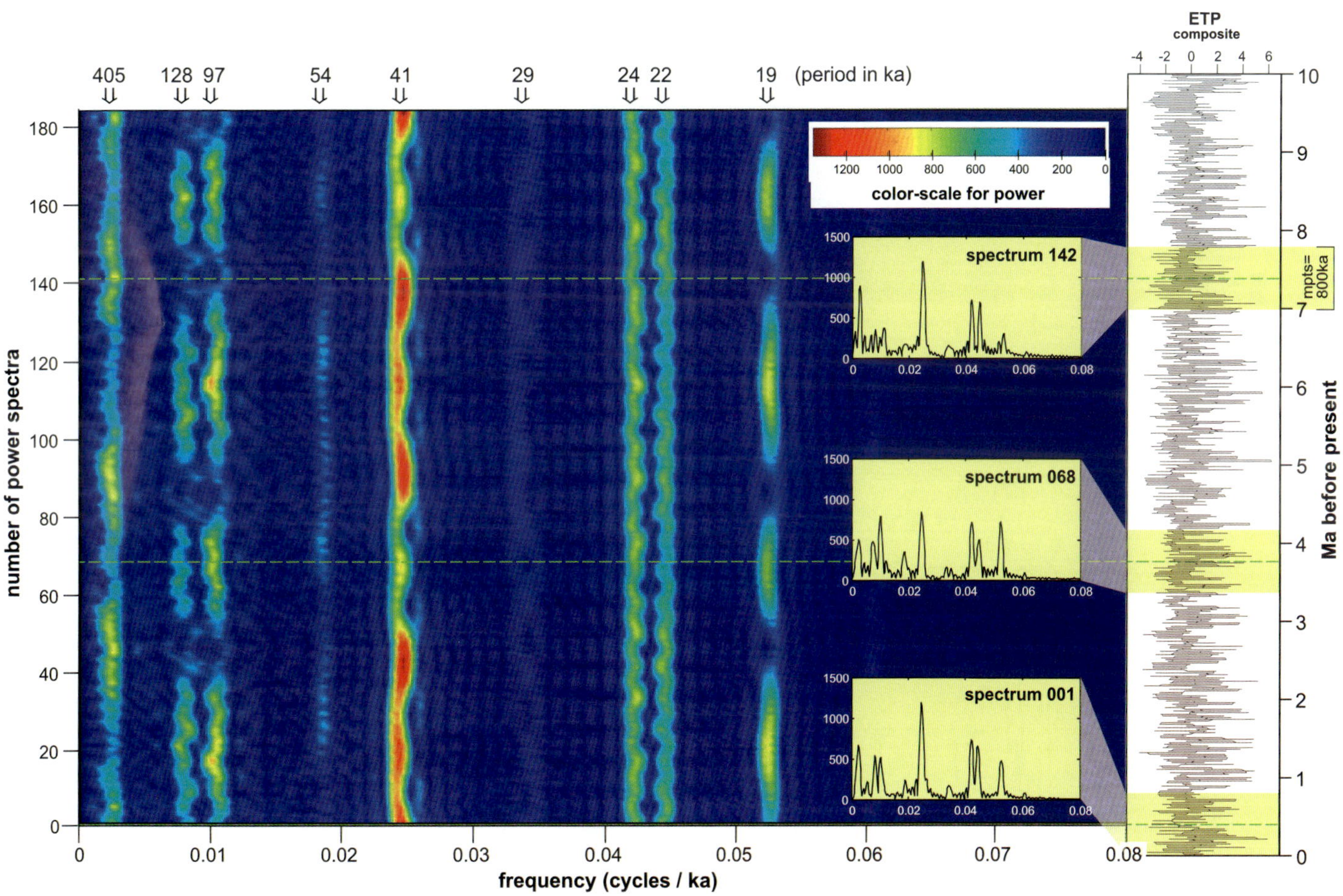

Fig. 5.7.— The spectrogram illustrates the change of the power spectrum with time. Example, based on Laskar et al. (1993), shows the changes of insolation in the last 10 My with characteristic shifts in power among the various orbital cycles. Spectrograms can also be constructed from measured stratigraphic sections where they offer clues to changes in sedimentation rates. Graph courtesy of L. Hinnov.

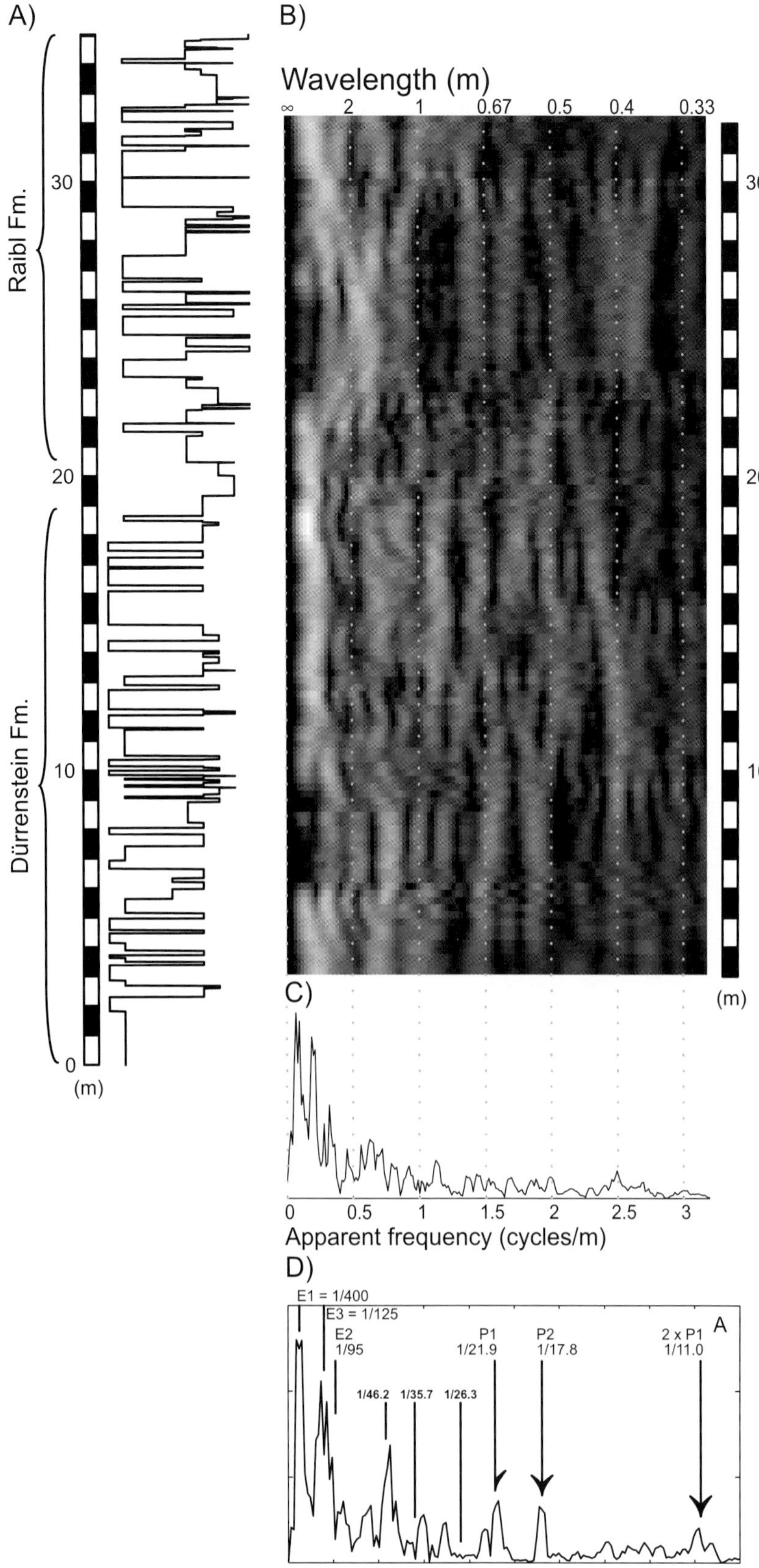

Fig. 5.8.— Analysis of bedding rhythms in search of orbital cycles - a case study from the Triassic of the Southern Alps. A) Measured section with lithofacies ranked according to presumed water depth (maximum depth on the right). B) Spectrogram of apparent frequency (cycles/m) throughout the section. The spectrogram was constructed from 121 overlapping line spectra such as the one shown in C. Lateral shifts of peaks are attributed to changes in sedimentation rate. These sedimentation-rate changes were compensated for by assuming that the spectral peak at 1.6 cycles/m represents the short precession cycle with an estimated Triassic period of 17.8 ky. D) Power-spectral analysis of the time series after tuning to the short precession cycle, P2. Shown is 1.5 π multitaper spectrum with arrows indicating peaks that passed the F test for significance at the 95% level or better. E1 – E3 = eccentricity cycles, P1 – P2 = precession cycles. Peaks in between may represent obliquity cycles. After Preto and Hinnov (2003), modified.

LONG OSCILLATIONS IN THE OCEAN-ATMOSPHERE SYSTEM

Autocycles as well as orbital rhythms can be studied in outcrop. In addition, the past few decades have produced convincing evidence for rhythms in the sediment record at time scales of 10^7 to 10^8 years. These rhythms are rarely obvious at the outcrop scale; they were discovered by intensive collaboration among paleoceanographers, paleoclimatologists, geochemists, sedimentologists and stratigraphers and tectonicists. For most of these rhythms it remains open if they are truly periodic. However, almost all of them can be shown to be true oscillations rather than undirectional trends with occasional incisions, such as the record of biotic evolution.

Environmental variables showing long oscillations that are particularly relevant for carbonate sedimentation and sequence stratigraphy are eustatic sea level, sea-surface temperatures, ice cover and the primary mineralogy of carbonates and evaporites. The causes invoked for these oscillations fall in three categories:
(1) oscillations of the rate of plate motion;
(2) celestial oscillations (solar system and beyond);
(3) autocycles in complex systems with numerous feedbacks.

The course of research on carbonate-relevant oscillations in the past three decades shows first a convergence of opinions followed by renewed discrepancies and divergence, despite growing data and insight (Fig. 5.9).

Long-term eustasy was the first variable to be quantified for the Phanerozoic by Hallam (1977) and Vail et al. (1977) (Fig. 5.9a). Both curves showed a little less than two cycles in the Phanerozoic with highs in the mid-Paleozoic and Cretaceous. In a bold synthesis, Fischer (1982) combined this eustatic curve with the new insights on the history of climate and ocean circulation obtained mainly by the Deep-Sea Drilling Program (Fig. 5.9b). He proposed that the Earth was oscillating between a greenhouse state with high sea level, equable climate and high levels of atmospheric CO_2, and an icehouse state of low sea level, large climatic gradients between poles and equator, and low CO_2 levels. Fischer (1982) estimated the ice-house-greenhouse cycle to be about 300 My long and be driven by changes in the rate of mantle convection and plate motion. Sandberg (1983) linked Fischer's icehouse-greenhouse cycles with changes in the mineralogy of abiotic carbonate precipitates in the ocean and proposed changes in atmospheric CO_2 content as the causal connection (Fig. 5.9d).

The contributions by Fischer (1982) and Sandberg (1983) were very influential. "Icehouse world" and "greenhouse world" became household terms and many aspects of the sediment record were tied to this long oscillation. In the past decade, the notion of long, CO_2-driven icehouse-greenhouse cycles was both significantly expanded and fundamentally challenged.

Hardie (1996) showed that the greenhouse phases of the Fischer-Sandberg wave correlate with times of marine evaporites rich in KCl, the icehouse phases with evaporites rich in $MgSO_4$ (Fig. 5.9c, d). He proposed that changes in seawater composition rather than fluctuations in atmospheric CO_2 levels had caused the alternation of aragonite seas and calcite seas in the past 600 My. Hardie (1996) also pointed out that hydrothermal alteration of young basalt at spreading ridges leads to net transfer of Mg and SO_4 from seawater to rock, and of K (and Ca) from rock to seawater. Thus, high rates of sea-floor spreading reduce the concentration of Mg and SO_4 in the sea and increase the concentration of K. The hydrothermal pump, therefore, offers an explanation for the abundance of KCl-evaporites at times of high sea level and fast spreading, and the abundance of $MgSO_4$ evaporites at times of slow spreading. Consequently, Hardie (1996), like Sandberg (1983), considered changes in the rate of plate tectonics the ultimate driver of the aragonite-calcite oscillations, albeit with a coupling mechanism in the form of sea-water chemistry rather than atmospheric CO_2. Stanley and Hardie (1998) expanded the concept of aragonite-calcite oscillations in the sea by demonstrating that it is not restricted to abiotic precipitates. Organisms with relatively weak internal control on the calcification process or organisms that calcify very rapidly ("hypercalcifiers") also follow the aragonite-calcite wave in the Phanerozoic. Dickson (2004) added important data gleaned from echinoderm remains.

Veizer et al. (2000) challenged the 300 My icehouse-greenhouse cycle with a large data set on temperature-sensitive oxygen-isotope ratios from pristine carbonate shells (Fig. 5.9e). They discovered persistent oscillations with an approximate period of 135±9 My. The strong correlation of this signal with the distribution of ice-rafted debris and other glacial indicators supports its climatic significance. Veizer et al. (2000) report a drastic mismatch with the 300 My icehouse-greenhouse cycle as well as with direct estimates of atmospheric CO_2 levels; they reject CO_2 variations as a possible driver of the oscillations. Shaviv and Veizer (2003) argue for an astronomic cause of the 135 My cycle (Fig. 5.9f). They point out that the flux of cosmic rays is modulated by the passage of the solar system through the spiral arms of the milky way, creating an oscillation of 143±10 My, very similar to the 135±9 My oscillation in oxygen-isotope ratios. High cosmic-ray flux leads to more low-level clouds, higher albedo and therefore cooler climate. Royer et al. (2004) reject the notion of strong celestial control on climate and maintain that the case for CO_2 as the principal driver is robust. They correct Veizer et al.'s (2000) isotope data for likely pH effects. This modification produces a better match with the CO_2 record and climate models but the 135 My oscillation remains.

This brief tour through the science of long oscillations in Phanerozoic climate, sea-level and carbonate chemistry shows that, at present, the situation is heaven for a scientist writing proposals for funding and hell for a book author trying to extract the eternal truth of the matter. This quandary notwithstanding, some important messages clearly emerge.

- There is strong evidence that ice-free greenhouse conditions were rare in the Phanerozoic. According to Frakes et al. (1992) and Crowell (2000), ice-rafted debris occurs in about 60% of Phanerozoic time if one uses a 5 My window. This implies that glacio-eustatic fluctuations are more of a rule than an exception in the Phanerozoic. Gibbs et al. 2000 conclude from modelling studies that significant polar ice caps may form at 10 - 14 times present CO_2 levels given suitable paleogeography.
- Evidence for ice and temperatures estimated from $\delta^{18}O$ oscillated with an estimated period of 135 My, i.e. less than half the period of the icehouse-greenhouse cycle of Fischer (1982), estimated as 300 My.
- The 300 My cycle is in phase with chemical changes, i.e. the alternation of KCl and $MgSO_4$ evaporites, of aragonite and calcite seas, and certain proxies for atmospheric CO_2 levels.

BIOTIC EVOLUTION

The topic of evolution takes us back to a point made in chapter 1: Biotic evolution is one of the most significant causes of change in the geologic record, rivaling the effects of plate tectonics and chemical cycling. Throughout the Late Proterozoic and Phanerozoic, the time interval with detailed sediment record, species and higher taxonomic units came and went. Particularly in the Phanerozoic, severe extinctions are followed by short intervals of rapid speciation and longer periods of slow change, in agreement with the concept of punctuated equilibria (Eldredge and Gould 1972). The profound effect of evolution on the sediment record is beyond doubt and provided a solid basis for biostratigraphy already 150 years ago.

The scope of this book, however, leads to a more specific question: what are the effects of evolution on the basic functioning of the carbonate system, on loci and rates of carbonate production, facies zonation, construction of platforms. In the Phanerozoic carbonate world, most evolutionary change falls in the category: the actors change but the play goes on, the play being the basic functioning of the system of carbonate production, deposition and early diagenesis. This statement may sound bold and off the mark in view of the drastic evolutionary changes in carbonate-secreting biota during the Phanerozoic. I consider it too conservative. The processes for the construction of flat-topped, rimmed platforms with a zonation analogous to Wilson's standard facies belts (chapter 4) evolved in the Archaean and the early part of the Proterozoic. The platform had a "modern" anatomy by Neoproterozoic times (Hoffman, 1974; Grotzinger and James, 2000). This does not mean, of course, that modern scleractinian reefs and Proterozoic stromatolite reefs are comparable ecosystems with similar food webs, carbonate precipitation mechanisms etc. At that level, the differences are enormous but the gross anatomy of the respective accumulations and their fundamental control by the rate of change in accommodation and the rate of sediment production are very similar. So similar, in fact, that we could probably not distinguish Proterozoic rimmed platforms from Neogene ones in seismic data.

Empty buckets may serve as a case in point. This characteristic geometry of raised platform rims and deeper lagoons is conspicuously developed on extant platforms with scleractinian reefs or oolite shoals forming the rim. Analogous geometries have been reported from various parts of the Phanerozoic as well as the Proterozoic (Fig. 3.23 3.24; Meyer, 1989; Wendte et al. 1992; Playford et al. 1989; Van Buchem et al., 2000; Grotzinger and James, 2000; Adams et al. 2004).

Perhaps the most conspicuous effects of Phanerozoic biotic evolution are relatively short disturbances caused by extinction events. The detailed record of the Phanerozoic shows that certain extinction events did affect the carbonate production systems (Fig. 5.10). Some events caused production to shift from the T factory to the M factory, but there were also events that caused a shift in the opposite direction and still other events shifted production from one metazoan group to another within the T factory. Hottinger (1989) argued that extinction of reef builders may take millions of years to repair because these organisms typically are K strategists with very long life cycles and therefore slow evolution. For instance, Hottinger estimates that it took 5 – 6 My to rebuild vigorous reef communities after the end-Cretaceous extinction.

Data on the number of Phanerozoic reefs (Fig. 5.11) show significant variations with some rapid declines related to extinctions. There is also some indication that the maximum carbonate production, a crude measure of the growth potential of reef communities, decreased in the wake of major extinctions (Bosscher and Schlager, 1993; Flügel and Kiessling, 2002, Fig.1).

Arguably the most drastic change in Phanerozoic carbonate sedimentation is the advent of calcareous plankton and nannoplankton. Estimates of carbonate sediment accumulation on continental margins and epeiric seas, on slopes and rises, and the pelagic centers of the ocean basins reveal a fascinating pattern: the overall rate of carbonate accumulation seems to have increased several-fold in the past 100 My and the locus of accumulation has shifted from continents to oceans (Fig. 5.12′. The onset of significant pelagic carbonate deposition on ocean crust about 100 My ago is the likely cause of both trends (Hay, 1985; Veizer and Mackenzie, 2004). In the Late Cretaceous and Cenozoic, planktonic foraminifers and coccolithophorids progressively replaced shallow-water benthos in precipitating carbonate from the ocean. This explains the shift from continents to oceans in carbonate accumulation. The increase in total carbonate accumulation probably relates to the fact that the pelagic sediment is mainly deposited on ocean crust and therefore more rapidly recycled than the carbonate rocks on the continents. The "carbonate mill" seems to be grinding faster since the changeover to pelagic sedimentation.

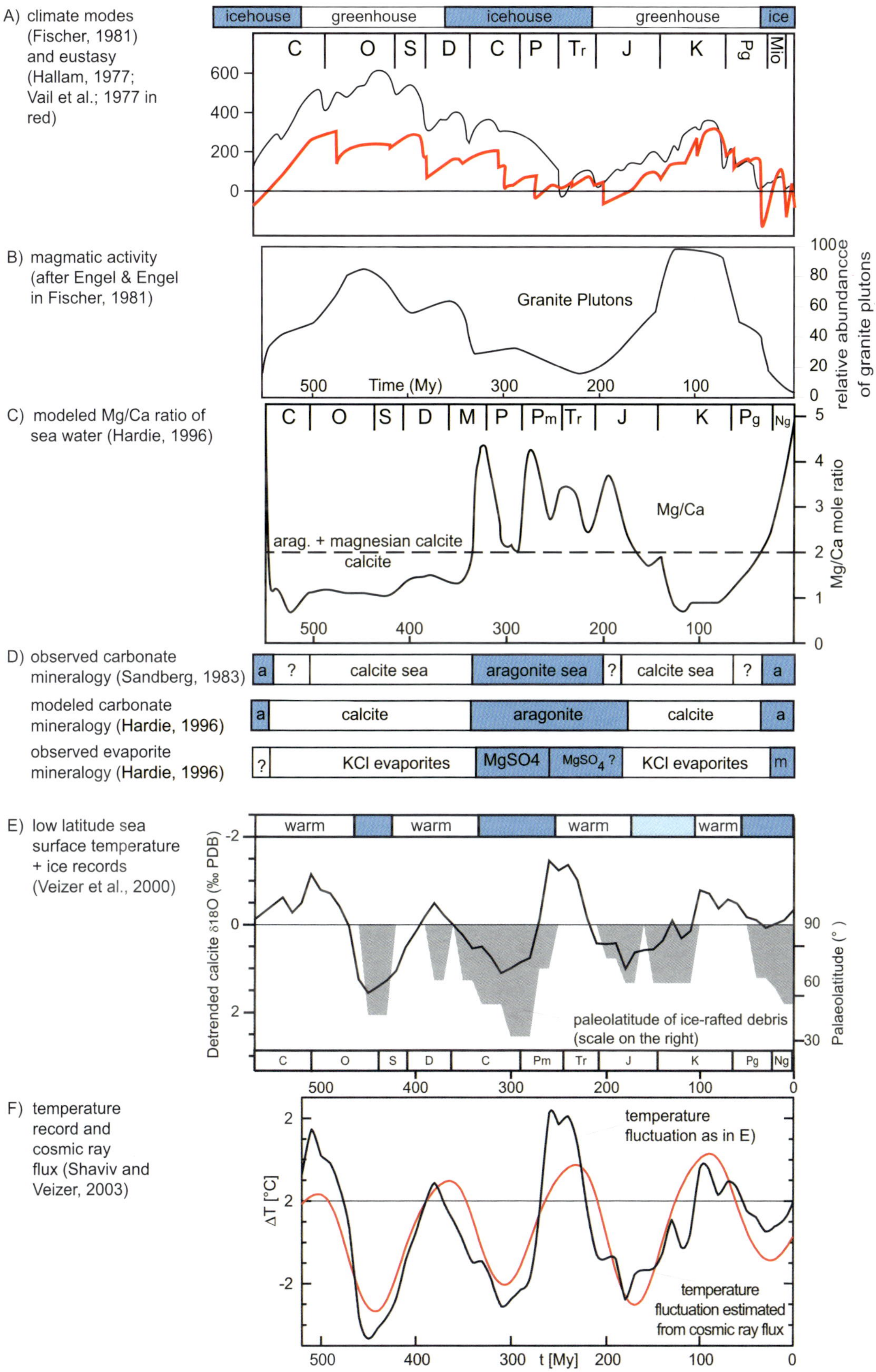

Fig. 5.9.— Long oscillations observed in the Phanerozoic sediment record and their proposed interpretations. Figs A - D present an internally consistent pattern of an oscillation with about 300 My period that affects climate, sea-level and the mineralogy of carbonates and evaporites; the oscillation is in phase with magmatic activity and with postulated changes in the rates of plate motion. Figs E – F indicate an oscillation of about 135 My period that approximately correlates with a celestial driver. Compiled from references shown in the left column.

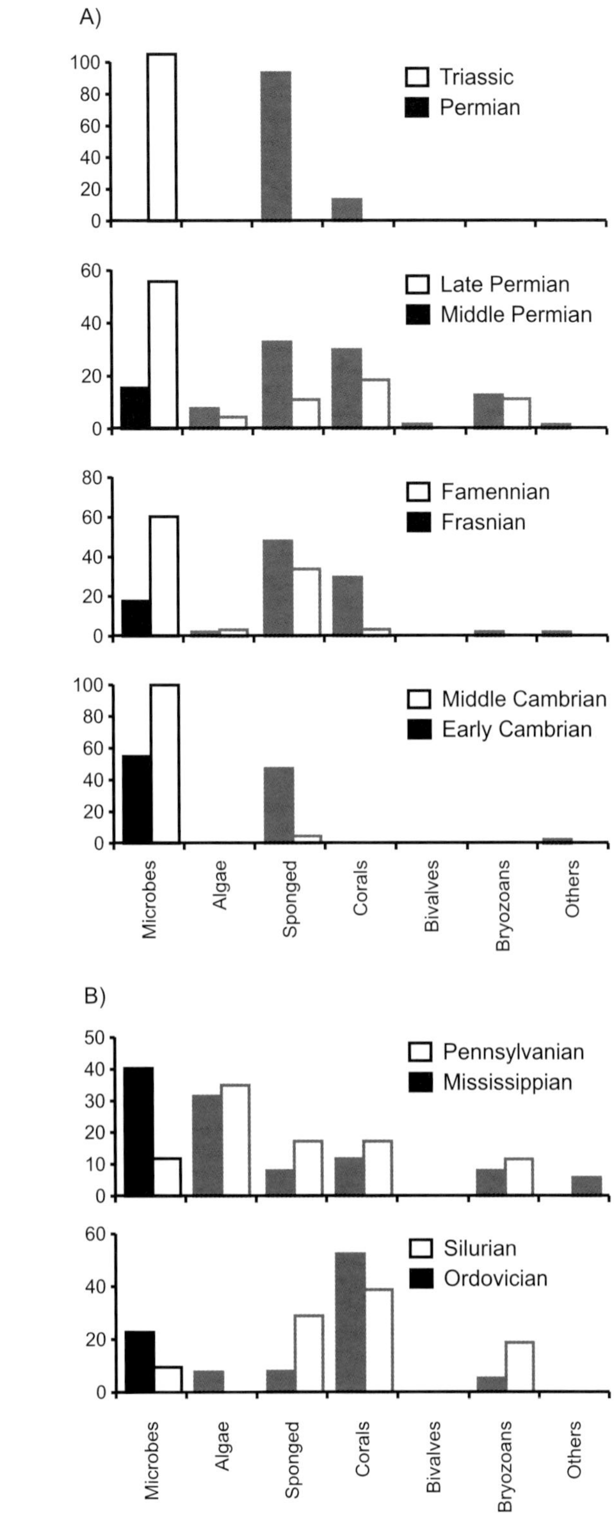

Fig. 5.10.— A) Phanerozoic extinction events that caused a shift from predominantly skeletal to microbial precipitation and from T to M factory. B) Extinction events that cause a shift from microbial to skeletal dominance. After Flügel and Kiessling (2002), modified.

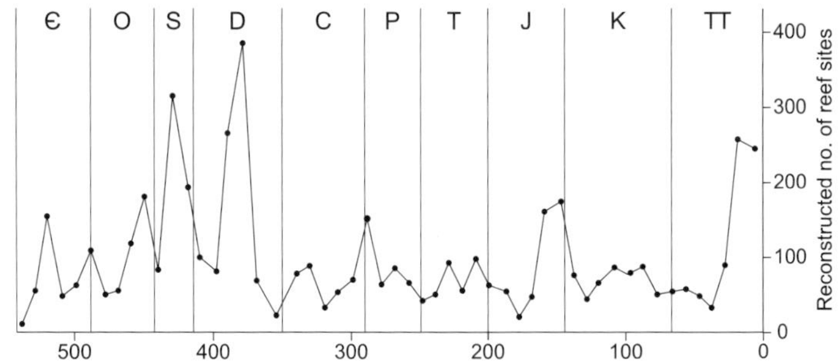

Fig. 5.11.— Number of documented Phanerozoic reefs fluctuates but the fluctuations lack a constant period. After Kiessling (2002), modified.

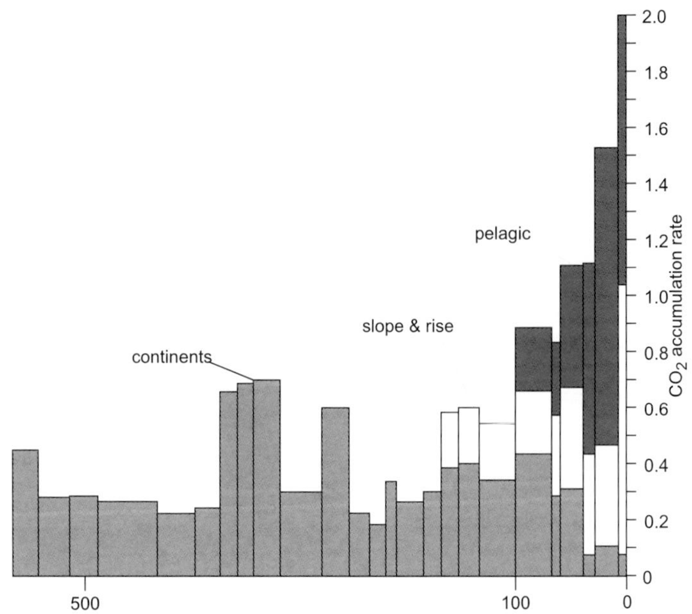

Fig. 5.12.— Carbonate accumulation rates in the Phanerozoic (expressed as units of 10^{21} g/My of CO_2). Drastic increase of accumulation on passive margins and in the pelagic domain since the mid-Cretaceous is attributed to the advent of calcareous plankton. After Hay (1985), modified.

CHAPTER 6

Fundamentals of sequence stratigraphy

INTRODUCTION

PRINCIPLES AND DEFINITIONS

Careful definitions and consistent terminology have been a hallmark of sequence stratigraphy from the outset. By now, however, the field suffers from an inflation of terms. I will introduce and use only the most essential ones. Others are explained elsewhere and can be recovered via the index.

The discussion in this chapter assumes that the following principles hold.

- Sedimentary bodies and thus all physical stratigraphic units are of finite extent - they are lenses in the most general sense of the term.
- Surfaces of erosion or non-deposition are also of limited extent. On the flat basin floors, the ultimate limits are set by the principle that siliciclastic sediment mass is conserved. Basins continually receive sediment and the water column cannot hold this material for any geologically significant time. As dissolution of siliciclastic sediment is negligible, the bulk must come rest on the basin floor. Thus, erosion in part of the basin must be compensated by increased sedimentation in others. For carbonate material, the conservation of mass does not hold for the abyssal sea floors but it remains a reasonable assumption for neritic and bathyal settings.
- The sediment record is dissected by hiatuses at all geologically relevant scales. The assumption of a "complete section" is an (often useful) idealization.
- The sediment record at any given locality reflects the history of *relative* sea-level movements. The history of eustasy requires either evidence of a world-wide phenomenon (such as orbital signals), or a global stack of records plus constraints from geodynamics. Geodynamic input is necessary because crust and mantle deform under the changing water load and therefore the same eustatic signal is recorded differently in different regions (Watts, 2001).

Sequence and sequence boundary

When Sloss (1963) formally introduced the concept of depositional sequences, he defined sequences as "rock stratigraphic units ... traceable over major areas of a continent and bounded by unconformities of interregional scope". Vail et al. (1977) adjusted this definition for work with smaller units and seismic data. These authors defined a depositional sequence as a "stratigraphic unit composed of a relatively conformable succession of genetically related strata and bounded at its top and base by unconformities or their correlative conformities". The definition by Vail et al. (1977) is in essence a geometric one and therefore well suited for seismic interpretation. It contains no statement on scale in time or space. The definition furthermore avoids any reference to the origin of a depositional sequence even though Vail et al. (1977) clearly expressed their conviction that sea level fluctuations were the dominant control.

Sequence boundaries were defined by Vail et al. (1977, p. 53) as "observable discordances ... that show evidence of erosion or nondeposition with obvious stratal terminations but in places they may be traced into less obvious paraconformities recognized by biostratigraphy or other methods". This definition, too, is essentially based on geometric relationships and again abstains from any statement on scale in time or space. Vail et al. (1977, p. 55) add that the hiatus at the unconformity generally is on the order of a million to hundreds of millions of years. The definition of sequence boundary also lacks a statement on the origin of the unconformity (except for the reference to erosion and non-deposition).

Van Wagoner et al. (1988, p. 39) added an important genetic qualification to the definition of sequence and sequence boundary. They follow Vail et al. (1977) in defining a depositional sequence as an unconformity-bounded succession of strata but then proceed to re-define unconformity as "a surface ... along which there is evidence of subaerial erosional truncation (and, in some areas, correlative submarine erosion) or subaerial exposure". This definition has been used in several key publications (e.g. Posamentier et al., 1988, p. 110; Emery, Myers et al.,1996, p. 24). Van Wagoner et al.'s definition makes for an internally consistent concept: all sequence boundaries are related to relative falls of sea level and thus sequences are predominantly controlled by sea-level fluctuations. In spite of this, I recommend not to use this unconformity definition in sequence stratigraphy because it is at odds with the traditional usage of the term unconformity and faces grave difficulties in practical application. Van Wagoner et al.'s definition of unconformity is much more restrictive than the generally accepted connotation of this old and widely used term (for instance Bates and Jackson, 1987). Unlike the geometric features used by Vail et al. (1977), the criterion of subaerial exposure is extremely difficult to verify in seismic data. Even in outcrop studies, demonstration of subaerial exposure in carbonates often requires extensive laboratory analyses (e.g. Esteban and Klappa, 1983; Goldstein et al., 1990; Immenhauser et al., 1999). In siliciclastics, the situation is further complicated by the fact that much of the record of subaerial exposure is washed away by the subsequent transgression.

The original definition of sequence boundary as an unconformity between two units of conformable, genetically related strata does not automatically imply sea-level control. It simply means that at this boundary the pattern of sediment input or sediment dispersal changed abruptly. Sea-level change commonly causes these shifts in input and dispersal because it reshuffles the sediment pathways on the shelf and slope. However, there is a significant number of examples where sea level interpretation is inadequate or outright inappropriate. An example from the Gulf of Mexico may serve to illustrate this point.

Figs 6.1 6.2 shows the mid-Cretaceous sequence boundary (MCSB, Buffler, 1991). It is a basin-wide seismic marker that

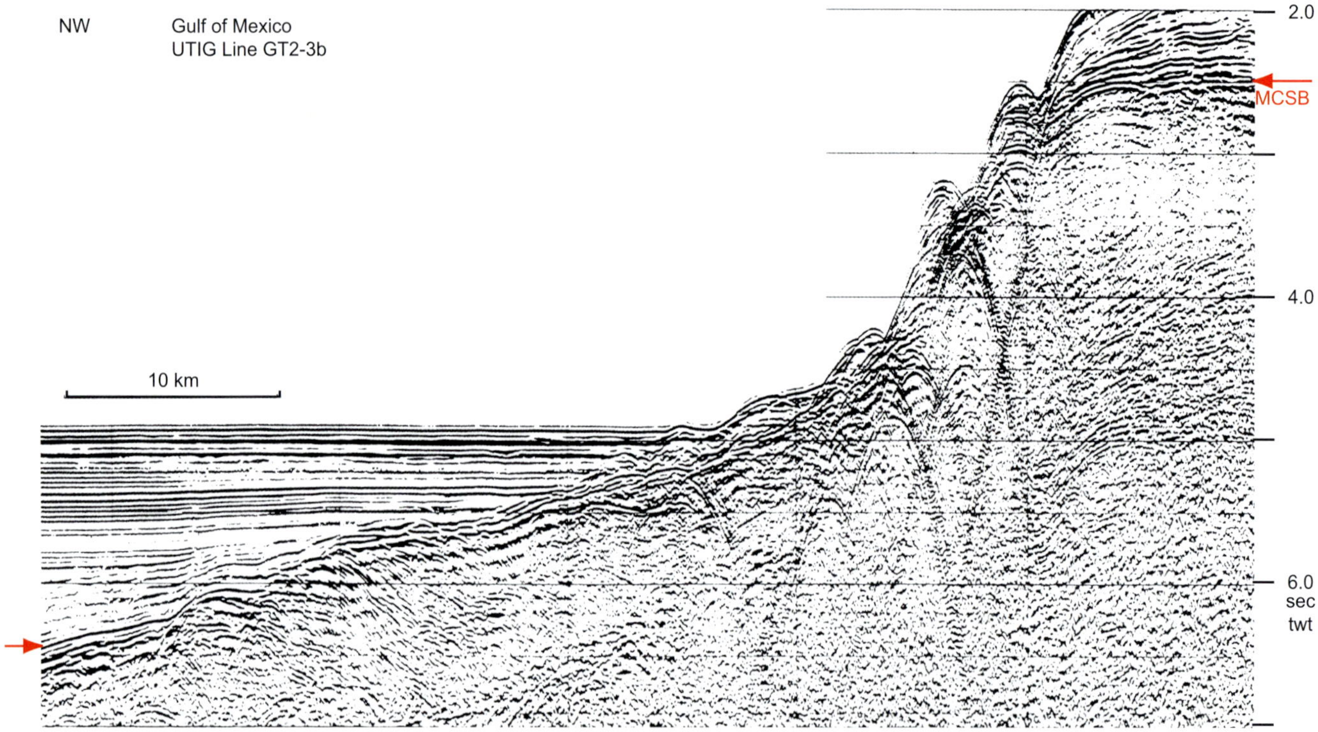

Fig. 6.1.— Change of sediment input at mid-Cretaceous sequence boundary (MCSB, marked by red arrows) on Campeche Bank, Gulf of Mexico. The carbonate platform on the right was drowned in the mid-Cretaceous and later covered with pelagics. As carbonate input ceased, the debris apron of the platform was gradually buried by clastic sediments transported at right angle to the profile. Seismic data courtesy of R.T. Buffler.

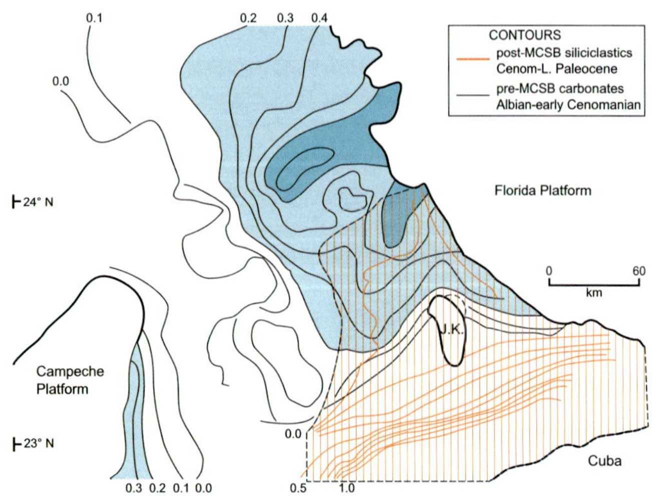

Fig. 6.2.— Sediment isopachs above and below the mid-Cretaceous sequence boundary of Fig. 6.1 show the change in sediment input at the unconformity. Pre-unconformity isopachs reflect the influx of carbonate debris from the platforms, post-unconformity isopachs reflects the influx of siliciclastic material from the advancing Cuban island arc. After Schlager (1989), modified.

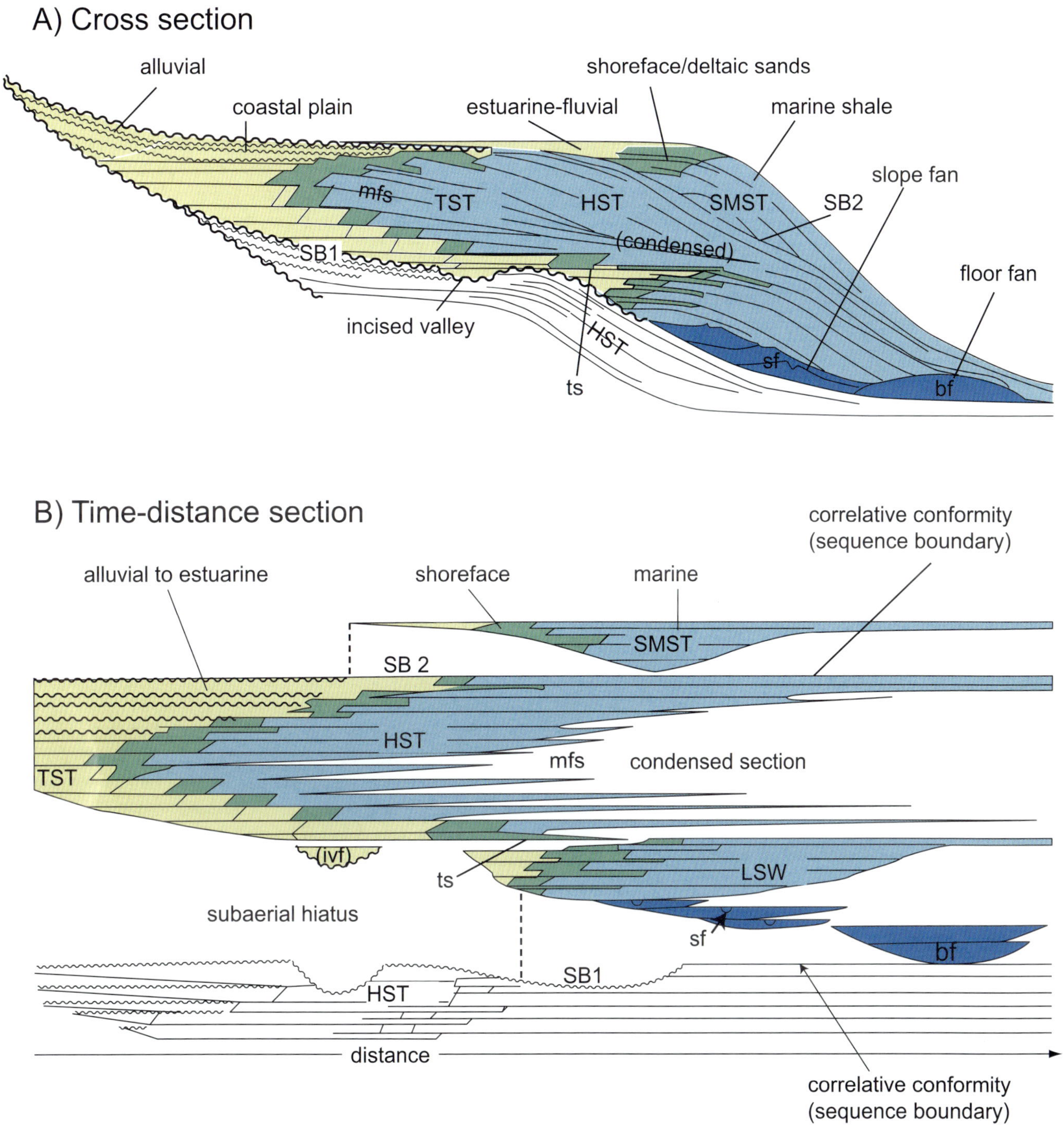

Fig. 6.3.— Stratigraphic sequences and their systems tracts . Abbreviations in alphabetic order: bf = basin-floor fan; HST = highstand systems tract; LSW = lowstand wedge; mfs = maximum flooding surface; SB1 = type-1 sequence boundary; SB2 = type-2 sequence boundary; sf = slope fan; SMST = shelf-margin systems tract; ts = transgressive surface; TST = transgressive systems tract. A) Schematic cross section of the standard model B) Section with distance on horizontal axis (same scale in A) and time on vertical axis ("Wheeler diagram"). After Vail (1987), modified.

has been dated as late Albian to Cenomanian and was tentatively correlated with the 94-My lowstand on the curve of Haq et al. (1987) (Winker and Buffler, 1988). However, at least two other lowstands, at 96 and 98 My are equally possible given the stratigraphic data. Furthermore, there is a mismatch between the prominence of the sequence boundary and the amplitude of the invoked sea-level events. The MCSB is arguably the most prominent sequence boundary in the Gulf whereas the amplitudes of the mid-Cretaceous sea-level falls are very modest. On the curve by Haq et al. (1987) they are predated by two pronounced Valanginian lowstands and postdated by a Turonian one. All of them have about twice the amplitude of the mid-Cretaceous events, yet none of them has a comparable seismic expression in the Gulf. Thus, the postulated correlation to eustatic events fails to explain the prominence of the MCSB. What sets the MCSB apart from all other sequence boundaries in the Gulf is the associated change in depositional regime: the mid-Cretaceous sequence boundary marks the termination of the rim of carbonate platforms around the Gulf and the spread of pelagic deposits and marine hardgrounds, later to be covered by Tertiary siliciclastics (Schlager et al., 1984; Buffler, 1991). The change in deposition goes hand in hand with a drastic change in the input and dispersal of sediment in the basin (Fig. 6.2). The unconformable nature of this boundary is accentuated by the fact that the platforms had high and steep flanks when they were drowned and this pronounced relief tends to amplify ocean currents (see chapter 5).

Systems tracts

The term "depositional system" was introduced by Fisher and McCowan (1967) for a three- dimensional assemblage of lithofacies genetically linked by a common set of depositional processes. Rivers, deltas and slopes are examples of depositional systems. Coeval systems are often linked by lateral transitions, for instance along a topographic gradient, to form systems tracts. The most common example of a systems tract is the succession of systems encountered in a traverse from basin margin to deep water. Such a transect may cross the systems river, delta, shelf, slope and basin floor.

Sequence stratigraphy has adopted and somewhat modified the concept of systems tracts. The standard model of sequence stratigraphy stipulates that the systems tract from basin margin to deep water varies in a systematic fashion during a sea-level cycle such that lowstand, transgressive and highstand systems tracts can be distinguished (Posamentier and Vail, 1988). Fig. 6.3 shows the standard model applicable to siliciclastics, tropical carbonate ramps and cool-water carbonates. For systems-tract definitions of rimmed platforms see chapter 7. The lowstand systems tract consists of the suite of depositional systems developed when relative sea level has fallen below an earlier shelf margin. The transgressive systems tract consists of the depositional systems developed when sea level rises from its lowstand position to an elevation above the old shelf margin and depositional environments shift landward. Finally, the highstand systems tract consists of the depositional systems developed then sea level stands above the old shelf margin and depositional environments and facies belts prograde seaward. The standard model postulates further that systems tracts follow each other in regular fashion. The lowstand systems tract immediately overlies the sequence boundary, the transgressive sytems tract occupies the middle, the highstand tract the top of a sequence. Systems tracts in sequence stratigraphy were originally defined by lap-out patterns at the base and top, internal bedding, stacking patterns and position within a sequence (Posamentier et al., 1988, 110; Van Wagoner et al., 1988, 42; Emery et al., 1996, p. 26). All these criteria are based on geometry. The characterization of sequence systems tracts in terms of facies are a later addition.

The standard model assumes that the fall of sea level from highstand to lowstand position does not leave a significant sediment record. Subsequent work on outcrops and cores has shown that this generalization is not justified. There exists a growing number of examples where the retreating sea has left a significant sediment accumulation that records the downward shift of the shoreline and shelf surface (Hunt and Tucker, 1992; Nummedal et al., 1995; Naish and Kamp, 1997; Belopolsky and Droxler, 2004). These observations agree with theoretical considerations (Nummedal et al., 1993) and insights from numerical modeling. Numerical models allow one to explore under what circumstances the falling sea may produce a sediment accumulation. It turns out that the parameter space for generating a falling-stage systems tract is quite large while the space for creating the geometry of the standard model is small. To produce the erosional unconformity of the standard model one has to assume either intensive terrestrial erosion or a highly asymmetric sea-level cycle with a rapid fall (Figs 6.4, 6.5). Extreme erosion rates, on the other hand, are incompatible with the very minor erosion shown in the classic diagram of the standard model. In summary, field observations and numerical models indicate that sea-level falls are likely to produce a sediment record. The situation shown in the standard model, i.e. only modest erosion of the highstand tract and no sedimentation during sea-level fall, requires a highly asymmetric sea-level cycle with a very rapid fall. Diagrams relating the standard model to a symmetric sine wave of sea level are misleading.

Several different names have been suggested for the sediment body formed during sea-level fall: Forced-regressive wedge systems tract (Hunt and Tucker, 1992), falling-stage systems tract (Nummedal et al., 1995) and regressive systems tract (Naish and Kamp, 1997) are commonly used. I prefer *"falling-stage systems tract (FST)"* because it refers to the critical process - the relative fall of sea level that can be directly deduced from the geometry or the facies pattern of the systems tract. The phrase "forced-regressive wedge systems tract" is a bit awkward and the term "regressive systems tract" does not distinguish between forced regression and depositional regression, for instance during highstand

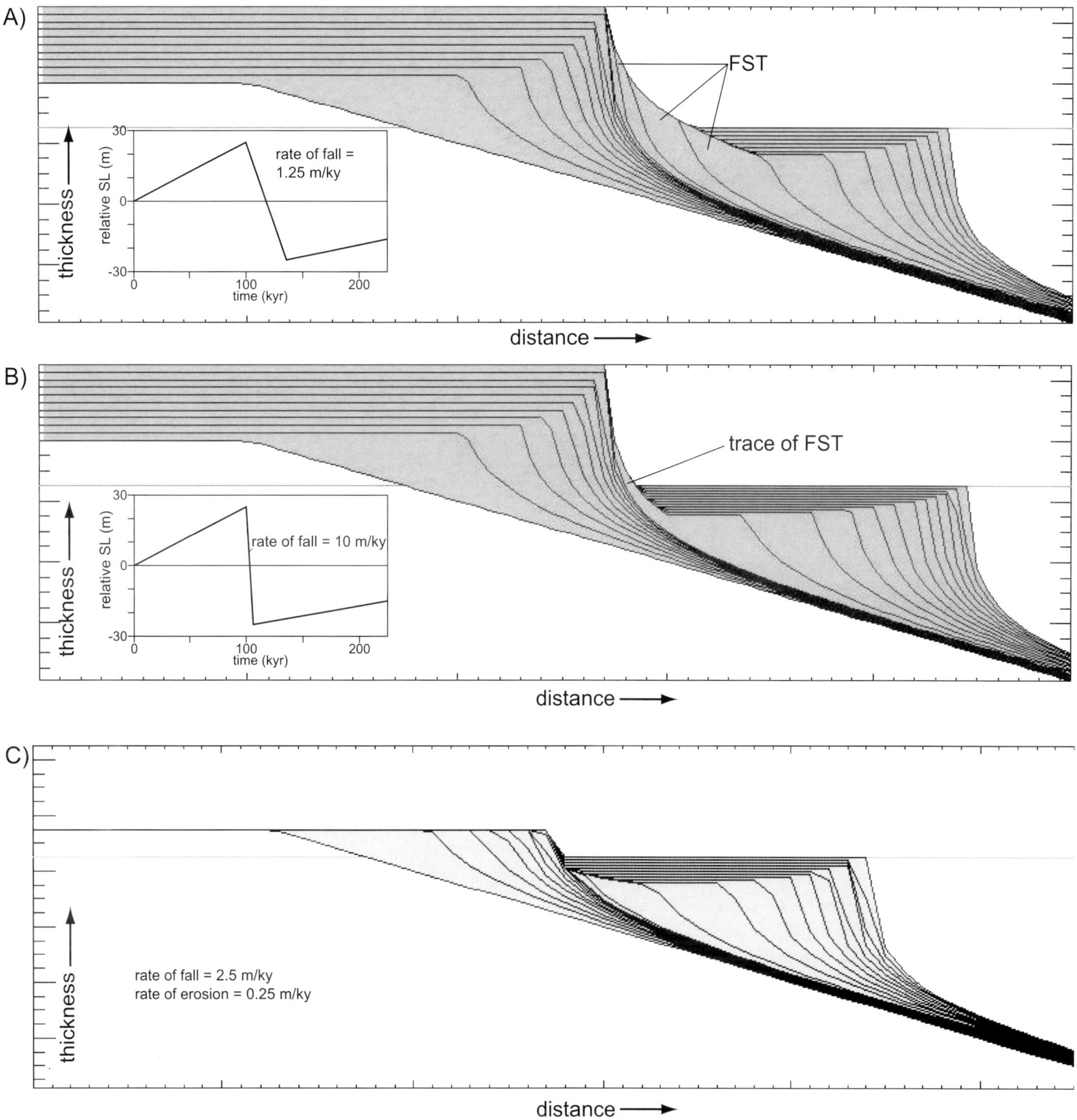

Fig. 6.4.— Modelling runs with CARBONATE 3D (appendix B) illustrating the difficulty of avoiding formation of a falling-stage systems tract in a system that resembles the standard model of Fig. 6.3. Forcing with symmetrical sea-level waves invariably produces a FST. Even an asymmetric fluctuation with rapid fall of 1.25 m/ky is insufficient to suppress FST. A record resembling the standard model requires a rate of fall of 10 m/ky as in B). The FST can also be eliminated if slower sea-level fall is coupled with subaerial erosion as in C). However, this setup severely truncates the preceding HST.

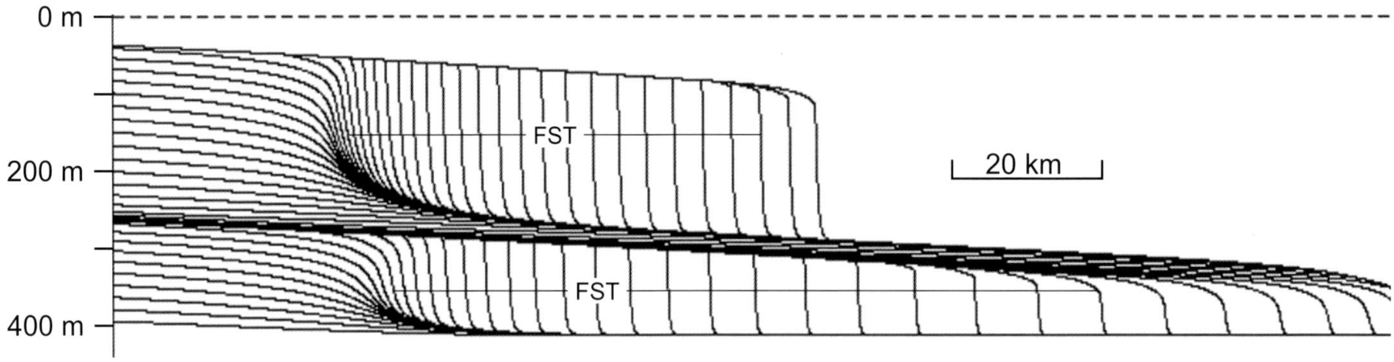

Fig. 6.5.— Model of siliciclastic falling-stage systems tract (STRATA program, appendix B). Shelf subsides uniformly, is supplied with material from left and is exposed to sinusoidal sea-level fluctuations. Falling limb of sea-level cycle produces extended FST. Standard LST forms during the last part of the falling limb and the earliest part of the rising limb.

progradation.

I consider the falling-stage systems tract an important addition to the systems-tract model but do not think that the category is as fundamental as lowstand, transgressive and highstand tracts. Systems tracts have been defined by geometry because the direct link to sea-level remains speculative (Posamentier and Vail, 1988). In chapter 7, geometry is used to define systems tracts on rimmed carbonate platforms. If one applies the same principles here, then the lowstand tract would be defined as a unit whose shoreface and shelf surface are lower than the respective surfaces of the preceding highstand tract. This definition puts the falling stage systems tract in the lowstand category - in agreement with the argumentation in Posamentier and Allen (1999). The falling-stage systems tract differs from the lowstand tract of the standard model by the downward shift of the shelf surface during deposition. Posamentier and Allen (1999) proposed to refer to falling-stage tract and standard lowstand tract as early and late lowstand tracts respectively. This is fine but the term falling-stage systems tract is useful as it describes the critical process. See Plint and Nummedal (2000) for examples and discussion.

The discussion around the falling-stage tract illustrates the importance of a reference profile when defining systems tracts by geometric criteria. If not stated otherwise, one should assume that the reference level is the top profile of the immediately preceding sequence of the succession.

The subdivision of sequences into systems tracts led to the recognition of two other bounding surfaces besides the sequence boundary.

The *transgressive surface* forms the boundary between lowstand and transgressive tract. It "... marks the initiation of transgression after regression ..." (Posamentier and Allen, 1999, p. 95) and "... represents the first significant flooding surface across the shelf within a sequence ..." (Van Wagoner et al., 1988, p. 44).

The *maximum flooding surface* constitutes the boundary between transgressive and highstand tract. It represents "... the surface that exists at the time of maximum transgression of the shelf ..." (Posamentier and Allen, 1999, p.95). The maximum flooding surface is also called "downlap surface" on seismic data because along it the clinoforms of the highstand tract downlap on the transgressive systems tract (Fig. 6.3). Maximum flooding surface seems a better term because downlap is also common at the sequence boundary where the clinoforms of the lowstand systems tract downlap on the distal parts of the preceding highstand tract.

In both instances, the term "surface" could just as well be replaced by "interval". The term surface has its justification in seismic interpretation. In seismic data one often can identify one reflection as the horizon of lapout. However, several authors have pointed out that in boreholes this apparent lapout frequently corresponds to transitional lithologic boundaries (e.g. Van Hinte, 1982; Posamentier and Allen, 1999, p. 97). The lithologic transitions indicate that sedimentation continued, albeit at a lower rate, and that there is no single surface of lapout.

"Condensed section" is another term in the systems-tract literature that merits discussion. In Fig. 6.3 we see that the areas of lapout in the cross section appear as condensed section in the Wheeler diagram. This reminds us again that the classic diagrams were derived from and designed for seismic interpretation. The condensed section in the dia-

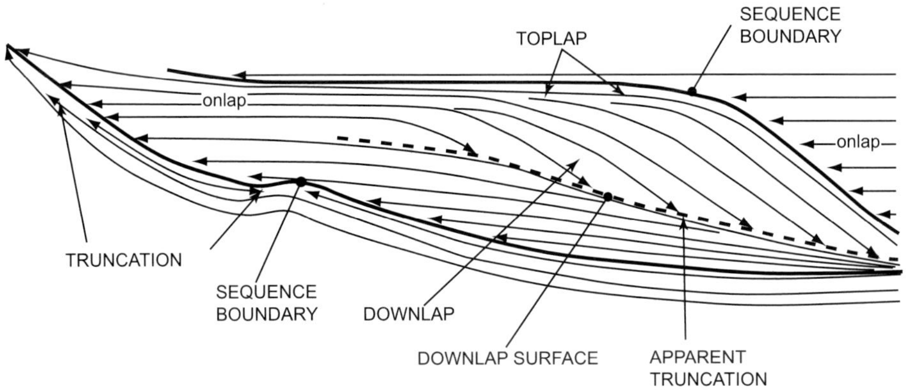

Fig. 6.6.— Lapout patterns in a sequence of the standard model. Note that besides the sequence boundaries there are downlap and toplap patterns, i.e. unconformities in common geologic language, within the sequence. After Vail (1987). (Reprinted by permission of the AAPG whose permission is required for further use).

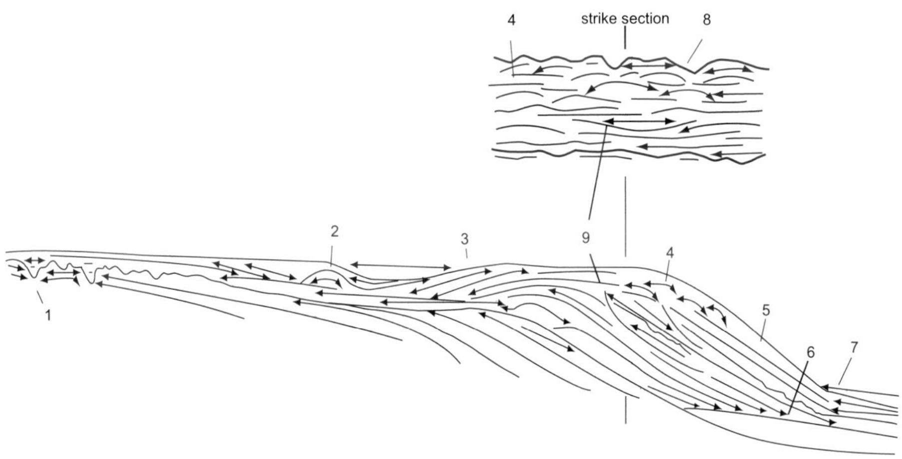

Fig. 6.7.— Lapout patterns in carbonate sequences are more varied than in siliciclastics particularly because there are many localised centers of high production, such as reefs, and localised areas of erosion, such as inter-reef channels. Particularly characteristic is the elevated platform margin that may simultaneously prograde landward and seaward. Numbers refer to characteristic situations: (1) karst buried by marine sediment, (2) lagoonal patch reefs or mounds, (3) prograding backreef apron, (4) bioherms at platform margin, (5) slope clinoforms, (6) slope clinoforms downlapping on basin floor, (7) onlapping basin-floor sediments, (8) incisement of shelf edge at sequence boundary, (9) shelf-margin incision within sequence. After Handford and Loucks (1993), modified.

gram represents pelagic or hemipelagic sediments whose sedimentation rate is sufficiently low that they frequently are not resolved by seismic data. It is important to note that this does not automatically imply that these sediments show evidence of biostratigraphic condensation, i.e. fossils of several biozones in one layer (Heim, 1934); nor does it necessarily imply field-geologic evidence of starved sedimentation such as hardgrounds or authigenic minerals in the form of Fe-Mn oxide crusts, phosphate crusts or glauconite grains (Flügel, 2004, p. 211-216).

For the sequence stratigrapher the term "condensation" is broader than for the biostratigrapher or the sedimentologist. The condensed facies in sequence stratigraphy includes the facies with evidence of biostratigraphic or sedimentologic condensation mentioned above but the term also includes normal pelagic or even hemipelagic facies of the ocean basins with sedimentation rates of 10 μ/y or higher (Loutit et al. 1988, p. 186).

The geometry of systems tracts leads to characteristic stratal patterns in seismic profiles and large outcrops (Fig. 6.6; Fig. 6.7). It should be noted that the lapout patterns indicated in this scheme refer to seismic lapout and that seismic lapout does not necessarily imply a genuine unconformity in outcrop or cores (see sections on unconformities, p. 90f and pseudo-unconformties p. 127 f).

Stratigraphic time lines and seismic reflections

Field geology of continental interiors was the cradle of sequence stratigraphy but what makes it such an important tool at this day and age is its immediate applicability in seismic interpretation. In fact, most of what here is called the standard model of sequence stratigraphy was introduced under the heading "seismic stratigraphy" by Vail et al. (1977). The connection between seismic reflections and time lines (or time surfaces) in the sediment record is of pivotal importance in this context and merits some comment.

Vail et al. (1977, p. 51) commence their classic expose with the following statements: "Primary seismic reflections are generated by physical surfaces in the rocks ... Therefore, primary seismic reflections parallel stratal surfaces and unconformitie", and "... the resulting seismic section is a record of the chronostratigraphic ... depositional and structural patterns and not a record of the time-transgressive lithostratigraphy ...".

I believe that these elegant, clear and simple statements are somewhat misleading because of their categoric style. In a questionaire that leaves only the options, right and wrong, the statement - seismic reflections are time lines - would have to be classified as right. However, the refined statement - seismic reflections always represent time lines - is wrong and a number of authors have indicated this (e.g. Christie-Blick et al., 1990; Tipper, 1993; Emery et al. 1996, p. 45). Sheriff and Geldart (1995, p. 150) succinctly summarize the situation "... overwhelming evidence indicates that reflections coincide with time boundaries except in occasional unusual circumstances". In fact, Vail et al. (1977) themselves provide an example of such "unusual circumstances", shown in Fig. 6.8. The task before us as sedimentologists and stratigraphers is to acknowledge the overwhelming evidence for time-parallel reflections without losing sight of the, often important, exceptions such as the ones presented in the next section.

Unconformities in outcrop and seismic data

The above discussion on sequences, time lines and seismic reflections automatically leads to the stratigraphic unconformity as a particularly important element in sequence-stratigraphic theory. As sequence stratigraphy is applied to both outcrops and seismic data, it is important to realize that outcrop unconformities and seismic unconformities do not always match. Ground truthing by bore holes and seismic modelling of outcrops provide evidence of this mismatch. Three different situations may develop.

1. Boundaries where both outcrop bedding and seismic reflections abut against a surface. This situation is unproblematic because the field geologist and the seismic interpreter will classify this feature as an unconformity.
2. There are outcrop unconformities that cannot be seen in seismics because their geometric expression consists of a microrelief on the scale of centimeters or decimeters while the large-scale bedding of the two units remains parallel. Common in deep-sea sediments.
3. Seismics shows unconformities that correspond to transitional boundaries in outcrop. Two situations must be distinguished:
   (a) Seismic unconformities where time lines converge into an interval of continuous but very slow sedimentation and where the seismic tool portrays this condensed interval as a lapout surface. In most instances this condensed interval meets the definition of unconformity by Vail et al. (1977) in so far as all deposits below the condensed unit are older than the oldest deposits above it. The difference to the classical situation is merely that "non-deposition" needs to be replaced by "slow deposition". For the seismic interpreter this is a minor difference, but one that should be kept in mind when seismics is tied to boreholes or outcrops. (Van Hinte, 1982)
   (b) Most disturbing are pseudo-unconformities where a rapid facies change occurs in each bed at a similar position and the seismic tool merges these points of change into one reflection. Time lines cross this reflection; it is therefore not an uncon-

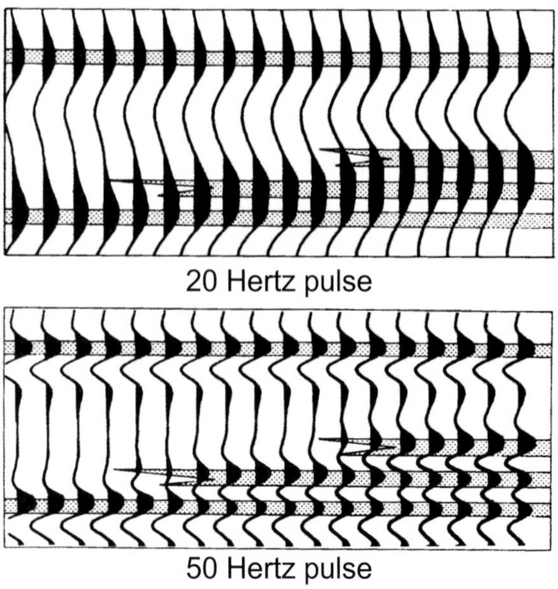

Fig. 6.8.— Seismic model of sandstone interfingering with shale. At 20-Hz frequency, the lower reflection runs oblique to bedding, approximately connecting the terminations of the sandstone tongues. At 50 Hz, the individual sandstone tongues are revealed by reflections parallel to bedding. After Vail et al. (1977). (Reprinted by permission of the AAPG whose permission is required for further use).

formity the sense of Vail et al. (1977) and needs to be recognized and eliminated before sequence analysis is done (Fig. 6.8, 6.9 and 6.10). Carbonate-siliciclastic transitions and sand-shale transitions are particularly prone to this effect.

Recognition of pseudo-unconformities still is in its infancy. Vail et al. (1977) spotted the phenomenon (Fig. 6.8) but did not think it was serious. I believe it deserves some attention. In seismic models of outcrops the true nature of the interfingering can often be revealed by increasing the wave frequency. Before the pattern is completely resolved, a series of short, en-echelon reflections appears in the transition zone. Complex-trace attributes such as instantaneous phase and reflection strength offer additional possibilities to recognize pseudo-unconformities (Bracco-Gartner and Schlager, 1999).

Parasequence and simple sequence

The parasequence was originaly defined as "... relatively conformable succession of genetically related beds ... bounded by marine flooding surfaces or their correlative surfaces" (Van Wagoner et al., 1988, p. 39). The marine flooding surface was defined by the same authors as "a surface ... across which there is evidence of an abrupt increase in water depth. This deepening is commonly associated by minor submarine erosion (but no subaerial erosion or basinward shift in facies) ... ".

The simple sequence was defined by Vail et al., 1991, p. 630) as a unit that "has the stratal and lithologic characteristics of a sequence, but its duration is that of a parasequence". The authors indicate that "simple sequences are picked where possible, but in most cases simple sequence boundaries are difficult to identify, therefore the more readily recognizable parasequence boundaries are used."

There can be little doubt that shoaling successions of the parasequence type are a common feature of the sediment record. Van Wagoner et al. (1988) point out that they are best developed in deposits of the coastal plain, the nearshore zone and the shelf. They are difficult to define in fluvial and deep-water deposits. This distribution points to the origin of the shoaling successions: they reflect sedimentation in settings with limited accommodation where the sedimentation pile gradually approaches the accommodation limit (in carbonates the upper supratidal zone). There is no significant fall of relative sea level and the sea transgresses the flats again by gentle inundation with only minor erosion. Shoaling cycles of this kind are a well-established pattern in shoal-water carbonates (chapter 3). However, shoaling cycles bounded by flooding surfaces are not the only pattern in shoal-water carbonates. Symmetrical shoaling-deepening cycles, deepening cycles or oscillating successions without distinct shoaling or deepening trends are quite common (e.g. Enos and Samankassou, 1998).

In contrast to the parasequence, the simple sequence requires falls of relative sea level as bounding events. This pattern, too, is common in the geologic record, particularly in epochs with pronounced glacio-eustasy.

Application of the concept of parasequence and simple sequence requires that a clear distinction be made between marine flooding surfaces reflecting abrupt change from shallow to deep, and exposure surfaces reflecting change from shallow marine to exposure to deep flooding. In carbonates, this can be quite cumbersome as the upper supratidal zone already develops fresh-water diagenetic overprints and terrestrial exposure surfaces need not appear as dramatic breaks in the record. As a consequence, this distinction has been largely ignored in the literature such that the Pleistocene glacial-interglacial cycles of the Bahama Banks and other extant platforms were classified as parasequences even though they are bounded by exposure unconformities during which sea level stood 100 m or more below the platform top (e.g. Kievman, 1998).

Sea level

Sea level is a crucial element in sequence stratigraphy. According to the standard model, it is the principal cause of stratigraphic sequences and it certainly represents one of the most important environmental boundaries in sedimentology. Sea level is easy to visualize but measuring the elevation of the sea surface and its change with time is a formidable task.

Relative sea level refers to sea level measured relative to a fixed point on land (e.g. Revelle et al., 1990). In ocean engineering, this fixed point is normally chosen at the land surface near the coast. Sequence stratigraphy choses a fixed point in the sediment pile, preferably near or at its base (Posamentier et al., 1988, p. 110; Emery et al. 1996, p. 16). Many stratigraphers and sedimentologists take, explicitely or by tacit assumption, the sea bottom as the fixed point for measuring relative sea level (Hallam, 1998). The difference between this common stratigraphic practice and the approach of sequence stratigraphy is fundamental. Relative sea-level changes determined by the sequence-stratigraphic approach represent the sum of eustatic and tectonic movements, relative sea-level changes measured from the sea floor yield the sum of eustasy, tectonics, sediment accumulation and sediment compaction. By sequence-stratigraphic standards, a 1,000 m-pile of shoalwater limestones deposited at sea level has recorded a 1,000-meter rise of relative sea level (probably largely caused by regional subsidence); by common stratigraphic standards, the limestone pile indicates no relative change of sea level (or only minute oscillations indicated by alternation of supratidal and subtidal facies).

Eustatic sea level. When he introduced the term, Suess (1888) called eustatic changes those that originated in the sea and contrasted them with sea-level changes that were caused by uplift or subsidence of the surface of the solid Earth. Nowadays, eustatic sea level is defined as the elevation of the

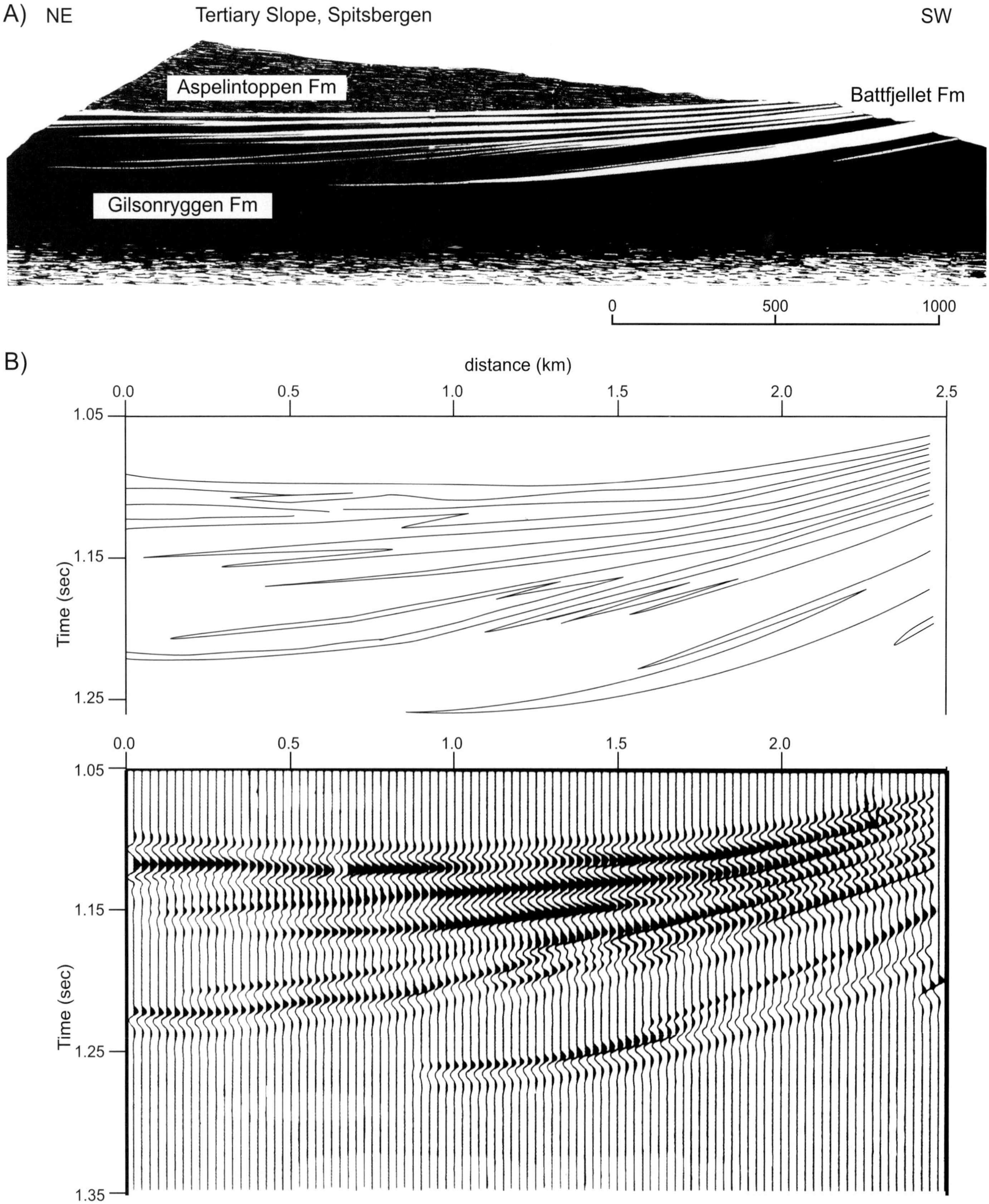

Fig. 6.9.— Pseudo-unconformities in seismic outcrop model, Tertiary, Spitsbergen . A) Outcrop geology shows clinoforms of sandstone (white) interfingering with shale (dark). B) Boundaries of impedance model (upper panel) and vertical-incidence seismic model (lower panel). Seismic model shows false lapout where resolution is insufficient to portray the interfingering in detail (e.g. 0.7 – 1.9 km/1.16 – 1.18 sec, 1.6 –2.2 km/1.19 – 1.24 sec). After Helland-Hansen et al. (1994), modified.

global sea surface relative to a fixed datum on the planet, such as the center of the Earth (Kendall and Lerche, 1988, p. 3). Estimating the eustatic sea level of past epochs is very difficult as it depends on the use of proxy indicators (Kendall and Lerche, 1988; Harrison, 1990).

Regression. The distinction between relative and eustatic sea level is not sufficient to properly extract the sea-level signals from the stratigraphic record. Yet another distinction needs to be made, the one between depositional and erosional regression (Grabau, 1924 and Curray, 1964) or normal regression and forced regression (Posamentier et al., 1992b). Depositional or normal regression develops where the rate of sediment supply to the coastal zone exceeds the rate of accommodation creation by relative sea-level rise. Erosional or forced regression is caused by a fall of relative sea level; under this condition the shoreline shifts seaward (and downward) irrespective of sediment supply. The progradation of the highstand tract produces normal regression, the downstepping from highstand to lowstand during formation of the sequence boundary or the downward shift within a falling-stage systems tract are examples of forced regression.

Forced regression plays a pivotal role in the construction of relative sea-level curves because it is clear evidence of a relative sea-level fall, whereas the progradation and retrogradation of highstand tracts and transgressive tracts may be caused by changes of sea level or sediment supply (p. 94f; Jervey, 1988; Schlager, 1993). Distinguishing normal and forced regression in siliciclastics relies on geometric criteria, such as downstepping of the shelf break or incised valleys, as well as facies patterns such as shoreface deposits with erosional base (Posamentier et al., 1992b; Naish and Kamp, 1997). On carbonate platforms, downstepping of the margin is a good criterion, particularly since the shelf break is often better defined than in siliciclastics (chapter 3). Lithologic evidence of exposure includes karst, soils, relicts of terrestrial plants etc. Freshwater diagenesis alone is not diagnostic because it may also develop during depositional regression when the system builds into the high supratidal zone, for example on tidal flats (e.g. Halley and Harris, 1979; Gebelein et al., 1980,p. 45).

Sea level from sequence anatomy. Stratigraphers are historians and, like most historians, are in danger of overinterpreting the documents at hand. Sequence stratigraphy is no exception. The literature contains numerous suggestions on how certain features of sequence anatomy correlate with the underlying sea-level curve and, conversely, how to construct a sea-level curve from sequence anatomy. Most of these techniques are heuristically valuable thought experiments. We should keep in mind, however, that usually we are dealing with an underdetermined system, i.e. we have more variables than equations. As a consequence, there are several ways to interpret the record. The road from sequence to sea-level history usually is paved with simplifying assumptions such as constancy of other environmental variables, near-sinusoidal shape of sea-level fluctuations etc.

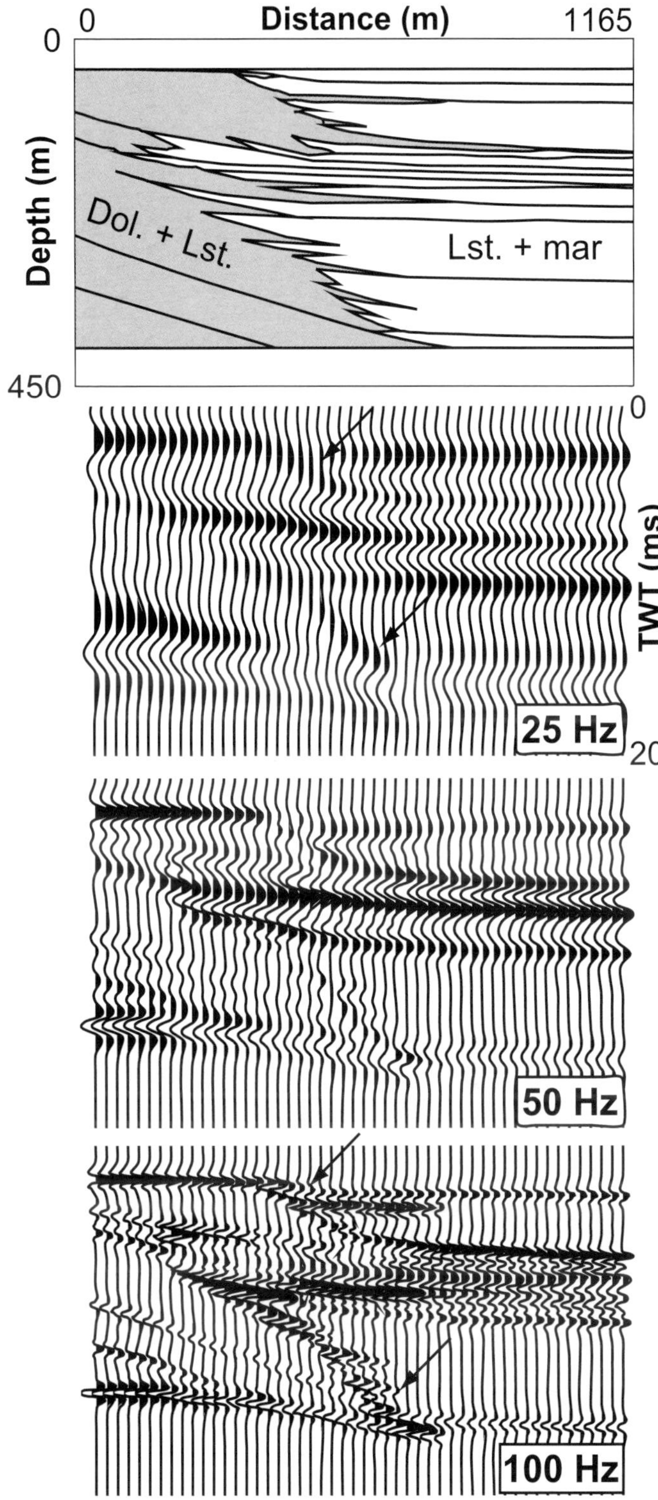

Fig. 6.10.— Increased seismic resolution can solve the problem of pseudo-unconformities . Uppermost panel shows slope-carbonates interfingering with argillaceous basin sediments. Seismic model (vertical incidence) at 25 Hz shows pseudo-onlap at two places (arrows). As frequency is increased from 25 to 100 Hz, the pseudo-onlap is correctly displayed as an interfingering pattern (arrows in 100 Hz panel). Particularly diagnostic are lens-shaped reflectors set en-echelon in the transition zone. Examples from Picco di Vallandro, Southern Alps. After Stafleu (1994), modified.

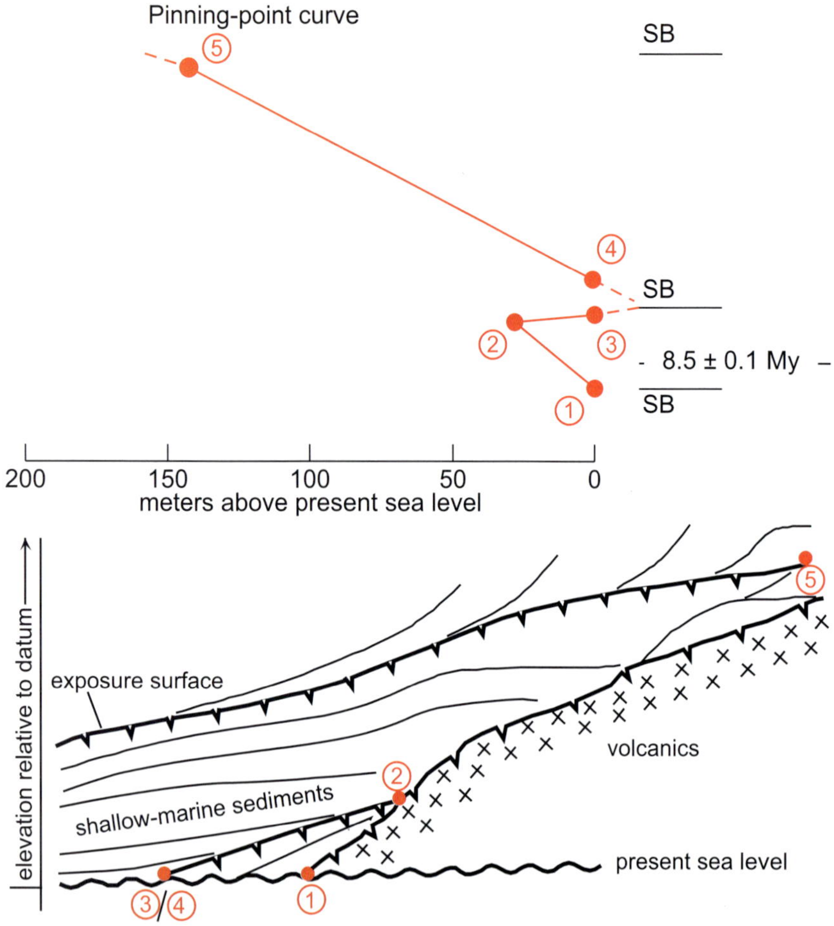

Fig. 6.11.— Pinning-point technique of reconstructing sea-level changes from well-dated sediment record. After Goldstein and Franseen (1995), modified. A) Schematic stratigraphy of Miocene marine strata onlapping a volcanic edifice. Volcanics are capped by terrestrial weathering surface, two more terrestrial exposure surfaces (= sequence boundaries) dissect the marine succession. Pinning points shown in red: (1) Terrestrial surface of volcanis intersects modern sea level, providing datum and starting point of sea-level curve shown in B). (2) Nearshore marine deposits onlap terrestrial surface of volcanics and are capped by another terrestrial exposure surface; thus, point 2 lies very close to an upper turnaround of the sea-level curve. (3)/(4) Terrestrial surface of point 2 dips down to modern sea level where it is overlain by younger marine deposits. This yields two pinning points, nr. 3 for the terrestrial surface and nr. 4 for the oldest marine sediments covering the surface at this location. B) Reconstructed sea-level curve (red) based on careful age dating and measuring of elevations in outcrop. Note that point 2 describes sea-level turnaround very closely. In contrast, points 3 and 4 are at the same location and have identical elevation but different ages; they occupy two points on the sea-level curve and it remains open how far sea level fell between points 3 and 4.

Separating firm data from speculative interpolations is facilitated by the pinning-point approach of Goldstein and Franseen (1995). The technique relies on the fact that while most sediments provide only a crude estimate of water depth and sea-level position, there are some features, the pinning points, where sea-level position is very narrowly constrained, for instance in backshore deposits or at the seaward end of an exposure surface. Fig. 6.11 shows a sea-level curve reconstructed with the pinning-point method. It has the advantage that fixed points and speculative interpolations are well differentiated.

ACCOMMODATION AND SEDIMENT SUPPLY - A DUAL CONTROL OF STRATIGRAPHIC SEQUENCES

A basic tenet of standard sequence stratigraphy holds that sequences and their systems tracts are essentially controlled by sea-level change (Vail et al., 1977; Van Wagoner et al., 1988; Posamentier et al., 1988; Emery et al., 1996). It is important to examine the ideas and assumptions that led to this statement.

Even a cursory examination of the standard model in Fig. 6.3 shows that the first-order patterns are governed by two variables - the rate of change in accommodation, and the rate of sediment supply. If we define S as the volume of sediment in the system and A as the accommodation space, i.e. the space available for sedimentation, then sequences and systems tracts are primarily controlled by two rates, A′ and S′.

A′ = dA/dt, represents the time rate of change in accommodation, i.e. the sum of the rates of subsidence by crustal cooling, sediment loading, sediment compaction, structural deformation, eustasy etc. The sign of A′ can be negative, indicating decrease of accommodation.

S′ = dS/dt, represents the time rate of sediment supply; again, the sign of S′ can be negative, indicating erosion and sediment withdrawal from the system under consideration.

Fig. 6.12 summarizes the relationship of A′ and S′ for the systems tracts and major surfaces of the classical sequence stratigraphic model as described by Vail (1987). The progradation and retrogradation of systems tracts, the development of the transgressive surface and the maximum flooding surface all are controlled by the interplay of A′ and S′. Only the downstepping of the shelf break from highstand to lowstand and the concomitant formation of a type-1 sequence boundary are controlled by A′ alone - they are the result of a negative change of accommodation. If accommodation decreases and baselevel falls, the downstep and the exposure unconformity must develop, irrespective of changes in the rate of sediment supply.

The term A′ is almost entirely a function of relative sea-level change. One caveat must be made, though: A′ as well as S′ represent changes of volume in time, the derivative dV/dt. Sequence stratigraphic models consider either the change in the vertical dimension only, the partial derivative $\partial z/\partial t$, or the change in two dimensions, $(\partial z/\partial t).(\partial x/\partial t)$. One tacitly assumes these partial derivatives to be good approximations of the change in accommodation volume - a premise that does not always hold. In fact, one limitation is immediately obvious. If slope height increases as the system progrades, the sediment volume required to maintain a constant slope also increases. This reduces the rate of progradation and eventually leads to retrogradation in siliciclastics (Jervey, 1988, p. 54). Rimmed carbonate platforms experience the same effect but respond somewhat differently by forming an elevated rim, the empty bucket, and finally drowning completely (Schlager, 1981).

The term S′, rate of sediment supply, is only remotely related to sea level and is a major control on sequences that may be completely independent of sea level. In siliciclastic systems, sediment supply is governed largely by conditions in the hinterland. The growth term, G′, of carbonates (p. 107f) is tied to the ocean environment and to organic evolution. The Holocene is particularly important in this respect because sea-level history is well constrained. Holocene coasts abound with examples of highstand and transgressive tracts developing side by side in response to differences in sediment supply. For instance, many major river deltas currently build prograding highstand tracts while the adjacent shorelines, undersupplied with sediment, retrograde and develop transgressive systems tracts.

The standard model of sequence stratigraphy treats sediment supply in a somewhat inconsistent way. On the one side, supply is being acknowledged as an important control on sequences (Vail, 1987; Jervey, 1988; Van Wagoner et al., 1988; Posamentier et al., 1988; Haq, 1991; Emery et al., 1996). On the other side stands the categoric statement that relative sea-level change dominates sequence stratigraphy (e.g. Vail et al., 1977; Vail, 1987, p. 3). The standard model of systems tract assumes that " ... sediment supply is constant ... " (Posamentier et al., 1988, p. 110) or that sediment supply, changing more slowly than sea level, has only a modifying effect on sequences (Vail, 1987). This assumption is also a prerequisite for correlating the transgressive and highstand tracts as well as the maximum flooding surface to specific parts of the relative sea-level curve (e.g. Posamentier et al., 1988; Vail et al., 1991; Handford and Loucks, 1993; Emery et al., 1996).

First principles of sedimentation provide little support for the notion that accommodation effects generally dominate over supply effects (with the exception of the exposure unconformities and downsteps, of course). Much of the spread in sedimentation rates is probably generated by supply variations in space. Schlager (1993) shows that the sediment yield of rivers in Italy varies by a factor of 5 and the yield of rivers of Taiwan varies by a factor of 40. Both regions are small and reside entirely within one climate belt. Thus, the figures certainly are relevant for sequence stratigraphy in the sense that the spatial scale, 10^1-10^2 km, is comparable to that of many sequence-stratigraphic studies. Limiting this comparison to regions within one climate belt increases their relevance for sequence stratigraphy: individual third-order sequences are likely to form within one climate belt as the characteristic time scale of sequences, 10^5-10^6 years, generally is too short for plate tectonics to move the area into another climate belt.

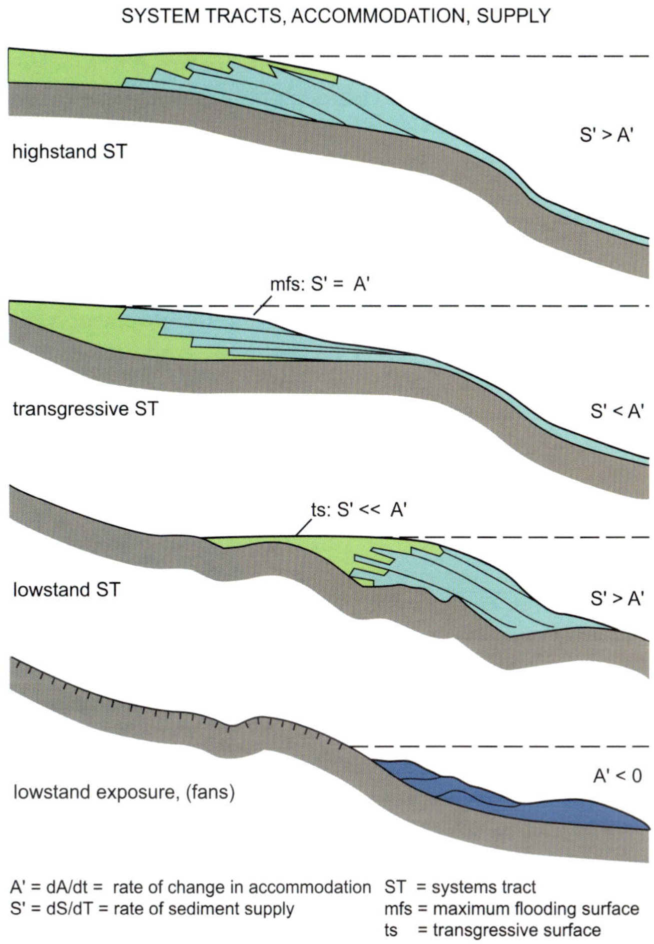

Fig. 6.12.— Sedimentologic interpretation of systems tracts in siliciclastics. They are generated by the interplay of the rate of change in accommodation, A′, and the rate of change in sediment supply, S′. Exposure surfaces that extend into formally marine areas are diagnostic of relative sea-level falls and cannot be generated by variations of supply. Colors: green – alluvial to littoral, light blue – neritic to bathyal, dark blue – turbidite fans. After Schlager (1992), modified.

The geologic record also shows that sediment supply to the ocean has varied through geologic history. For instance, sedimentation rates of the latest Cenozoic are sigificantly higher than in the earlier Cenozoic - perhaps as a consequence of the extensive glaciations. Longer term variation in supply must have resulted from the changes in the global rates of sea-floor spreading and subduction. It can be expected that this cyclic change in sediment input to the ocean is correlated with the first-order cycle of Haq et al. (1987) and Hallam (1977).

In summary, the assumption of standard sequence stratigraphy that the sequence pattern is dominated by sea-level related changes in accommodation as opposed to supply variations is not rooted in some basic principle of sedimentary geology but based on case studies and their interpretation. Consequently, this assumption should not be accepted a priori but tested wherever possible. Progradation and retrogradation of shelf breaks are unreliable guides to sea level. The reconstruction of sea-level changes from depositional sequences should proceed from two other sources of information: first, the timing and extent of exposure unconformities and downsteps to determine the sea-level falls; second, the rates of vertical aggradation of shoal-water systems whose facies indicate that water depth during sedimentation did not deviate significantly from sea level (such as certain carbonate platforms); such sections provide reasonable estimates on timing and amount of sea level rises.

The balance of the rate of accommodation change and the rate of sediment supply not only controls progradation and retrogradation of depositional systems. It also profoundly affects small- scale geometry and facies patterns by variation of water depth. We will examine this effect for carbonate systems in chapter 7.

ORDERS VERSUS FRACTALS IN THE SEQUENCE RECORD

The concept of "orders" of sequences already appeared in the first monographic publication on sequence stratigraphy. Vail et al., (1977, p. 86) described the sequence record as a hierarchy of cycles that were ranked by their duration as first, second and third order. Soon after their introduction as categories in time, sequence orders were also characterized by depositional architecture. An important step was the recognition of parasequences - building blocks of classical sequences that were made up of shoaling successions bounded by flooding surfaces rather than exposure surfaces (Van Wagoner et al. 1988; Van Wagoner et al. 1990). On the other hand, it was also observed that in many instances the building blocks of standard 3rd order sequences were "simple sequences", i.e. units bounded by exposure surfaces and built like the standard 3rd order sequence (p. 91; Vail et al. 1991, p. 630; Mitchum and Van Wagoner, 1991).

Comprehensive classifications that assigned a specific anatomy to each sequence order were advanced by Vail et al. (1991) and Duval et al. (1998), see Fig. 6.13. These authors proposed that the cycles of fourth order and shorter have the anatomy of parasequences, third-order cycles follow the standard model of sequence stratigraphy with units bounded by exposure surfaces and composed of lowstand, transgressive and highstand systems tracts. This standard sequence is distinctly asymmetric. Second and first order cycles were assumed to be progressively more symmetric. First-order cycles in particular were viewed as nearly symmetric patterns of flooding and exposure of the continental interiors and were called "continental encroachment cycles" by Duval et al. (1998).

Critique of the concept of orders in the standard model

While the principle of defining orders by duration has been almost universally followed, the actual figures in the definitions scatter widely. Two data sets may illustrate this point. Fig. 6.14 plots the durations of sequence cycles of 2nd and 3rd order determined in one publication: the sea-level curve of Haq et al. (1987). The two categories clearly differ in their modes but broadly overlap in range. Fig. 6.15 shows the definitions used in key publications since the introduction of the concept. In the range of 2nd to 4th order, the discrepancies are about one order on either side. For the shorter categories the discrepancies are even larger. Moreover, the values do not seem to converge with time and growing amounts of data. Finally, one notes that many authors choose durations that are constant on a logarithmic scale, often coinciding with full powers of ten on a year scale. This practice, too, suggests that the orders are subdivisions of convenience.

In view of these difficulties, Hardenbol et al. (1998) abandoned the subdivision into 2nd and 3rd order cycles. They argue that better understanding of mechanisms is required to justify this classification. I certainly agree that the way sequence orders were defined and these definitions later modified leaves one with the impression that sequence orders are subdivisions of convenience, not an expression of natural structure.

This impression is strengthened if one looks at the characterization of sequence orders by sediment anatomy. The papers by Vail et al. (1991) and Duval et al. (1998) presented concepts and examples but no statistical data on the postulated correlation of sequence duration and depositional anatomy. And to this day, dearth of statistical data characterizes the discussion on this topic. Qualitative observation show no clear correlation of duration and anatomy of sequences. Sequences of 10^4 - 10^5 y duration are often of the parasequence type, but there are also many sequences of the standard type, for instance in the Neogene. Among sequences in the 10^6 y domain there is a significant number of units bounded by flooding surfaces and without recognizable lowstand tract. Finally, at all time scales we observe some sequences whose anatomy is more or less symmetrical, showing deepening followed by shoaling, or fining followed by coarsening trends. Thus, the notion that long cycles are symmetric, 3rd order and shorter cycles asymmetric needs to be examined quantitatively. The fraction of symmetric cycles among long cycles may be higher, but they certainly occur also in "third-order" and shorter domains (see chapter 7 for examples).

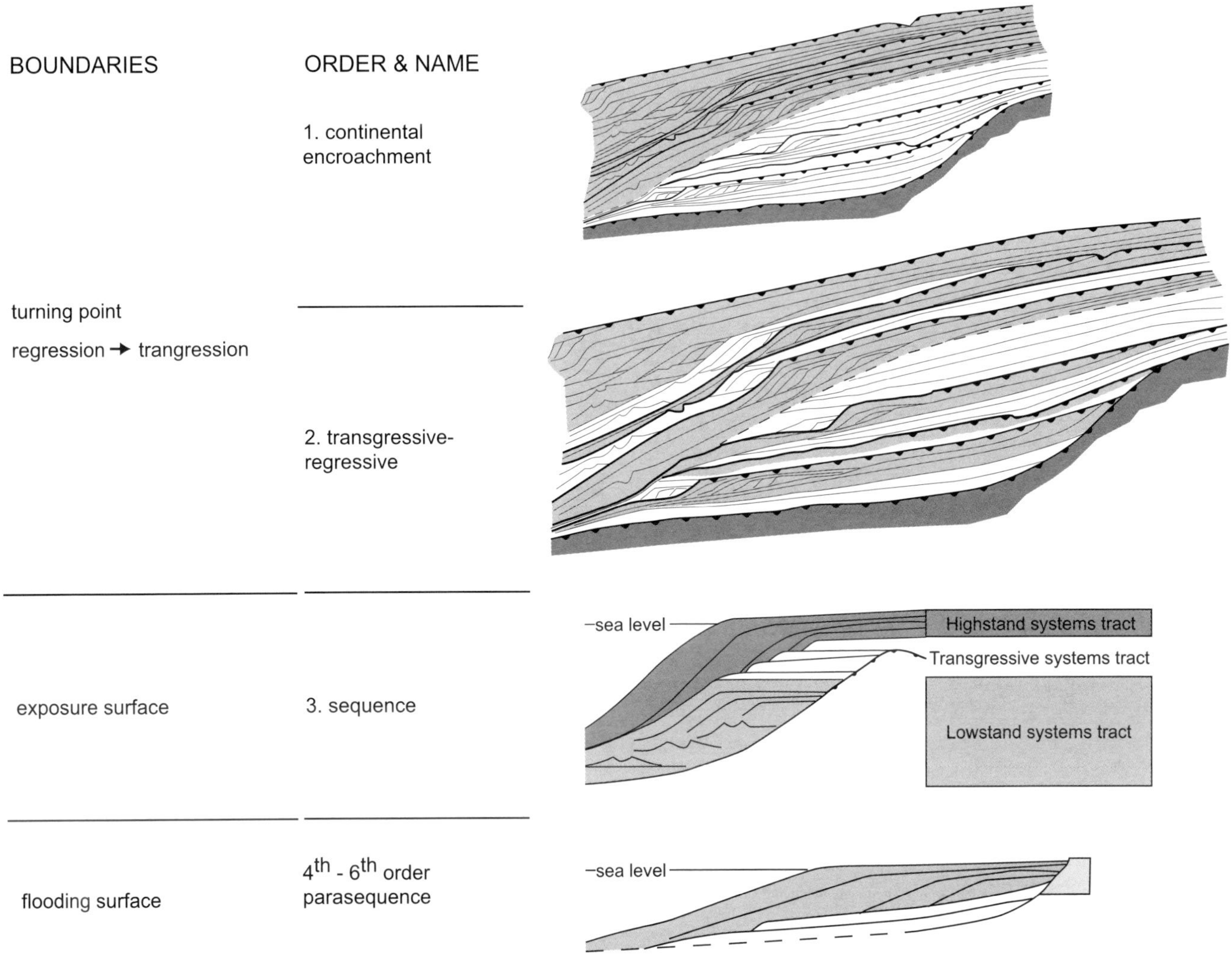

Fig. 6.13.— Orders of sequences and sequence anatomy after Duval et al. (1998), modified. Orders are defined by duration but also characterized by their depositional anatomy and the kind of sequence boundary. See text for discussion.

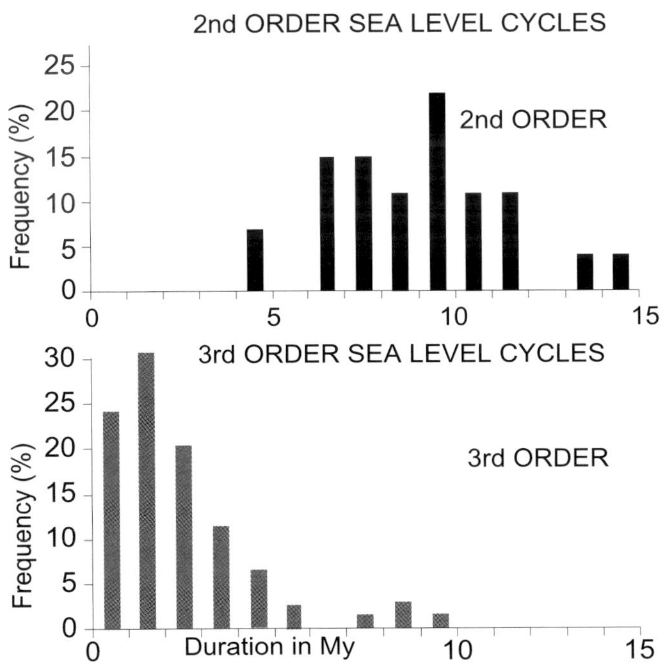

Fig. 6.14.— Duration of sea-level cycles of third and second order of the eustatic curve of Haq et al. (1987). The two orders broadly overlap even though their modes are different. After Schlager (2004).

Fractals - an alternative to orders in sequence stratigraphy

Work on the sequence record in the domain of 10^3 to 10^6 years has gone far enough to formulate an alternative to the concept of orders in sequence stratigraphy.

Examination of sediment anatomy indicates that many patterns relevant for the sequence model are invariant in a wide range of scales. Moreover, a strong case has been made that the two basic controls of sequences - changes in sea-level and sediment supply - are fractals in the time domain.

Based on these observations, I proposed the following conceptual model (Schlager, 2004): The pattern of sequences and systems tracts is scale invariant and a statistical fractal. In the arrangement of systems tracts and the nature of sequence boundaries two types are about equally likely in the validity range of the model (Fig. 6.16, 6.17):

- *Standard sequence*, or S sequence , shows the succession: highstand tract - sequence boundary - lowstand tract - transgressive tract - highstand tract. The sequence boundary is a type-1 or type-2 boundary caused by relative sea-level lowering (forced regression).
- *Parasequence*, or P sequence , shows the succession: highstand tract - sequence boundary - transgressive tract - highstand tract. The sequence boundary is a type-3 boundary, i.e. a flooding surface that overlies marine deposits without demonstrable evidence of terrestrial exposure or forced regression.

The crucial distinction between the two types is that standard sequences are bounded by surfaces of forced regression that require a relative fall in sea level. Parasequences are bounded by type-3 sequence boundaries, i.e. flooding surfaces that lack evidence of forced regression. Consequently, parasequences and their type-3 boundaries may be formed by fluctuations in sediment supply or by relative changes of sea level. The type-3 sequence boundary was already recognized as an important type of unconformity by Vail and Todd (1981, Fig. 3c) and by Schlager (1999b) who proposed the name.

As the terms parasequence and standard sequence had been defined by both the length of time represented by the unit and the type of bounding surface, Schlager (2004) introduced the terms S sequence and P sequence for sequences defined solely by the nature of their boundaries. The terms can be avoided and replaced by "standard sequence" and "parasequence" if one defines the two sequence types strictly by the nature of their boundaries and disregards the length of time they represent: Parasequences are sequences bounded by flooding surfaces, standard sequences are bounded by exposure surfaces. Throughout this book, these definitions will be applied. This usage is common (e.g. Nummedal et al. 1993,p. 55; J.F. Sarg, written communication), but not unanimous.

Conservative estimates of the validity range of the model are 10^3 – 10^6 y in the time domain, 10^1 – 10^5 m in the horizontal directions of space and 10^0 – 10^3 m in the vertical. Justification of these ranges is given below.

The time range of the model can be justified as follows. The statistics on flooding versus exposure boundaries is based on cycles with durations of 10^4 – 10^5 y. Physical processes operate in similar fashion also on much shorter time scales. However, the formation of recognizable exposure surfaces depends on soil processes whose rates are considerably lower. Accumulation of organic matter, one of the fastest soil processes, takes thousands of years to advance

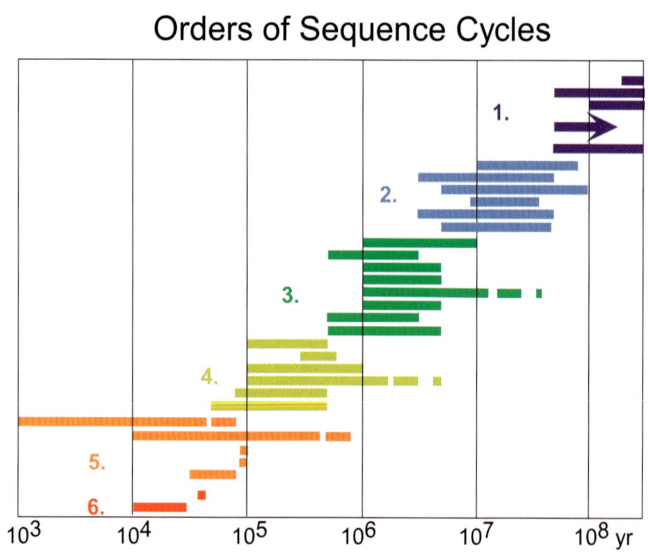

Fig. 6.15.— Orders of stratigraphic sequences as defined by various authors since 1977. In each category, the oldest publication is on top. Differences are about 1/2 order at each boundary, in the 4th - 6th orders even larger; opinions do not seem to converge with time. After Schlager (2004), modified.

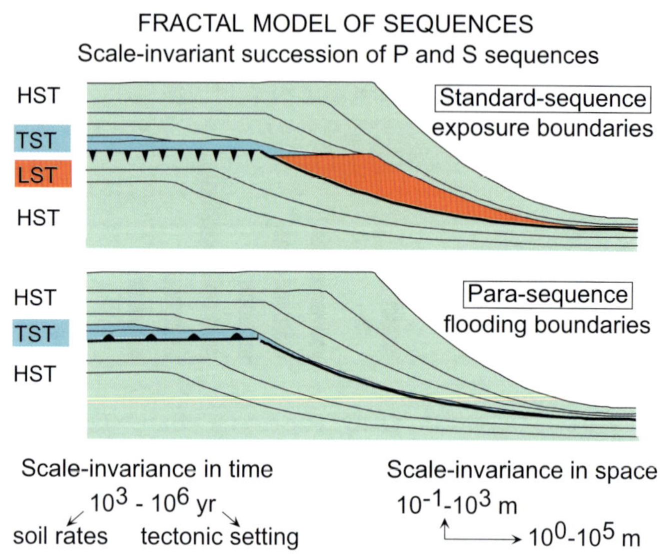

Fig. 6.16.— Scale-invariant model of stratigraphic sequences. Standard sequences, i.e. sequences bounded by exposure surfaces, and parasequences bounded by flooding surfaces, are assumed to occur in about equal proportions in the range of 10^3 to 10^6 y. After Schlager (2004), modified.

significantly (e.g. Birkeland, 1999, p. 215). Similarly, the easily preserved calcrete crusts form at rates of several millimeters per 10^3 y (e.g. Robbin and Stipp, 1979). Finally, modelling of rock-water interaction suggests that significant decrease in bulk-rock $\delta^{13}C$ requires $10^3 - 10^4$ y (Yang, 2001). Based on these indications, I assume that the formation of a preservable exposure record, even in carbonates, takes about 10^3 year. This number is taken as the short time limit of the model because it assumes that the occurrence of parasequences and standard sequences is about equally likely. Among sequences shorter than 10^3 y, parasequence-cycles are likely to dominate simply because there was insufficient time to develop recognizable exposure surfaces.

One question that still needs to be sorted out is the different preservation potential of exposure events in carbonates and siliciclastics. Hard soil crusts are uncommon in siliciclastics and sediment lithification is much slower than in carbonates. Thus, soils and other vestiges of subaerial conditions may easily be washed away during the subsequent transgression. The situation is different during long exposure where deeply incised valleys may form and preserve the record of exposure by their morphology and by the presence of soils in protected pockets.

The upper time limit for the fractal model is set by lack of data. As Fig. 6.17 includes no cycles longer than ten million years, it seems prudent to limit the validity range to sequences shorter than 10^7 y. Beyond this limit set by the data base, tectonics may set yet another limit. Few margins of basins or continents exist for 100 My or more without significant episodes of deformation or magmatism. Consequently, the record of cycles in the 10^8 y-range is largely derived from flooding and exposure of continental interiors whereas data on shorter cycles mainly are from ocean margins. This difference in setting may explain the characteristics of first-order cycles reported by Vail et al. (1991) and Duval et al. (1998).

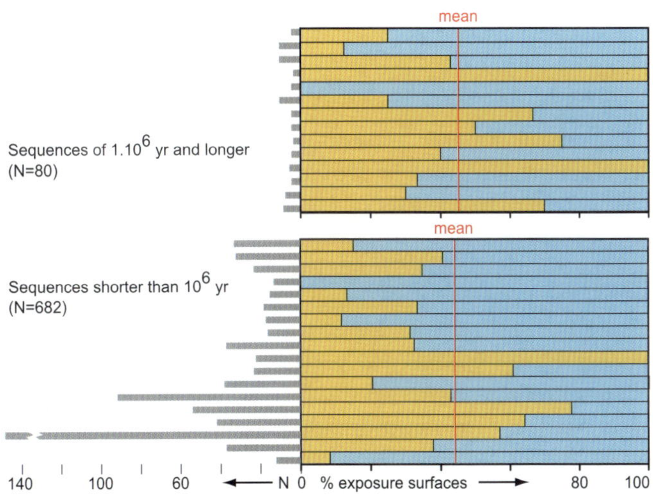

Fig. 6.17.— Standard sequence boundaries (orange) and parasequence boundaries (blue) in well-documented carbonate rocks. Each bar represents a measured section with number of analysed surfaces on the left. Top panel: cycles longer than 1 My. Bottom panel: cycles shorter than 1 My. Sources: Buchbinder et al. (2000); D'Argenio et al. (1997); D'Argenio et al. (1999); Egenhoff et al. (1999); Föllmi et al. (1994); Hillgärtner (1999); Immenhauser et al. (2001); Immenhauser et al. (2004); Minero (1988); Saller et al. (1993); Strasser and Hillgärtner (1998); Van Buchem et al. (2000); Van Buchem et al. (1996); Wendte et al., (1992); Wendte and Muir, (1995). After Schlager (2004), modified.

The spatial range of the model is essentially a consequence of the time limits. In the range of 10^3 - 10^6 y one typically observes sequence packages that range in thickness from 10^0 - 10^3 m and extend horizontally from 10^1 - 10^5 m. The larger horizontal extent is a characteristic of nearly all sedimentary bodies and reflects the fact that at the surface of the Earth vertical gradients generally exceed horizontal ones by several orders of magnitude.

Support for the fractal nature of sequences

The conceptual model introduced above draws support from three areas - sediment anatomy, the nature of sediment supply and sea-level fluctuations, and the distribution of parasequences and standard sequences in time.

Sediment anatomy, i.e. the external geometry and internal depositional structures of sediment bodies, is scale invariant in a wide range of scales in time and space. (In fact many features are scale-invariant in a much wider range of scales than considered here). Without characteristic objects for scale, we could not tell if the delta-canyon system in Fig. 6.18 is tens of centimeters or hundred kilometers wide, if it formed within hours or hundreds of thousands of years. Similar statements can be made for avalanche foresets, braided and meandering streams and many other depositional patterns. Thorne (1995) illustrated the scale-invariance of prograding clinoforms and developed a scale-invariant quantitative model for them. Van Wagoner et al. (2003) proposed that delta-shaped accumulations fed by a

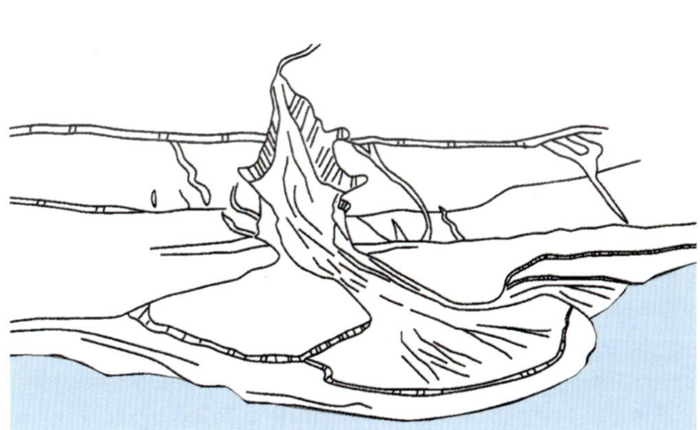

Fig. 6.18.— Scale-invariance of depositional anatomy is elegantly demonstrated by this valley-delta system described by Posamentier et al. (1992a). As objects of characteristic size are absent, the dimensions of the system cannot be estimated. (The delta is about 2 meters across).

point source are scale-invariant, energy-dissipation patterns governed by the principles of thermodynamics. Posamentier and Allen (1999) offered numerous examples of characteristic sequence anatomy that develop at temporal and spatial scales far smaller than those of stratigraphic sequences but nonetheless in completely analogous fashion. Finally, the trajectories of shelf edges that prograde and move up and down in step with sea level have fractal characteristics (Fig. 6.19). The last example is important because it allows one to estimate fractal dimensions - an important prerequisite for using the fractal model in subsurface prediction.

Changes in sea-level and sediment supply were identified in the previous section as the two fundamental controls on sequences. Both are known to have fractal characteristics. The rates of sediment supply decrease systematically as the time span of observation increases (Fig. 6.20). It is generally accepted that this persistent pattern is caused by the occurrence of hiatuses at all scales. The fact that the decrease follows a power law strongly suggests that the distribution of hiatuses in the stratigraphic record has fractal characteristics and Plotnick (1986) proposed that stratigraphic successions are random Cantor sets with gaps at all scales.

Sea-level fluctuations have been studied at time scales of seconds to hundreds of millions of years. For many data sets it was found that power laws govern the relationship between frequency and wave power. Harrison (2002) combined data covering 12 orders of magnitude in frequency and concluded that at time scales of years to hundreds of millions of years, – the geologically relevant range –, the first order trend of sea-level fluctuations has fractal character (Fig. 6.21). There are islands of order that break this trend, for instance the sea-level history of the recent past with its strong orbital control. The trend breaks down in the domain of high-frequency sea waves. Hsui et al. (1993) found fractal characteristics in the sea-level curve of Haq et al. (1987) using rescaled-range analysis. Fluegeman and Snow (1989) observed the same in the Neogene oxygen-isotope record (a proxy for sea level) of deep-sea sediments.

The distribution of parasequences and standard sequences in the sediment record does not support the subdivision into a domain of standard sequences at the million-year scale and a domain of parasequences at shorter time scales. Rather, it was observed that the two types co-exist and often alternate. Vail et al. (1991) reported the occurrence of parasequences and standard sequences among short cycles. Furthermore, there is a growing number of sequences in the million-year domain that are terminated by drowning unconformities, i.e. flooding surfaces, without prior exposure (see chapter 7). Finally, Fig. 6.17 summarizes over 700 data from detailed observations on sequences boundaries in carbonates. Important criteria for data selection in Fig. 6.17 were that the authors

(1) presented detailed and specific observation on the nature of each discontinuity surface and

(2) accepted at least the possibility that a flooding surface could be a sequence boundary.

The last criterion was important for the long cycles in Fig. 6.17 because many authors would follow Van Wagoner et al. (1988) and accept only exposure surfaces as sequence boundaries in the million-year domain.

In siliciclastics, the record of exposure may have a strong preservational bias. As lithification at the sediment surface is largely absent, soils formed during short exposure events are likely to be washed away by the subsequent transgression (ravinement surface). Long exposure frequently is preserved in incised-valley fills.

Fractals and the impression of ordered hierarchy

In many sequence data, the impression of a hierarchy of cycles is very strong. The model does not imply that this impression is false. It is characteristic of fractals that the same pattern is repeated at finer and finer scales. Consequently, any snapshot of the fractal taken at a certain resolution will show a superposition of coarser and finer patterns. The crucial difference to an ordered hierarchy of cycles is the lack of characteristic scales. Moreover, the apparent hierarchies in statistical fractals – and this is what the model entails –, will be different for each random sample drawn from a population.

The fractal model proposed here predicts that the sequence record, like many other natural time series, has the characteristics of noise with variable persistence and thus variable predictability (see Turcotte, 1997; Hergarten, 2002, for examples). The model also predicts that in the sequence record the effect of ordered oscillations, such as the orbital cycles, generally is subtle and becomes dominant only in special circumstances (e.g. "orbital" data in fig. 6.21).

Purpose and scope of the fractal model

The model is meant as a conceptual framework to steer future data analysis and to provide a basis for statistical characterization of sequences. A logical next step is to determine fractal dimensions of important features of sequences and explore the limits of the fractal domain. This insight, in turn, can be used for interpolating between data sets and extrapolating beyond technical limits of observation, for instance beyond the resolution of seismic data.

The difference between the fractal model and the orders model with regard to prediction is best illustrated by an example. Let us assume that seismic data of a passive margin reveal systems tracts and unconformable sequence boundaries. Crude estimates indicate that the boundaries are spaced at intervals of a few million years. The model of Duval et al. (1998) would predict that the seismically recognizable sequences are of the third order and therefore standard sequences bounded by exposure surfaces and lowstand systems tracts. The building blocks of these sequences, usually beyond seismic resolution, are predicted to be parasequences bounded by flooding surfaces. The fractal model predicts that the seismically visible sequences as well as their building blocks consist of a mix of parasequences and standard sequences in about equal proportions. The model is open for refinement on the exact proportions of the two types depending on expanding statistical data bases. It is likely that the P:S ratio varies with tectonic setting, long-term eustatic trends, sediment composition (e.g. siliciclastics or carbonates) and other environmental factors. In the absence of specific constraints, the model assumes a P:S ratio of 1. Similarly, if regional or other specific constraints are lacking, the model would assume that the shelf-edge trajectory would be a fractal with a fractal dimension, D, in the range observed in Fig. 6.19 or other, more extensive data bases.

The fractal model delivers estimates of sedimentation rates or sea-level fluctuations in analogous fashion to sediment anatomy. In the absence of case-specific constraints, the model assumes sedimentation rates, S′, to decrease proportional to the inverse of the square root of the duration of sequences, i.e.

$$S' \propto t^{-0.5}$$

This formula is based on the first-order trend found by Sadler (1981, 1999) for the decrease of sedimentation rates with increasing observation span. The same approach holds for the prediction of sea level. In the absence of specific constraints, the model assumes that the power of sea-level fluctuations, i.e. the square of the amplitude, decreases with increasing frequency, f, according to the global trend found by Harrison (2002)

$$P = 10^{-4} \cdot f^{-2}$$

where f is in cycles per year (cpy), and P is the power in m^2 per cpy. It is of the utmost importance to realize that the model is open for more detailed quantification of any kind as long as it is based on statistically valid data and sound arguments.

Can the model be applied in the absence of unconformities? The standard model of sequence stratigraphy already states that the sequence-bounding unconformities may pass laterally into correlative conformities . Cycle-stacking patterns and facies analysis (e.g. Goldhammer et al., 1990; Homewood and Eberli, 2000; Van Buchem et al., 2000; Hillgärtner, 1998) have demonstrated that sequence boundaries, maximum flooding surfaces etc., may be zones of rapid transition rather than distinct surfaces. The scale-invariant models (Fig. 6.16) are applicable in these situations, too. Marine flooding surfaces, in particular, are prone to change into flooding intervals because it is impossible for the ocean to erode the seabed at the same time everywhere.

The type-3 sequence boundary in carbonates (p.121) is a logical consequence of the co-existence of standard sequence architecture and parasequence architecture (Fig. 6.16). This type of boundary was already recognized by Vail and Todd (1981, Fig. 3c) and Schlager (1999b). Drowning unconformities are a special case of type-3 boundaries where the carbonate factory is shut down.

Origin of scale-invariant fractal pattern

Scale-invariant fractals often appear as the result of other fractals (Hergarten, 2002). At a first, surficial level, stratigraphic sequences are essentially shaped by the interplay of rates of change in accommodation and rates of sediment supply (Fig. 6.11; Jervey, 1988; Schlager, 1993). Both variables show fractal properties: accommodation is largely governed by relative sea-level change and many sea-level curves of the past have been shown to possess fractal properties (Fluegemann and Snow, 1989; Hsui et al., 1993). The scale-invariant, fractal nature of sediment supply is indicated by the long-term decrease of sedimentation rates with increase in time interval (Sadler, 1981; 1999). At a more fundamental level, most geologically important processes of sedimentation and erosion are turbulent and the chaotic nature of turbulence may be one reason for the dominance of scale-invariant, fractal patterns in stratigraphic sequences and the sediment record in general. Another possible reason is the tendency of sedimentary systems to evolve to a self-organized critical state (Bak et al., 1987; Hergarten, 2002; Rankey et al., 2002).

ORIGIN OF SEQUENCES

The authors of the standard model of sequence stratigraphy offered not only a conceptual framework for the stratigraphy of unconformity-bounded units but also rather firm statements about the origin of stratigraphic sequences. Vail et al. (1977) argue on sedimentologic grounds that stratigraphic sequences are essentially caused by relative sea-level changes; they argue furthermore that most sequences are globally synchronous, thus eustasy must be the main cause of sequences. Vail (1987, p.1) lists subsidence, climate and sediment supply as major controls but concludes that "fundamental control of depositional sequences, is, we believe, short-term eustatic changes of sea level superimposed on longer-term tectonic changes." According to Van Wagoner et al. (1988, p. 39) "Sequences ... are interpreted to form in response to the interaction between the rates of eustasy, subsidence, and sediment supply." Overall, the authors of the standard model clearly assumed that sequences are shaped by the interplay of eustasy, local tectonic movements and sediment supply. However, they also postulate that eustasy dominates the sequence record. The main argument for this assumption is the success in global correlation of sequences and the similarity of the estimates in sea-level movements (e.g. Vail et al. 1977, p. 84-91; Haq et al., 1987; Hardenbol et al., 1998; Billups and Schrag, 2002, reproduced in Fig. 6.22). For the pre-Neogene, however, the sequence-based sea-level curve contains so many events that their spacing is close to the resolution of the best dating techniques. An elegant experiment by Miall (1992) illustrates the pitfalls of correlating stratigraphic events under these circumstances (Fig. 6.23). Miall's (1992) results are particularly relevant if one considers that the best global correlation is found in deep-water sediments whereas the sequence

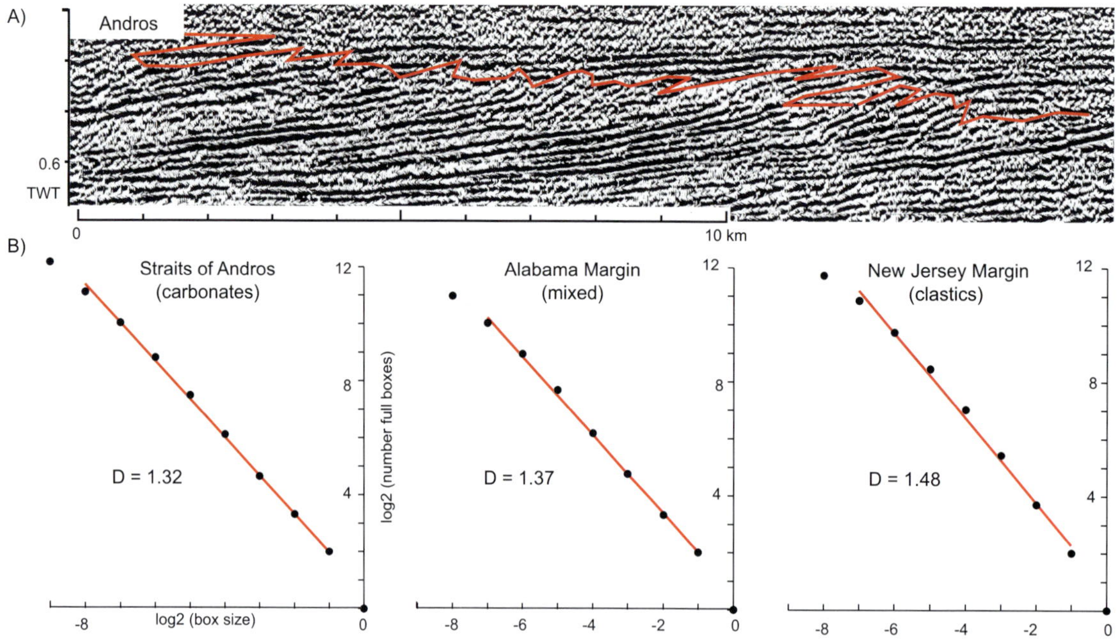

Fig. 6.19.— Trajectories of migrating shelf breaks, a pivotal feature of sequence anatomy, show fractal characteristics. Data from Cenozoic of Great Bahama Bank (Eberli and Ginsburg, 1988) and offshore New Jersey (Greenlee, 1988). (A) Part of the migration path of the Bahama shelf break as determined by the author using the seismic line by Western Geophysical described in Eberli and Ginsburg (1988). (B) Fractal analysis of shelf-break migration paths box-counting (see appendix). Straight-line trends for box sizes 2^2 to 2^6 indicate a power law relationship between box size and number of boxes required to cover the curve. This in turn suggests that the paths of shelf-break migration have fractal characteristics in these limits. The trends are somewhat irregular on the right because of finite-size effects and break down on the left because of limited seismic resolution.

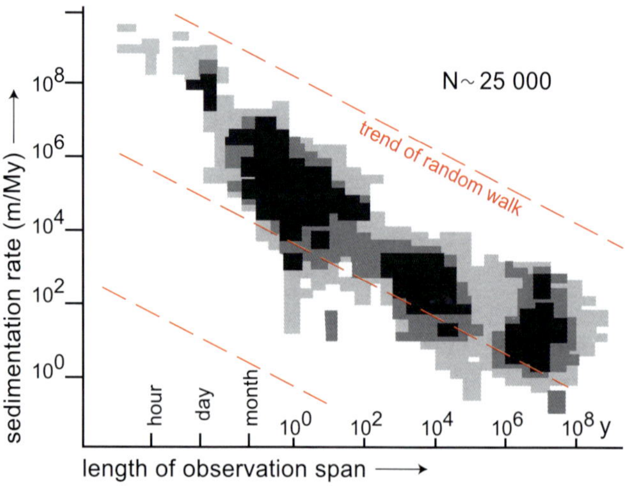

Fig. 6.20.— Log-log plot ofsedimentation rates vs. length of the time window of observation. Rates vary considerably in each time window but the overall trend is linear and close to that of a random walk. After Sadler (1981), modified.

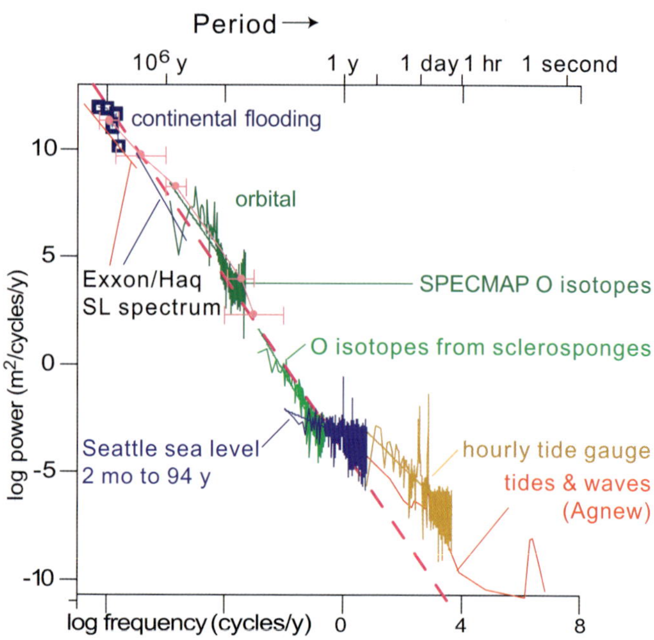

Fig. 6.21.— Power spectra of sea-level fluctuations covering 15 orders of magnitude of frequency. In the stratigraphically relevant range of 10^0 to 10^8 y, the first-order trend is a power law indicating that the spectrum of sea-level fluctuations is a fractal whose dimension is close to that of white noise. The trend breaks down for high frequencies. Records dominated by orbital rhythms appear as islands of order that break the trend. After Harrison (2002), modified.

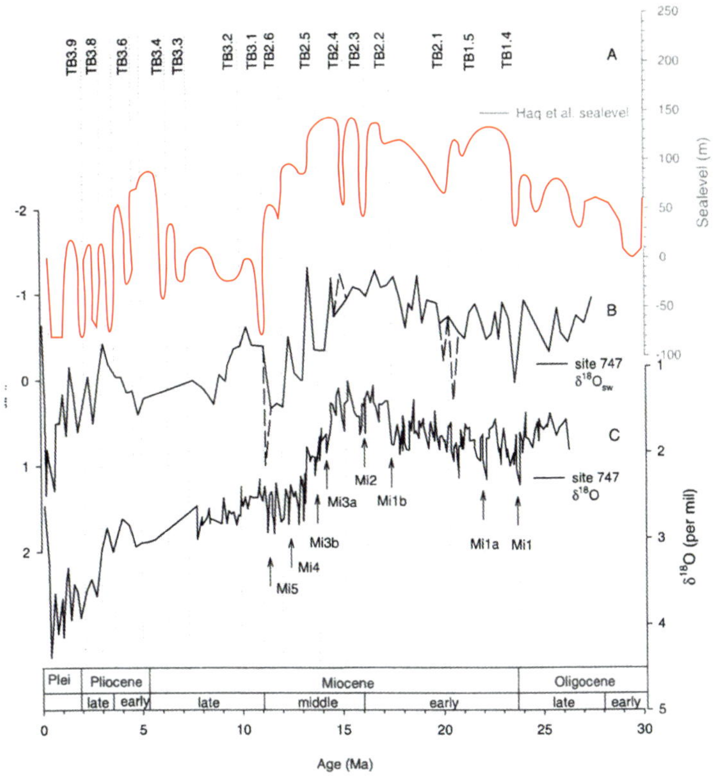

Fig. 6.22.— Comparison of the sequence-stratigraphic sea-level curve with paleoceanographic indicators of temperature and ice volume derived from ODP site 747. A) Red: sea-level curve of Haq et al. (1987). B) $\delta^{18}O$ seawater a proxy indicator of global sea level derived from paired measurements of Mg/Ca ratios and $\delta^{18}O$ of benthic foraminifera. C) $\delta^{18}O$ curve measured directly at site 747. Overall similarity of the curves as well as coincidence of many single events suggest that glacio-eustatic sea-level fluctuations were a dominant control of the sequence record in the past 30 My. After Billups and Schrag (2002).

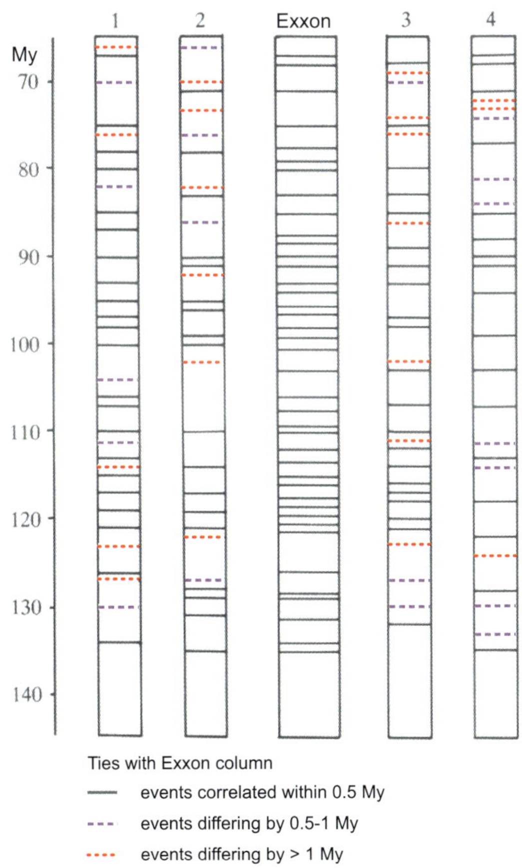

Fig. 6.23.— Correlation between the Exxon sea-level chart (Haq et al., 1987) and four synthetic stratigraphic columns constructed by random-number generation. Horizontal lines represent boundaries of sea-level cycles drawn at the time of most rapid fall. Degree of correlation is high; even in the worst case, i.e. column 3, 77% of the cycles correlate with the Exxon column by a margin of less than 1 My. After Miall (1992).

record shows its optimal differentiation in the shoal-water deposits.

The notion that eustasy is the principal control of the sediment record was quickly challenged by tectonicists (Pitman, 1978; Watts et al., 1982; Cloetingh, 1988) and the debate continues to this date (Miall, 1997; Watts, 2001). The issue is of critical importance for global correlation but it is somewhat beyond the sedimentologic scope of this book. The balance between eustasy and regional tectonics in sequence stratigraphy is not a central issue in this context because the sediment record at any given location does not distinguish between eustatic and regional tectonic sea-level change.

Two serious tests for eustasy are (1) to globally stack sea-level records and demonstrate the (near)-synchrony of events and (2) to tie records of relative sea-level changes to a globally operating cause such as the perturbations of the Earth's orbit. Both techniques make high demands on stratigraphy; they require high stratigraphic resolution in time as well as reliable correlation around the globe and across facies boundaries. These conditions are less frequently met than is generally assumed. Consequently, eustatic control of sequences is more often postulated than demonstrated.

The question of eustasy notwithstanding, sequence stratigraphy is a powerful and practically useful concept for analysing the sediment record. With regard to carbonates, the sedimentologic principles outlined in this chapter indicate that *a priori* sequences should be viewed as the result of the interplay of relative sea level movements (i.e. eustasy plus regional tectonics) and sediment supply. The next chapters will show that sediment supply in carbonates is strongly controlled by the ocean environment. Thus, the factor "climate" in Vail's (1987) list of controls gains added importance in carbonate systems.

CHAPTER 7

Sequence stratigraphy of the T factory

CARBONATE FACTORIES AND THE PRINCIPLE OF DEPOSITIONAL BIAS

Depositional systems resemble newspapers that all report on the events of the day but each with a different editorial bias. It behooves the reader to learn about the editorial bias of his paper. Similarly, the geologist ought to know about the bias of depositional systems in recording sea level and other environmental factors.

The three carbonate factories each have their own bias in recording sequence-stratigraphic events and all three differ to varying degrees from the siliciclastic standard model. This is not to say, however, that sequence stratigraphy does not apply to these systems. On the contrary, comparative sedimentology of depositional systems clearly shows that the basic features of the standard model are shared by all depositional systems, showing once again the power of sequence stratigraphy as a unifying concept.

In this chapter, we base our discussion of carbonate sequence stratigraphy primarily on the deposits of the T factory. They are volumetrically dominant in the geologic record and their sequence stratigraphy is best known. The sequence stratigraphy of the C and M factories is developed in chapter 8 by comparison with the T factory and the standard model.

T FACTORY – KEY ATTRIBUTES FOR SEQUENCE STRATIGRAPHY

The T factory was characterized and compared with the other factories in chapter 2. Below is a summary of attributes that are crucial for sequence stratigraphy.

- The T factory is highly productive but the depth window of production is very narrow, usually the uppermost 100 m (or less) of the water column.
- The factory can build wave-resistant structures in the depositional environment either by organic framebuilding or by cementation of sands on the sea floor or during brief exposures. These structures are independent of shoreline position. They may form close to shore or far offshore. Once formed, they tend to stay in place and build upward rather than migrating laterally.
- Because of the ability to form rigid structures, the classical seaward-sloping shelf in equilibrium with sediment supply, sediment caliber and wave energy is a transient condition rather than a stable equilibrium as in siliciclastics or cool-water carbonates. In the T factory, the dipping shelf - called ramp by carbonate sedimentologists - evolves to a rimmed platform where an offshore belt of reefs or sand shoals protects a lagoon. Large parts of this lagoon may fill with sediment right up to sea level because large oceanic waves are filtered out by the rim and cannot affect the lagoon floor.
- Rimmed platforms are flatter than siliciclastic shelves or may show a reversed dip from the rim into the lagoon. If the rim is drowned during sea-level rise, facies belts usually jump landward by a considerable distance.
- T carbonates are rich in metastable minerals. Lithification by dissolution of these metastable components and precipitation of stable calcite is rapid. Once lithified, the limestones develop karst surfaces where mechanical denudation is minimal and chemical denudation is relatively slow because much rainwater percolates through the young, porous rock rather than flowing over its surface. As a consequence, the highstand systems tracts of T carbonates are often better preserved than in siliciclastics or cool-water carbonates where they get washed down as sea level falls.
- Large amounts of particulate material from clay- to granule size are shed down the slope and into the basin because accommodation on the very shallow T platforms is limited and the factory normally produces far more than can be stored on the platform top.

T FACTORY - SEQUENCE ANATOMY

Importance of platform rims

The ability of tropical carbonate platforms to develop rigid, wave-resistant structures at the shelf break normally leads to elevated margins at sea level that protect a slightly deeper lagoon that gently rises to the littoral zone farther landward (see chapter 3). These "defended" platform margins are probably the most important feature in the sequence anatomy of tropical carbonate accumulations. When these rims are overwhelmed by sea level, they cause facies belts to jump and interrupt the gradual shift in onlap. This jump must not be confused with a sea-level fall. Furthermore, elevated rims have a strong tendency to stack vertically, putting reef on reef, sand shoal on sand shoal, because the environmental conditions favor the establishment of a new reef on top of an old reef, an oolite shoal on top of an older shoal. Fig. 7.1 shows the combined effects of rim-building

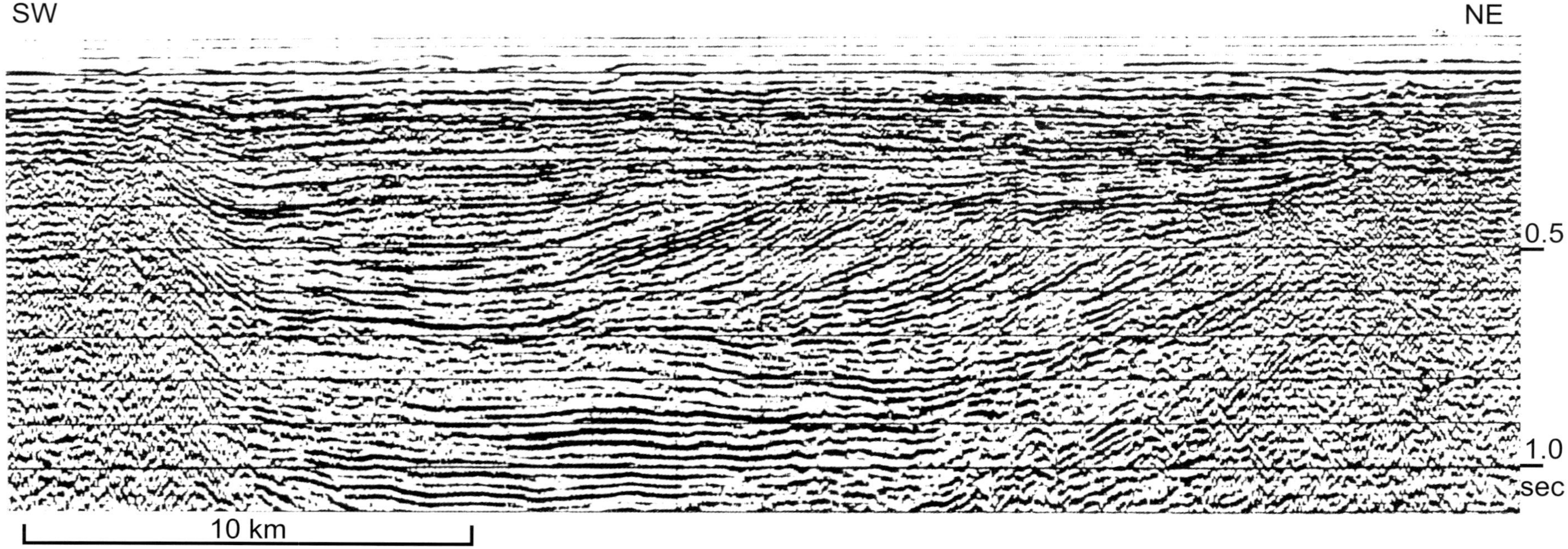

Fig. 7.1.— Seismic profile across southwestern Great Bahama Bank. Two Cenozoic platforms enclose a basin that is gradually filled by the prograding platform on the right. The platform margin on the left only aggrades. The seismically incoherent patterns of the left margin and its tendency to rise above the adjacent platform suggest that it consists of stacked reefs or sand shoals. Recognizing sequences and systems tracts is difficult in this setting; it is much easier on the prograding margin on the right. Continuous reflections extending across the basin indicate that there were no synsedimentary tectonic movements between the two platforms. Both platforms experienced the same sea-level history and their differences are caused by sediment supply: the right margin faces downwind and receives abundant sediment from the adjacent platform; the left margin faces upwind and is therefore stripped of sediment. After Eberli and Ginsburg (1988, modified).

and wind on the large-scale anatomy of platforms in the Bahamas. On the left, windward, margin the rims stack vertically and the gradual lateral migration of systems tracts in response to sea level is interrupted. The prograding leeward margin on the right in Fig. 7.1 more closely resembles the classical sequence model because loose sediment from offbank transport plays a bigger role. However, rim building by reefs or lithified sand shoals is important even in prograding margins: the constructional rims tend to occur intermittently and as lenses; they are not resolved by the low-frequency seismics in Fig. 7.1 but can be seen in high-resolution data of the Holocene (Fig. 7.2). The fluctuating shelf-margins of these prograding platforms with their buried rims seem to yield the most reliable sea-level record in carbonate seismic stratigraphy.

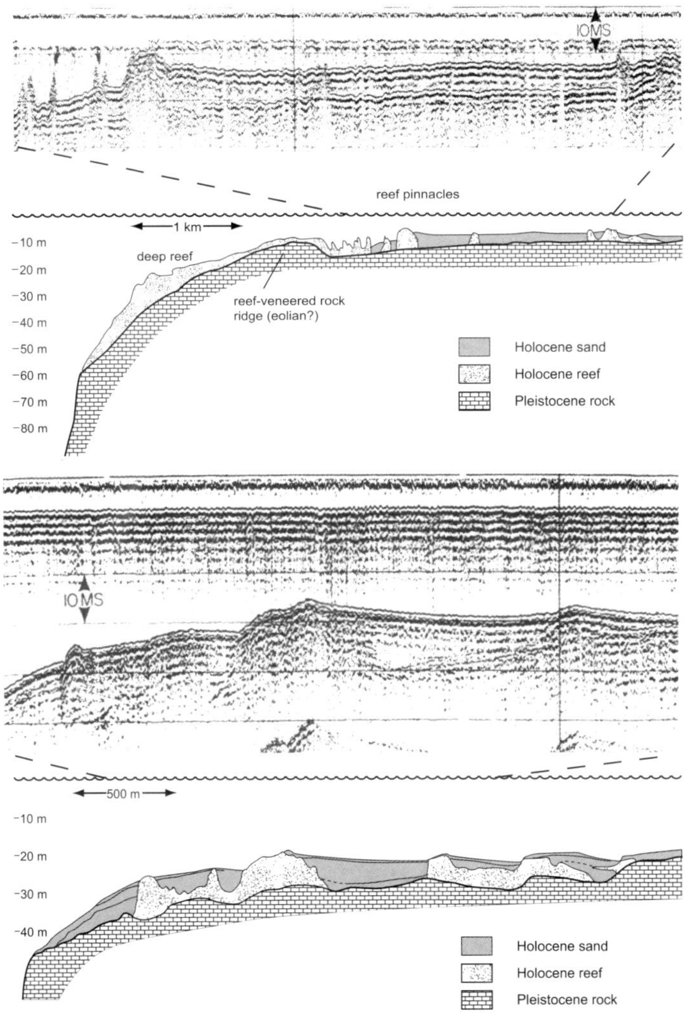

Fig. 7.2.— Windward and leeward Holocene platform margins in the Bahamas. Reefs are actively growing on the windward margin (upper panel) and mostly buried in sediment on the leeward margin because of offbank sediment transport (lower panel). After Hine and Neumann (1977). (Reprinted by permission of the AAPG whose permission is required for further use).

Systems tracts in T carbonates and their control by accommodation and production

Systems tracts of the standard model were defined geometrically and subsequently interpreted in terms of relative sea-level changes (see chapter 6). Very similar definitions can be applied to tropical carbonates (Fig. 7.3) . One significant difference is that well-rimmed tropical platforms have essentially horizontal tops rather than seaward dipping shelf profiles.

In chapter 6, the systems tracts of the standard model were interpreted as the balance of the rate of change of accommodation and the rate of sediment supply (Fig. 6.12). In carbonates, the situation is analogous except that outside sediment supply has to be replaced by G, the *in-situ* growth and production of carbonate material. Sediment geometry and systems tracts are again controlled by two a priori independent rates: A′ = dA/dt, the rate of change in accommodation as defined in chapter 6, and G′ = dG/dt, the rate at which a platform produces sediment and builds wave-resistant structures. The maximum rate of growth that the system can sustain, the growth potential, varies across the platform. At the very least one must distinguish between G_r'= the growth rate of the platform rim, and G_p'= the growth rate of the platform interior.

Based on the relationship between A′ and G′, five characteristic patterns can be distinguished (Fig. 7.4). The first three situations correspond to the three systems tracts of the standard model in Fig. 6.12. The last two situations are typical for rimmed tropical carbonates. Empty bucket and drowned platform tops are particularly important because they yield diagnostic patterns for recognizing tropical platforms in seismic data.

The importance of sediment supply on sequence anatomy is illustrated by Fig. 7.1. The entire area forms a stable

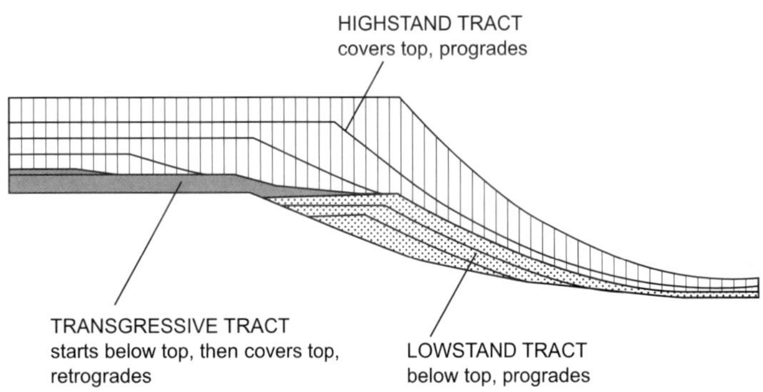

Fig. 7.3.— Terminology of systems tracts on rimmed carbonate platforms. The platform margin of the preceding cycle serves as a reference level. A systems tract whose top is lower than the top of the preceding platform is a lowstand tract, systems tracts that cover the top of the preceding platform are either transgressive tracts or highstand tracts, depending on whether they retrograde or prograde. The tops of rimmed platforms tend to be very flat and flooding of the rim often leads to abrupt backstepping of the platform margin. After Schlager et al. (1994).

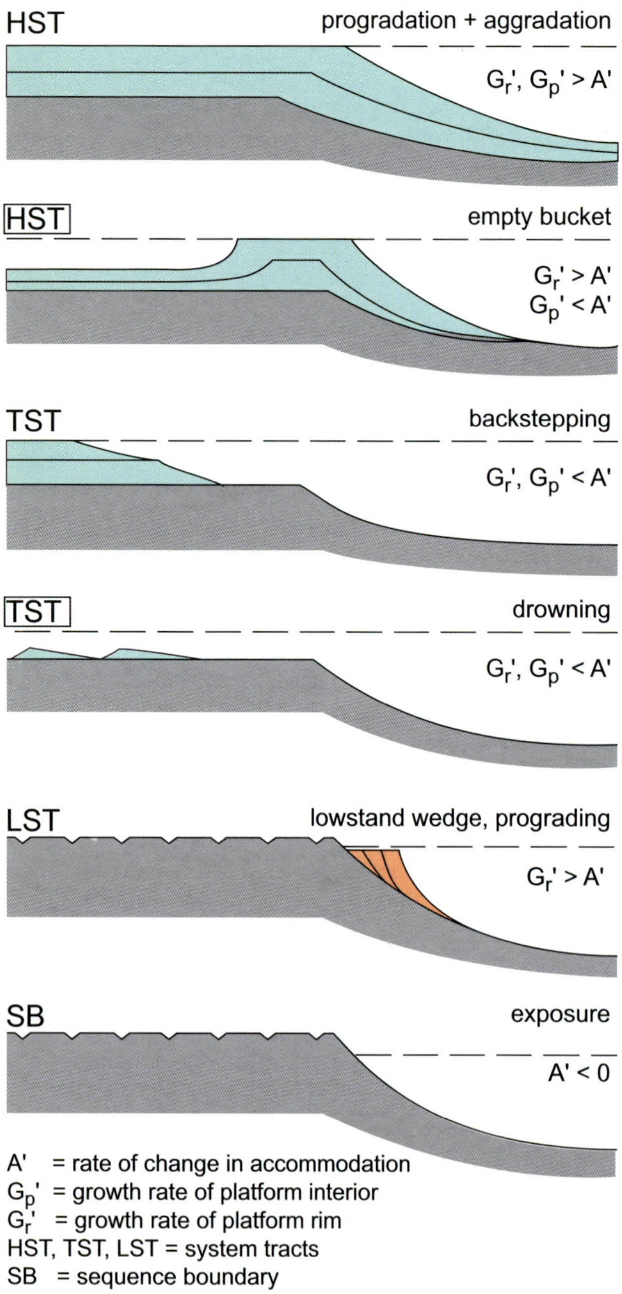

Fig. 7.4.— Basic geometries of tropical platforms and their interpretation in terms of rate of change in accommodation, A', and rate of sediment supply, G'. The letter G stands for carbonate growth, indicating that most material is produced within the depositional environment even though lateral transport may be significant. Panels 1,3,5 and 6 are strictly analogous to the systems tracts of the standard model (Fig. 6.12). Panels 2 and 4 labelled in red, are specific to tropical carbonates. Empty bucket and complete drowning illustrate the importance of the growth potential of the various elements of the system. Submergence below the euphotic zone means that parts or all of the production system are shut down.

block and therefore experienced the same relative sea-level changes. The pronounced difference between windward and leeward platform margin must be entirely due to differences in carbonate production and supply.

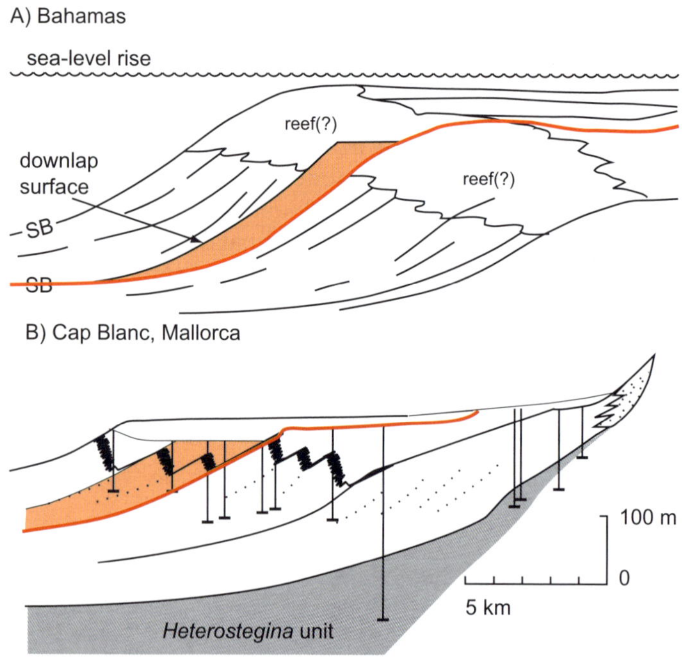

Fig. 7.5.— Downstepping platform margins are one of the best geometric indicators of sea-level fluctuations in tropical carbonates. In both examples, red shading indicates the minimum extent of the lowstand systems tract that can be defined with respect to the red reference profiles. A) Bahamas, Cenozoic, based on seismic data and some bore holes. After Eberli and Ginsburg, (1988), modified. B) Miocene, Mallorca, based on continuous outcrops and borings. Upstepping and downstepping of the reef belt (black) reveals a hierarchy of rhythms. After Pomar (1993), modified.

Sea-level movements deduced from seismic images of carbonate platforms

With their flat tops built to sea level and their biota very sensitive to water depth, carbonate platforms are one of the most reliable dip sticks in the ocean. This quality is enhanced by the resistance to erosion of exposed platforms. Reefs are "born" as rock-hard structures, other platform deposits frequently lithify within a few thousand years when exposed. Subsequent erosion is largely chemical and operates within the rock rather than at its surface (see above). Surface denudation is generally less than in siliciclastics and a reasonable sea-level record can be gleaned from platforms by determining overall subsidence and measuring the thickness of marine intervals plus position and timing of exposure horizons (see Fig. 7.4 and Ludwig et al., 1988; McNeill et al., 1988).

The combination of defended margins and enhanced resistance to erosion creates some special opportunities for sequence stratigraphy. Rapidly prograding platform margins tend to preserve the original elevations of the shelf surfaces particularly well, including the very important lowstand systems tracts. Eberli and Ginsburg (1988), Eberli et al. (2001), Sarg (1988; 1989) and Pomar (1993) have contributed excellent examples of sea-level curves gleaned from carbonates using the technique of the fluctuating shelf surface (Fig.

Fig. 7.6.— Satellite image of Andros Island and surroundings on Great Bahama Bank. Despite uniform recent sea-level history, conditions for all three systems tracts are realized simultaneously in the vicinity of Andros. This is due to the interplay of inherited topography, variations of sediment production and lateral sediment transport. Lowstand conditions (L) exist on the narrow shelf between Andros and the deep-water trough of Tongue of the Ocean. Transgressive conditions (T) are well documented on the NW side of the island where the narrow belt of tidal flats is being eroded on the seaward side and transgresses the island on the landward side. Highstand conditions (H) prevail on the SW-facing tidal flats and on the Joulters Cay oolite shoal N of Andros. The platform E of Tongue of the Ocean is mostly in transgressive conditions. After Harris and Kowalik (1994), modified. (Reprinted by permission of the AAPG whose permission is required for further use).

7.5). See also Purdy et al. (2003, p. 562) for a seismic example from the Pleistocene.

SHOAL-WATER FACIES OF T-FACTORY SYSTEMS TRACTS

The geometric definition notwithstanding, many authors have suggested that systems tracts also display characteristic facies patterns (see Posamentier et al., 1988 for siliciclastics; Handford and Loucks, 1993, for carbonates).

A connection between geometry of sediment bodies and their facies is indeed very likely because the geometry reflects conditions that must also affect the depositional environment. With respect to carbonate platforms, the geometry of the three systems tracts immediately leads to several conclusions about the depositional environments and thus the facies of the systems tracts. For instance, lowstand tracts can be expected to be rather narrow, normal marine and devoid of the vast shallow lagoons of mature platforms. Transgressive tracts and highstand tracts, on the other hand, inherit the vast shallow environments of the platform top from the preceding highstand. In the transgressive tract, the retrograding platform margin implies that sediment supply lags behind the rate of accommodation creation by relative sea-level rise; as a consequence, water depth increases and open-marine conditions expand on the platform top during the transgressive stage. Conversely, progradation of the highstand tract indicates that the rate of supply exceeds the rate of change in accommodation; water depth decreases and circulation on the platform top tends to become restricted. Progradation of the highstand tract may be bi-directional, filling the empty lagoon and expanding the platform seaward (see Fig. 3.19; Saller et al., 1993).

Systems-tract facies of modern Bahamas and Florida

In order to test these tentative conlusions and examine the facies in more detail, we turn to modern environments of the Florida-Bahama platforms. If one follows the definitions on Fig. 7.3, all three systems tracts are currently developed on the platforms (Figs 7.6, 7.7). This may seem surprising at first because eustatic sea level movement has been practically the same in the entire area. However, the effects of inherited topography and variations in sediment supply suffice to differentiate the record.

Lowstand tract (Figs 7.6, 7.7 7.8). What I consider analogues of lowstand tracts are narrow carbonate shelves that rim the exposed limestones of the last interglacial highstand tracts. The width of these shelves varies from about 100 m to over 2,000 m. The sediment cover consists of the coralgal lithofacies of Bathurst, (1971, p. 108): skeletal sands, hardgrounds and reefs (fringing reefs, barrier reefs protecting narrow lagoons, and patch reefs in the lagoons). Most of the "outlier reefs" mapped by Lidz et al. (2003) also fall in this category. The biota is dominated by algae, corals, molluscs and foraminifers, derived from the reefs and from the sea-grass communities on the flat sea floors. Very fine sand and mud amount to ca. 10%, pellets to 25% (Bathurst, 1971). Ooids are rare. A possible exception are certain beaches on Caicos Bank (Wanless and Dravis, 1989), where precipitation seems to be so rapid that ooids form on the shore face before they are ejected from this narrow belt of favorable conditions. Chemical conditions have not been studied; however, it is likely that the process depends on backflow of hypersaline lagoon waters and thus ultimately requires a flooded platform top - a condition not met during genuine lowstands of sea level. Most of the Bahamian coralgal shelves are sediment-starved. However, the environment is highly productive and will quickly catch up with sea level and start to prograde. Progradation can be expected to be slow because the system faces a high, steep slope.

Transgressive tract (Figs 7.6, 7.7, 7.9a). In most parts of the Bahamas and southern Florida, sea level has flooded the platform top and sedimentation clearly trails the rising sea. Along the platform margin, reefs have stepped back repeatedly in the past 10,000 yr and the present barriers lie 200 - 1,000 m bankward of the marginal escarpment (Hine and Neumann, 1977). These barriers are discontinuous with only a fraction of the platform perimeter occupied by shallow reefs (rim index <0.25). Sand shoals, mainly composed of oolite, are widespread. Much like the reefs, these shoals are rather ineffective barriers; typical are tidal-bar belts that extend 20 – 30 km onto the platform but whose seaward tips do not reach the platform edge. Channels between the sand bars usually are wider than the sand bars themselves. The sediment covering the vast interior of the Bahama Banks is dominated by pellets, ooids and grapestone lumps; skeletal grains generally are less than 15%; mud is less than 10%, higher concentrations occur only in the lee of islands (Bathurst, 1971). Hardgrounds are common. In large areas, the Pleistocene surface lies bare or is thinly covered by corals and sponges, and sometimes studded with patch reefs ("transgressive surface"). Other hardgrounds occur within the Holocene section where sedimentation was interrupted for extended periods by currents and waves (e.g. on the eastern lobe of Great Bahama Bank). In the sediment cover, the facies succession is deepening upward, with mangrove peat and muddy deposits of restricted lagoons at the base and open-marine, winnowed sand on top (Enos, 1977 for Florida area).

Highstand tract (Figs 7.6, 7.7, 7.9a, 7.10). Even though the Holocene eustatic rise of sea level must have been rather uniform across the stable Florida-Bahama area, the sediment cover does not uniformly exhibit transgressive characteristics. Where currents and waves sweep the sediment together or where carbonate production is exceptionally high, sedimentation has caught up with the rising sea. In these areas, sediments have filled the available accommodation space and started to prograde laterally. Geometrically these parts of the Holocene cover are highstand systems tracts. Examples include the oolite shoal of Joulters Cays and the tidal flats SW of Andros.

The shoals of Joulters Cays differ significantly from their counterparts in a transgressive setting. At Joulters, the sand shoal forms a nearly continuous, margin-parallel barrier whose crest is several kilometers wide and reaches in the intertidal zone, locally topped by beach-dune islands (Harris, 1979). Tidal channels are narrow, widely spaced and frequently choked by migrating sand on either end. Cores show that the intertidal shoal has been established on top of shallow-marine pellet sand. In the last 1,000 y, islands formed and prograded seaward. Similar progradation of beach-dune complexes was described from other Bahamian platforms (Wanless and Dravis, 1989).

More significant progradation took place on the tidal flats of SW Andros. The belt S of Williams Island prograded 15 – 20 km in the past 6,000 y while the flats N of Williams Island

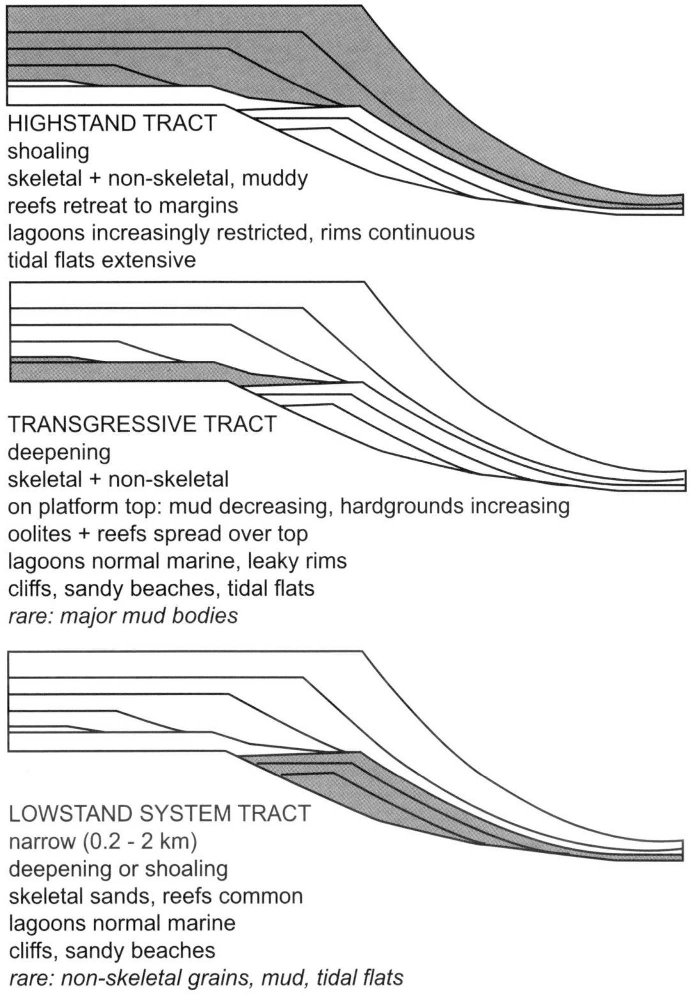

Fig. 7.7.— Facies characteristics of carbonate systems tracts based on the Holocene record of the Bahamas and Florida. After Schlager (2002).

were retreating (Figs 7.6, Hardie and Shinn, 1986). Progradation occurred by rapid accumulation of shallow-marine muds and muddy tidal flats. Narrow tidal channels, closed off at their seaward end, as well as shore-parallel, inactive beach ridges attest to this stepwise progradation.

Shoaling and prograding Holocene systems tracts are common on other carbonate platforms, too. The tidal flats and barrier islands of Qatar in the Persian Gulf have built a 15 m thick wedge that has prograded 5 – 10 km in the Holocene (Hardie and Shinn, 1986). Reefs and reef aprons in the Caribbean and the SW Pacific have built to sea level and prograded both seaward as well as landward (examples in James and Macintyre, 1985). This bi-directional progradation of platform margins is common during the early phases of highstand deposition when rapid sea-level rise left the lagoons empty.

Significance of shoaling and deepening trends. Much information in sedimentary geology becomes available in the form of vertical stratigraphic sections. In such sections, carbonate lowstand tracts appear as exposure surfaces on the platform top (or carbonate shelf) and as downstepped shoal-water deposits on the slope. Transgressive and highstand tracts both are represented by marine deposits on the platform top. They differ by the change in depositional environment observed in vertical section: deepening upward (and increasingly open marine) in transgressive tracts, shoaling (and increasingly restricted) in highstand tracts.

The above examination of Holocene systems tracts shows once more that in carbonate sections only the lowstand tract with its exposure surface on marine deposits is a reliable indicator of sea-level change. The change from transgressive to highstand tracts and vice versa is an ambiguous record. It may be produced by a change of relative sea level

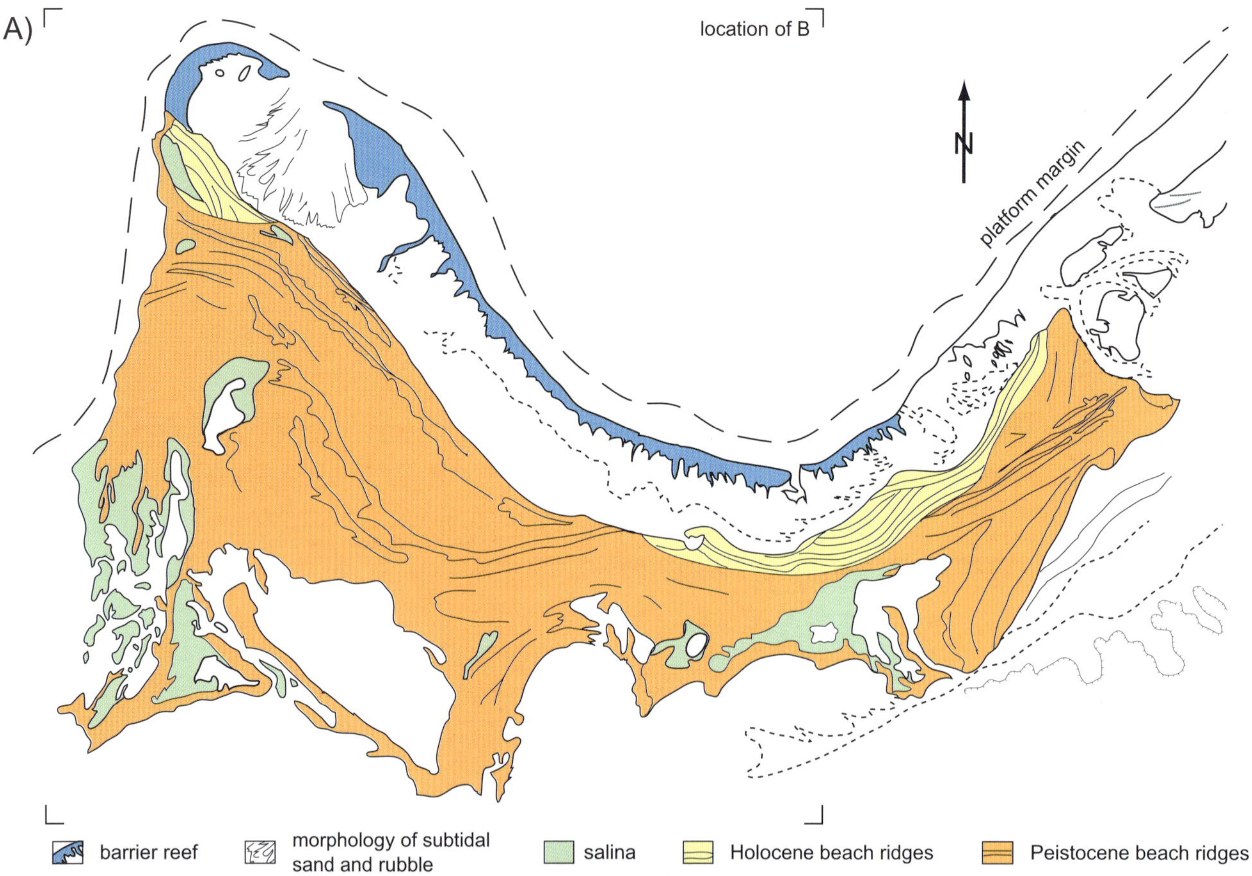

Fig. 7.8.— A) Part of Caicos Bank in the southeastern Bahamas, after Wanless and Dravis(1989), modified. The platform shows the complex interplay of sea level, antecedent topography and sediment supply. The narrow shelf W and N of the island still is analogous to a lowstand tract, abutting against the cliffed edge of the Pleistocene island. The remainder of the platform is in transgressive or highstand condition. B) Satellite image of left part of A. "L" indicates lowstand conditions. After Harris and Kowalik (1994). (Reprinted by permission of the AAPG whose permission is required for further use).

Fig. 7.9.— Bahamian oolite shoals in transgressive and highstand condition. A) Cat Cay - transgressive shoal. Sand bars are thin and anchored behind Pleistocene islands; tidal channels (c) are wide. B) Joulters Cay highstand shoal. Tidal channels are few and narrow, sand bars have coalesced to one continuous belt. C) Prograding beach ridges (r) and eolian dunes on Joulters shoal.

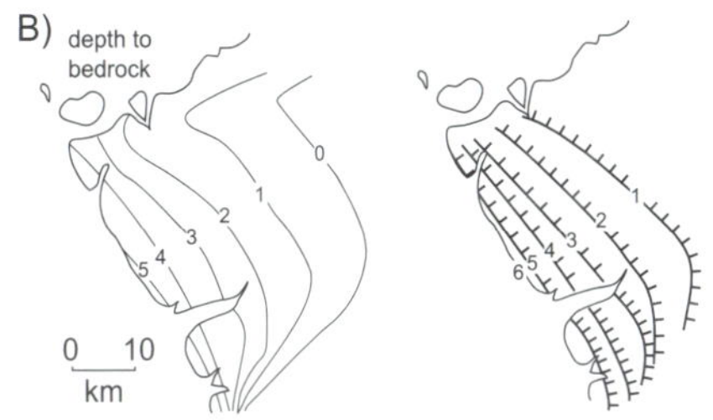

Fig. 7.10.— Tidal flats west of Andros Island, Bahamas. A) Satellite image showing NW tidal flats in transgressive conditions (T), SW flats in highstand condition (H). On transgressive flats, seaward beaches are ragged and erosional, channels deep and well connected. Channel network efficiently floods and drains the standing water of the inner flats. On the highstand flats, seaward beaches are smooth and accreting (older beach ridges indicated by x). Open channels are scarce, most have been choked by prograding beach ridges. After Harris and Kowalik (1994), reprinted by permission of the AAPG whose permission is required for further use. B) Ancient shore line positions and thickness of the highstand wedge of the SW-facing flats. After Hardie and Shinn (1986), modified.

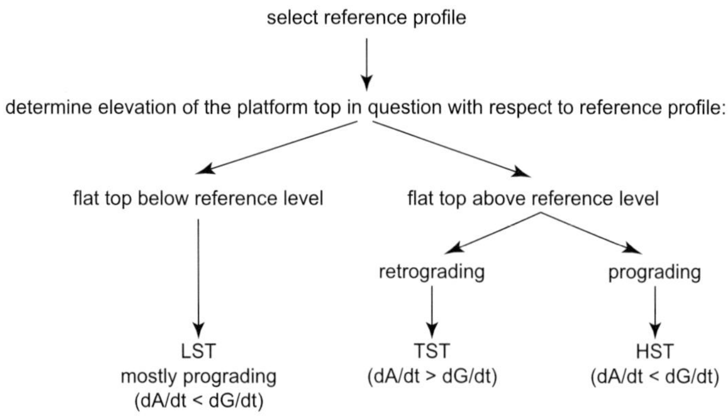

Fig. 7.11.— Flow chart for identification of systems tracts when a cross section of the carbonate platform is available, in particular the position of the platform margin.

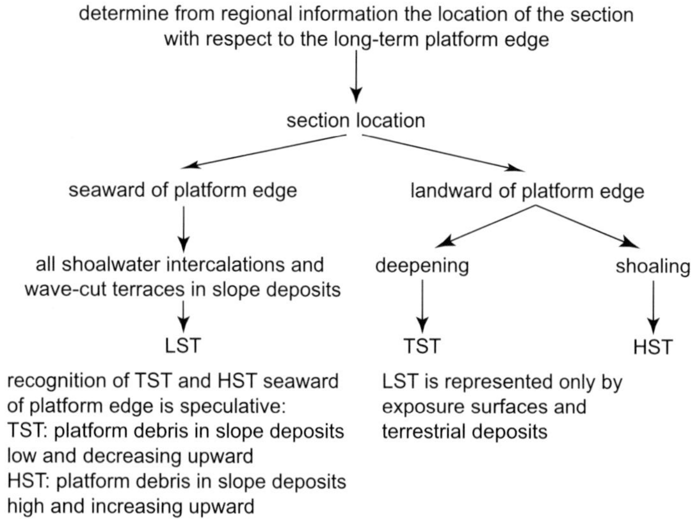

Fig. 7.12.— Identification of systems tracts in a single section or borehole. This technique is more speculative than the cross-section approach, relying on proxy indicators such as deepening and shoaling trends. The position of the section with respect to the long-term margin is required as input.

or by a change of sediment supply (or by a combination of both, of course). This conclusion is inevitable if one defines relative sea-level change not simply as a change in water depth but as a change in the distance between the sea surface and some deep stratigraphic reference level, such as the top of basement or a deep stratigraphic marker (p.91). With good reason, sequence stratigraphers have insisted on this definition in several strategic papers (e.g. Vail et al., 1977; Posamentier et al., 1988; Jervey, 1988). I recommend to accept this definition with its implications. Shoaling/deepening trends should be mapped and correlated where possible, but their relation to sea-level change should be left open until independent evidence removes the ambiguity. Such evidence may consist of correlating the shoaling or deepening trends with distinct exposure rhythms or with orbital cycles that have the additional advantage that their sea-level component is by definition eustatic.

Flow charts for identification of carbonate systems tracts are shown in Figs 7.11 and 7.12. Fig. 7.11 is based on the geometric definitions of Fig. 7.3. This procedure is applicable if one has information on the shore-to-basin cross section of the platform, in particular the position of the platform edge. Seismic profiles, large outcrops or well-correlated series of boreholes or outcrops may provide this information. However, there is great demand to recognize sequences and systems tracts also in single boreholes or outcrop sections. Under these circumstances, the diagnostic criteria for systems tracts must be inferred from circumstantial evidence. Fig. 7.12 offers a procedure for these situations. However, identification of systems tracts in single sections remains more speculative and less reliable than the cross-section method.

The testimony of ancient systems tracts

Handford and Loucks (1993) offer an extensive compilation on the facies patterns of systems tracts, relying on ancient examples and principles of carbonate sedimentology. Their results agree well with the observations on the Holocene summarized above. Handford and Loucks' (1993) lowstand tracts are narrow, favorable for reefs and carbonate sands; transgressive tracts are well flushed, reefs again thrive and spread over the platform top, tidal flats are narrow or absent; in highstand tracts, the platform interiors tend to become restricted and more muddy, tidal flats expand. On the slopes, slumps and sediment gravity flows occur throughout a sequence cycle as long as the slope is suffiently steep. For mixed carbonate-siliciclastic systems, Handford and Loucks (1993) predict that in the lowstand tract, incised valleys breach the narrow carbonate belt and funnel siliciclastics to the basin. In the highstand tract, the landward part of the platform is filled by siliciclastics forming a coastal plain under humid conditions or a salina under arid conditions.

Homewood (1996) approached systems-tract anatomy and facies from a theoretical point of view. He argued for strong coupling among the rate of accommodation creation, ecologic conditions and carbonate production. Such feedbacks are very likely but I doubt that they are simple enough for reliable predictions. For instance, Homewood (1996, p. 711-714) predicts that transgressive tracts will be dominated by r-strategists among the biota, highstand tracts by K-strategists such as reef builders (p. 9). The analysis of Holocene systems tracts given above indicates very nearly the opposite pattern: reefs thrive during the late part of the transgressive tract and their domain shrinks during the formation of the highstand tract when platform environments become increasingly restricted and prograding bodies of loose sediment bury many reefs.

Ecologic reef, geologic reef, seismic reef

Chapters 2 and 3 emphasized the importance of reefs at platform margins for the creation of flat-topped platforms so typical of tropical carbonates. The question of what is a reef continues to fuel heated discussions among geologists. Dunham (1970) introduced a novel perspective in this debate when he proposed to qualify the term by distinguishing between ecologic reefs, built and bound by organisms, and stratigraphic reefs where the binding may also be done by cementation of loose sediment. This distinction has proved very useful but the increasing use of seismics to identify reefs has complicated matters again.

The "seismic reefs" defined by geometry and reflection character do not correspond straighforwardly to either ecologic or stratigraphic reefs. Seismics tends to overlook small reefs and, in other circumstances, adds non-reef deposits to the reef. Living parts of reefs are small by seismic standards, easily destroyed and quickly buried by their own debris. The diagnostic criteria for identification of reefs in outcrop are far below normal seismic resolution (see Fig. 4.6). The seismic tool reveals reefs only where many generations of reef growth were stacked to thicknesses of tens to hundreds of meters. Where the sites of reef growth shift laterally, reefs do not stack to seismically recognizable structures but form small lenses embedded in detrital sediment. These small constructions are readily recognized in the field but remain largely hidden in standard seismic data.

Equally important are situations where the seismic tool shows more "reef" than is actually there in a stratigraphic or ecologic sense. Reefs defined by geometry and reflection character normally include, beside ecologic reefs, also the reef aprons on the landward side as well as the coarse, poorly stratified talus on the fore-reef side (Fig. 7.13. In addition, sand shoals interfingering with reefs may be included in the seismic reef unit.

Rules of thumb on systems tracts and facies

The facies characteristics of recent and ancient systems tracts are rooted in two principles of carbonate deposition: the balance beween rate of sediment supply and rate of accommodation creation, and the tendency of carbonate systems to form flat tops and steep slopes. The shifting balance of rates explains the differences between transgressive and highstand systems tract, the flat-top-steep-slope morphology sets the lowstand tract apart from transgressive and highstand tracts. Below follow some rules of thumb on sytems tract facies. Most of them are directly related to the balance of rates and to morphology.

- The shoalwater belts of lowstand tracts are much narrower than transgressive and highstand tracts. The limited width is a geometric consequence of the intersection of the photic zone with a steep slope. For a platform with a top dipping at 0.1° and a slope dipping at 10°, the widths of the lowstand and highstand tracts differ by about two orders of magnitude. Consequently, the lowstand tract tends to be well flushed and normal marine, rich in reefs, hardgrounds and rocky shores; sediment tends to be coarse and predominantly skeletal; ooids tend to be scarce because tidal currents are weak on the narrow shelf. The facies succession may be shoaling or deepening upward and the margin prograding or backstepping. There is no indication that carbonate lowstand tracs are only prograding, nor is there a theoretical basis for such a postulate.
- Transgressive systems tracts are characterized by the rate of accommodation creation exceeding the rate of sediment supply. Consequently, facies are deepening upward. Reef rims and sand shoals are narrow and breached by wide passages, the platform interior is generally well flushed and normal marine; patch reefs spread far landward. In extreme situations, the rim may be lacking altogether such that the platform temporarily represents a ramp that steepens at the distal end to merge with the profile of the old highstand slope. Mud is rare in the shallow-marine environments because of the scarcity of protecting barriers; however, deep flooding may increase water depth to below wave base such that deeper-water muds accumulate on top of winnowed early TST deposits.
- Highstand systems tracts are characterized by the rate of sediment supply exceeding the rate of accommodation creation. As a consequence, sand shoals and reefs at the margin become wider and more continuous. Facies successions are shoaling upward; the degree of restriction increases in the platform interior and mud may start to accumulate in the lagoon and in expanding tidal flats. Lagoonal patch reefs tend to decrease upward.
- Non-skeletal grains (ooids, peloids) tend to be more abundant in transgressive and highstand tracts than in lowstand tracts because they benefit from the amplification of tidal currents and the extensive winnowing by waves on the vast, flooded platform top. This distribution pattern holds as long as carbonate precipitation rates are in the same range as those of the modern Bahamas or the Persian Gulf. More rapid rates of precipitation may lead to formation of ooids and hardened peloids on narrower platforms.
- The deposits of slope and basin commonly permit a differentiation into times of bank-top flooding and times of bank-top exposure, i.e. highstand and transgressive tracts on the one side, and lowstand tracts on the other. The key to this differentiation is grain composition for instance abundance of ooids and platform mud. In this instance, facies analysis is a better tool than the geome-

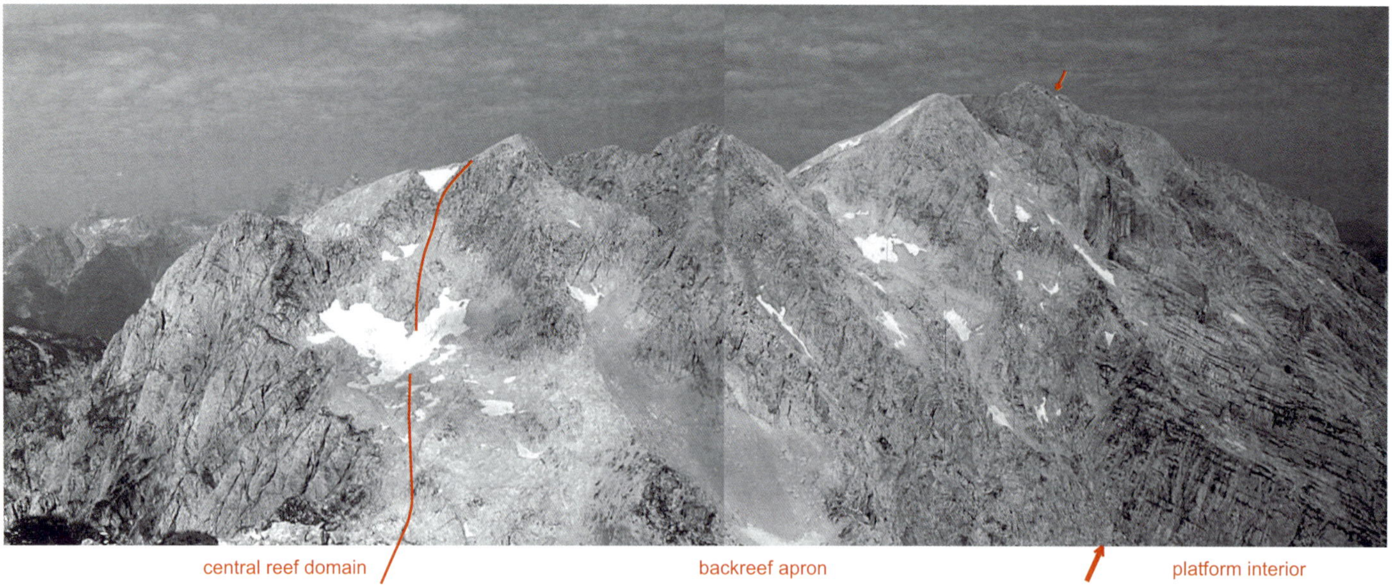

Fig. 7.13.— Reef-lagoon transition in Triassic Dachstein Fm. of the Northern Calcareous Alps. Massive limestones on the left gradually pass into well-bedded limestones of the platform interior on the right (bedding tilted by Alpine tectonics). Red line in massive limestones indicates the approximate boundary between in-situ reef belt on far left and rubble and sand of the backreef apron in the center; red arrows indicate boundary between reef apron and bedded lagoon facies (based on map by Zankl, 1969). Seismic data would almost certainly show reef and apron deposits as one zone of incoherent reflections and pick up the transition to the rhythmically bedded limestones as a facies change.

try of sediment bodies, which tends to be dominated by differences in slope angle.

The rules of thumb are based on first principles of carbonate sedimentation as well as observations on recent and ancient carbonate systems tracts. However, the link between geometrically defined systems tracts and carbonate facies remains an indirect and tenuous one that is easily perturbed by other effects on facies. Predicting facies from systems tract geometry will remain a blend of art and science for some time to come.

T SEQUENCES IN DEEPER-WATER

Periplatform environment - part of the platform system

Platform, slope and debris aprons on the basin floor are one connected system in the T factory. In this respect the T factory resembles siliciclastic systems where sediment supply from land and downslope transport in the marine domain create one interconnected system of depositional environments and facies. In the T factory, the sediment feeding the systems is produced in the photic zone at the platform top. Because of this shallow location, production is very sensitive to sea-level changes that expose and flood the platform top. The effects of changing production propagate downslope through the system and make themselves felt as changes of facies and sedimentation rates on the slopes and basin floors surrounding the platforms. Two important topics related to sea-level effects on deeper-water environments are highstand shedding and the origin of megabreccias. They are discussed below.

Highstand shedding

It is a well-established fact that in the Pleistocene, siliciclastic sediment supply to the deep sea was at its maximum during glacial lowstands of sea level. The insight that rimmed carbonate platforms were in antiphase to this rhythm developed first in the Bahamas: Kier and Pilkey (1971) and Lynts et al. (1973) showed that sedimentation rates in the interplatform basins peaked during the interglacials when large volumes of aragonite mud were swept off the platforms. Schlager and Chermak (1979) observed that turbidite input, too, was high during the Holocene and low in the last glacial. Mullins (1983) first emphasized this "carbonate way" of responding to sea level for which Droxler and Schlager (1985) coined the term "highstand shedding". It indicates that carbonate platforms produced and exported most sediment during interglacial highstands when the platform tops were flooded. The pattern is best documented for the Bahama Banks (Fig. 7.14; Droxler et al., 1983; Mullins, 1983; Reijmer et al., 1988; Spezzaferri et al. 2002; Rendle and Reijmer, 2002). However, the same trend has been observed on the platforms of the Caribbean, the Indian Ocean and the Great Barrier Reef (Fig. 7.15; Droxler et al., 1990; Davies et al., 1989; Schlager et al., 1994; Andresen et al., 2003).

Highstand shedding is pronounced on tropical carbonate platforms because of the combined effect of sediment production and diagenesis. Sediment production of a platform increases with its size, and the production area of a plat-

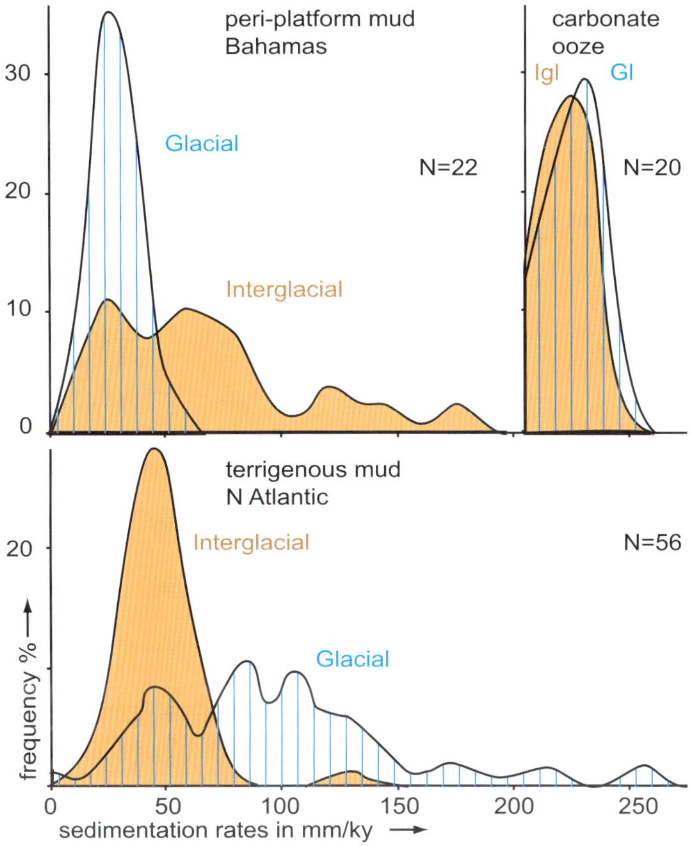

Fig. 7.14.— Sedimentation rates during highstands and lowstands of sea level in various depositional systems. In terrigenous muds from the continental rise, rates are low and uniform during interglacial highstands and high and variable (turbidite-controlled) during glacial lowstands ("lowstand shedding"). In periplatform muds, the pattern is reversed ("highstand shedding"). In pelagic carbonate ooze, rates remain low and uniform in both highstands and lowstands. After Droxler and Schlager (1985), modified.

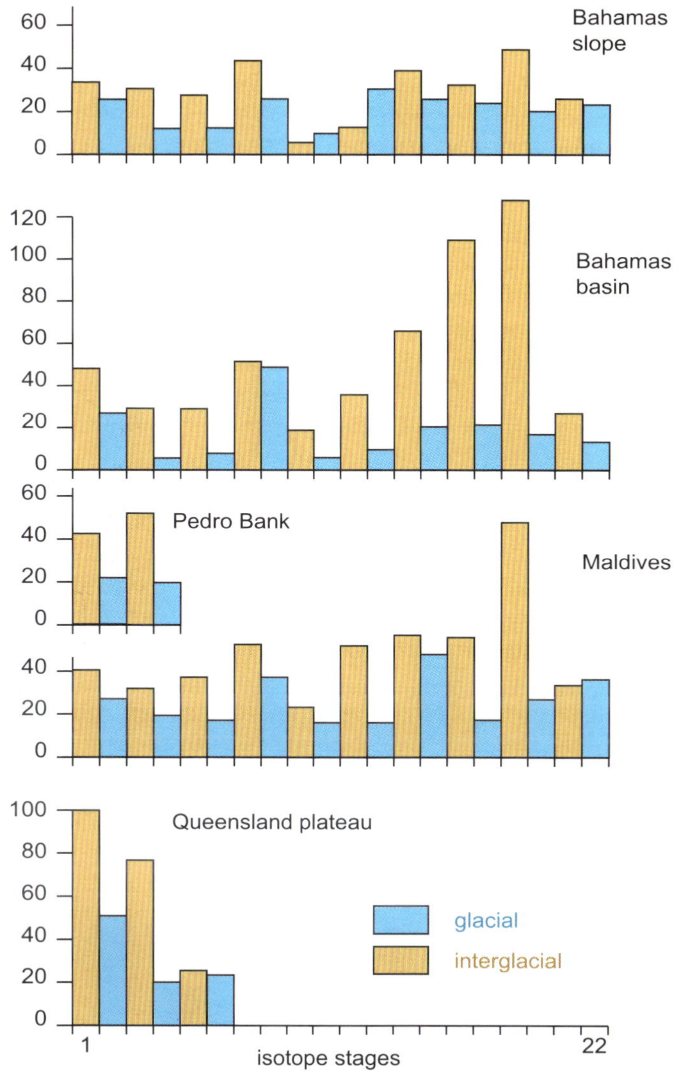

Fig. 7.15.— Highstand and lowstand sedimentation rates during the late Quaternary in the Bahamas, Pedro Bank in the Caribbean, Australia's Queensland Plateau and the Maldives (Indian Ocean). In all instances, rates during glacial lowstands (shaded) are generally lower than during the interglacial highstands of sea level. Other trends modulate this pattern (e.g. upward-decreasing rates in Bahamas and Maldives) but they do not erase it. Compiled from Reijmer et al. (1988), Droxler et al. (1990), Schlager et al. (1994).

form with its top is exposed is normally smaller than the production area of a flooded platform. In the Bahamas, the flooded top is one order of magnitude larger than the belt of shallow-water production during lowstands (Droxler et al., 1983; Schlager et al., 1994). The effect of increased highstand production is enhanced by the rapid lithification of carbonates during lowstands. Siliciclastics owe part of the high sediment input during lowstands to erosion of the preceding highstand tract. Carbonate diagenesis largely eliminates this effect. In the marine environment, there is a strong tendency for the sea bed to lithify and develop hardgrounds wherever waves and currents interrupt continuous sedimentation (Schlager et al., 1994). Thus, widespread hardgrounds protect the highstand deposits when sea level starts to fall and wave base is lowered. When sediments finally become exposed, they tend to develop an armour of lithified material within hundreds to a few thousands of years as borne out by the numerous rocky Holocene islands on extant platforms (e.g. Halley and Harris, 1979). Subsequent erosion acts largely through chemical dissolution, enlarging porosity and creating cave networks but keeping surface denudation lower than in siliciclastics (e.g. Fig. 2.28)

Turbidites are more frequent in highstand intervals, forming highstand bundles, (Fig. 7.16). Furthermore, the high sediment export by the flooded platforms leads to highstand wedges of sediment on the slopes, displayed in exemplary fashion by the Holocene sediment of extant carbonate platforms.

Platforms are not completely shut down during lowstands. Sediments continue to be produced from a narrow belt of sands and fringing reefs, and from eroding sea cliffs that cut into the exposed platform. There is no evidence, however, that the rates of lowstand input reach those of the highstand input with the possible exception of unusual breccia bodies (see below). The postulated apron-building around Pacific atolls during lowstands of sea level has not withstood close scrutiny (Thiede, 1981; Dolan, 1989). Lowstand input is compositionally different and can be recognized by petrographic analysis (see below).

The limitations of highstand shedding can be deduced from its causes. The difference between highstand and lowstand production depends on the hypsography of the platform. Flat-topped, rimmed platforms with steep slopes show more pronounced highstand shedding than platforms with gentle slopes (Fig. 7.17). Besides platform morphology, the duration of sea-level cycles is important. When sea-level cycles are long (e.g. millions of years) and flanks gentle, the platform can build a lowstand wedge that partly substitutes for the production area lost on the platform top (Fig. 7.17). Finally, lithification and resistance against lowstand erosion vary with latitude and possibly with age. Cool-water carbonates are less prone to submarine lithification than their tropical counterparts (Opdyke and Wilkinson, 1990). Lithification upon exposure, too, is reduced because of the low content of metastable aragonite and magnesian calcite. It is possible that Paleozoic carbonates with their generally lower content of metastable minerals are also more prone to lowstand flushing. Lowstand shedding of Cenozoic cool-water carbonates has been described by Driscoll et al. (1991).

The antagonistic behavior of carbonates and siliciclastics commonly results in reciprocal sedimentation with highstand bodies of carbonates and lowstand wedges of siliciclastics (Delaware Basin – Wilson, 1975; Canning Basin - Southgate et al., 1993).

Recently, sea-level studies of platform tops have been complemented by compositional analysis of platform-derived turbidites on the platform flanks. The depositional environment of these calciturbidites is below the range of sea-level fluctuations such that sedimentation is not interrupted during lowstands. Highstand turbidites differ not only in abundance but also in composition from their lowstand counterparts as they contain more ooids, pellets, and grapestones - grains that require flooded bank tops for their formation. This has been well documented for Pleistocene turbidites in the Bahamas (Fig. 7.16). Reijmer et al. (1991; 1994) report on a Triassic example, Everts (1991) on a Tertiary one.

Carbonate petrography reveals not only changes in the spectrum of contemporary grains. Lithoclasts derived from erosion of older, lithified parts of the platform are easily rec-

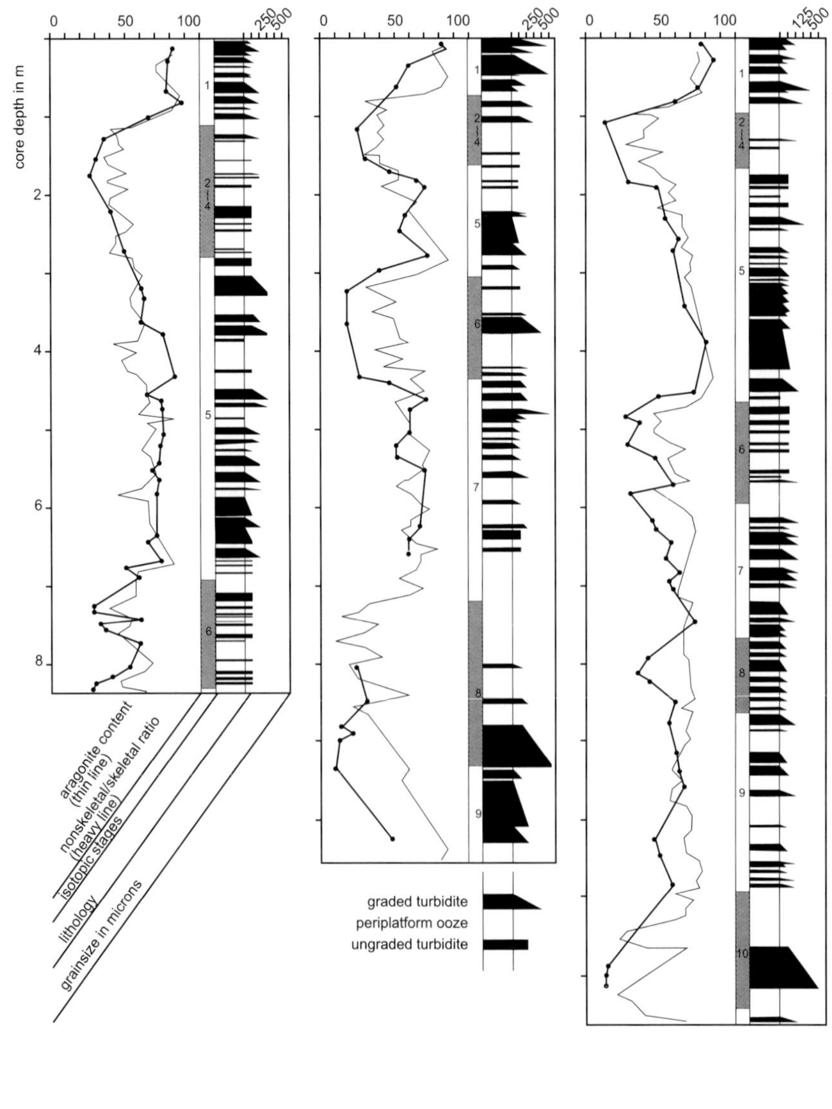

Fig. 7.16.— Highstand bundling and compositional signals in calciturbidites. Quaternary cores from Tongue of the Ocean, a basin surrounded by the Great Bahama Bank. Turbidites are most abundant during interglacial highstands, when bank tops are flooded and produce sediment; turbidites are thin and scarce during glacial lowstands. Turbidites also vary in composition: Highstand layers are rich in pellets and ooids - i.e. grains that form on the shallow banks by the interaction of tidal currents and winnowing from waves. Lowstand turbidites consist mainly of skeletal material, including reef detritus because fringing reefs and skeletal sand can migrate downslope with falling sea level. Glacial-interglacial stratigraphy is provided by variation in aragonite/calcite ratio, a property that is closely correlated with the oxygen isotope curve. After Haak and Schlager (1989).

ognized. However, lithoclasts do not prove lowstand conditions at the platform top because erosion may cut into older material on the flanks at any time during platform history (Grammer et al., 1993). Collapse of the Bahama margin in Exuma Sound produced high admixture of lithoclasts in a debris flow from the Sangamonian highstand (Crevello and Schlager, 1980).

Limitations of the compositional tool are the same as for highstand shedding. A platform that builds an extensive lowstand wedge during a long sea-level cycle may well export what looks compositionally like highstand material if the platform is large enough to develop a protected interior zone with its own characteristic sediment.

The model of systems tracts assumes that sea level controls the presence or absence of submarine fans as well as their position higher or lower on the toe-of-slope. A view at modern oceans as well as hydrodynamic theory indicate that fan development is largely a function of the transport capacity and competence of turbidity currents. These parameters, in turn, depend on sediment supply and topography (slope angle, size of valleys to channel the flows etc.). Sea level can influence sediment supply and, to a lesser extent, topography but in doing so it must compete with other processes.

For carbonate platforms, Schlager and Camber (1986) have proposed a model of slope evolution as a function of the height of the platform (Fig. 3.12). This model predicts that the depocenter will shift basinward as the slope angle increases. Steep slopes can be bypassed and even eroded by turbidity currents and fans may onlap these slopes irrespective of sea-level position. The position and shape of the gravity-displaced sediment bodies is largely a function of topography. Carbonate sediments, especially reefs, are known for their ability to rapidly build steep relief. Depocenters of gravity-displaced sediments will shift equally rapidly in response to the changes in sea-floor relief. Thus, fans and submarine canyons in carbonate terrains are rather unreliable indicators of sea-level movements.

Megabreccias and sea level

In chapter 4, megabreccias were introduced as common and conspicuous features on the slopes and basins of the T and M factories. They merit further discussion here because their specific position in sea-level controlled sequences is controversial.

The standard model as summarized by Vail (1987) and Van Wagoner et al. (1987) makes no specific statements regarding carbonate rocks. Sarg (1988), who contributed the pioneer paper on carbonate sequence stratigraphy, closely followed the standard model with regard to gravity-displaced sediments. He concluded that most carbonate material, including megabreccias, is shed during lowstands of sea level from the collapsing, oversteepened platform margins. Similar conclusions were reached by Vail et al. (1991, p. 656) and Jacquin et al. (1991). Hunt and Tucker (1993) note that megabreccias may be particularly abundant during lowstands but are "not specific to times of falling or lowstands of relative sea level" (1993, p. 314). Spence and Tucker (1997) also emphasize that margin failure and concomitant deposition of megabreccias may take place during highstands as well as lowstands. However, they also conclude that lowstands may favor slope failure and megabreccia formation by excess pore-water pressure in confined aquifers and by stress increase in the rock fabric. Below follows a closer look at these mechanisms.

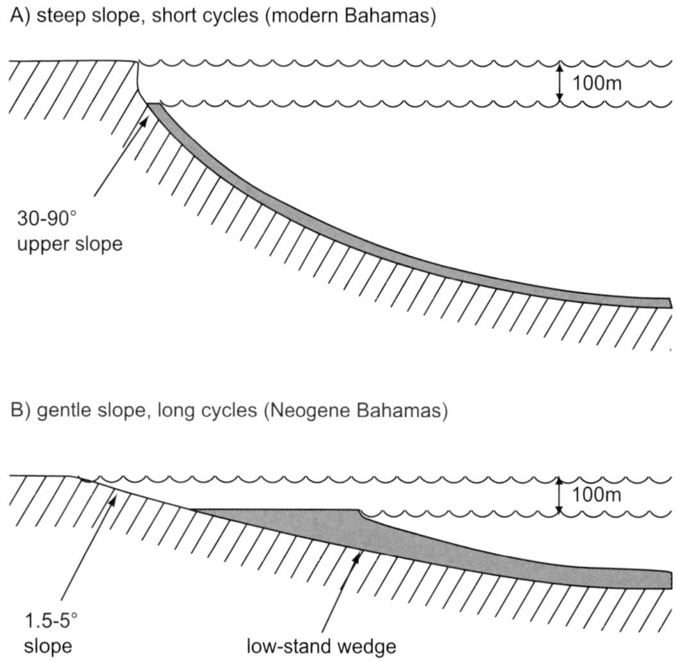

Fig. 7.17.— A) Highstand shedding is expected to be most pronounced around steep-sided platforms and during short sea-level cycles. B) Gentle flanks and long cycles lead to growth of a broad lowstand wedge that may produce bank-top sediment and dampen the effect of highstand shedding. After Schlager (1992), modified.

Failure by overpressured pore fluids. Slope sediments in which overpressures are generated during sea-level lowering will have less shear strength and thus be prone to failure. Confined aquifers seem favorable for such overpressure generation and failure because pore pressures in these aquifers are controlled by high water tables in the platform while local hydrostatic pressure is markedly reduced by reduction of the overlying column of sea water. (Fig. 7.18 The question is: under what circumstances will significant overpressures develop in young carbonate rocks, considering their high intrinsic permeability and the likely addition of cavernous karst systems during lowstands?

The Pleistocene/Neogene of Great Bahama Banks may serve as a case in point because the depositional anatomy and physical properties are fairly fairly well known (see p. 133ff). The Bahamian shoal-water facies have a mean matrix permeability of about 10 millidarcy (Melim et al. 2001) plus a well-connected network of karst conduits that supports long-distance circulation (Whitaker and Smart, 1990).

A quantitative flow model by H. Kooi (Vrije Univ. Amsterdam) indicates that under these circumstances, platform rocks are very unlikely to preserve significant overpressures during sea-level falls. Efficient flushing of the Pleistocene platform rocks of the Bahamas is confirmed by the observation that borehole Unda, 10 km inboard of the platform margin, presently is charged with well-mixed seawater down to about 200 m (Swart et al., 2001).

Slope deposits of Great Bahama Bank present a somewhat different picture. They have permeabilities of > 100 millidarcy separated by cemented layers of 0.01 millidarcy or less (Kenter et al., 2001; Melim et al., 2001). This configuration offers possibilities for preserving overpressure if the tight layers are extensive and laterally continuous, something that is not known at present. Local sealing potential of deeper-water carbonates was recently confirmed by extremely low permeabilities measured on North-Sea chalks (Mallon et al., 2005). However, it is also well established that the same chalks generally are in hydrostatic equilibrium in the upper 1,000 m or more of the rock column (Scholle et al., 1983, p. 619).

Stress increase in the rock fabric. Spence and Tucker (1997) propose that stress increase by loss of buoyancy during sea-level falls may lead to slope failure at the platform margins. This is a distinct possibility. Whether it is a common phenomenon remains to be seen. The large number of near-vertical sea cliffs in Quaternary limestones suggests that in many instances meteoric lithification strengthens the rock faster than the increase of apparent weight by sea-level fall can crush it.

Another mechanism of stress increase has gained support from recent studies of carbonate slopes - added stress by sediment loading. Spence and Tucker (1997) rightly point out that in detrital sediments at the angle of repose stress will not increase by addition of sediment. The situation is different, however, for *in-situ* growth of rigid bodies such as reefs or bodies of automicrite (see chapter 2). Several authors have observed that substantial volumes of automicrite in the form of layers, lenses or mounds have been added to the slopes of the T and M factories by *in-situ* precipitation (Playford, 1980; Brandner et al., 1991; Blendinger, 1994; Wood, 1999; Blendinger, 2001; Keim and Schlager, 2001). These bodies are rigid upon formation and represent localized loads that may be capable of triggering slides and slumps (Fig. 7.19). The common occurrence of automicrite and marine cement in megabreccia boulders certainly suggests a connection.

In summary, the evaluation of processes that can trigger large slides, slumps and debris flows does not, at this point, provide convincing arguments for sea-level falls as the prime driver of megabreccia formation. Without further information, megabreccias should not be considered as indicative of a particular sea-level position. Whether there is a specific connection in a particular instance, needs to be determined in a case study. Ineson and Surlyk (2000) have contributed an important example of this kind. Based on strong circumstantial evidence the megabreccias of their Cambrian example were associated with highstands and lowstands but the largest and most extensive ones occurred during lowstands.

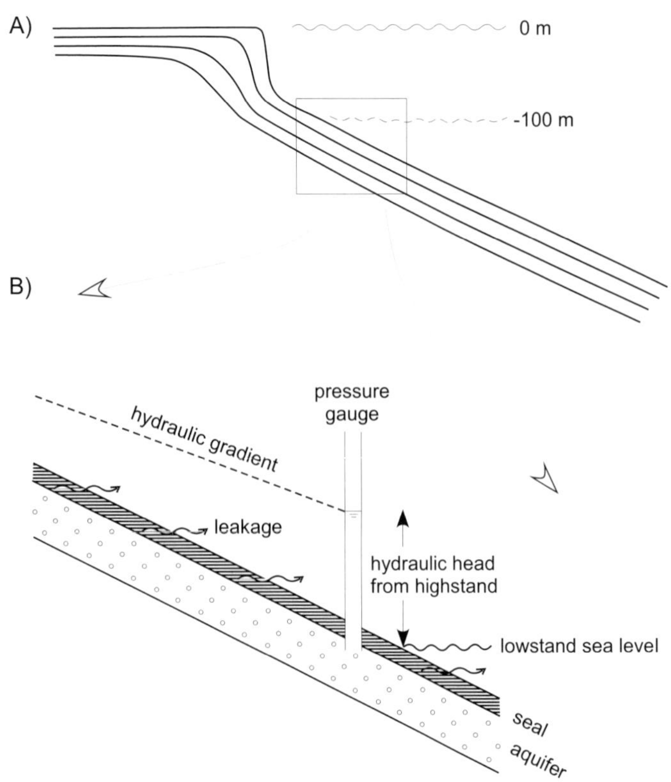

Fig. 7.18.— Possible setting of confined, overpressured aquifers during lowstands of sea level. A) Margin of rimmed platform experiencing sea-level cycles of 100 m amplitude. B) Situation of a confined aquifer on the slope, e.g. a layer of sand sealed by a layer of cemented mud. Hydraulic gradient is the difference in hydraulic head divided by the horizontal distance. Hydraulic gradient in a sealed aquifer is created by sea-level fall and dissipated by leakage across the seal. Sketch approximates the situation after the last interglacial-to-glacial transition. The process makes high demands on continuity and quality of the seal. See text for discussion.

BOUNDING SURFACES

Sequence boundary

Vail et al.'s (1977) definition of the sequence boundary as an unconformity that separates conformable successions of strata is broad and leaves room for further specification. In a pioneering paper, Vail and Todd (1981) presented six ways of producing unconformities by subaerial or marine erosion, three of them were deemed suitable to serve as sequence boundaries. Two of the three became widely accepted. I recently proposed to add a third type (not identical with the type 3 of Vail and Todd, 1981). The three types are briefly characterized below and subsequently evaluated in more detail.

Type-1 sequence boundary forms when relative sea-level falls below the shelf break of the preceding sequence. The new sequence starts with a lowstand tract whose flat top is distinctly lower than the youngest shelf surface of the underlying highstand tract.

Type-2 sequence boundary forms when relative sea-level falls to somewhere between the old shoreline and the shelf break. Consequently, only the inner shelf becomes exposed and the new sequence starts with a "shelf-margin wedge". This wedge is a lowstand tract whose shelf break is slightly lower than the preceding shelf break but whose flat top extends over the preceding shelf break.

Type-3 sequence boundary implies no fall of relative sea level. It forms when sea level rises faster than the system can aggrade such that a transgressive systems tract directly overlies the preceding highstand tract, often with a significant marine hiatus (Saller et al., 1993; Schlager, 1998; Schlager, 1999b). Marine erosion frequently accentuates this sequence boundary, particularly on drowned carbonate platforms.

From a sedimentologic perspective, the three types have rather different qualities. The type-1 unconformity is in many ways the ideal sequence boundary. It has a good geometric expression because of the downstepping of the shelf break and, in carbonates, the karst morphology that is sometimes visible even in seismic data. Type-1 unconformities also have a distinct lithologic signature in the form of terrestrial overprint on marine deposits (*Microcodium*, plant roots, calcrete soils). Finally, the type-1 unconformity is unambiguous evidence of a significant relative sea-level fall.

Type-2 boundaries are similar to type-1 boundaries but less pronounced. The geometric expression is subdued because the difference in elevation between the old highstand

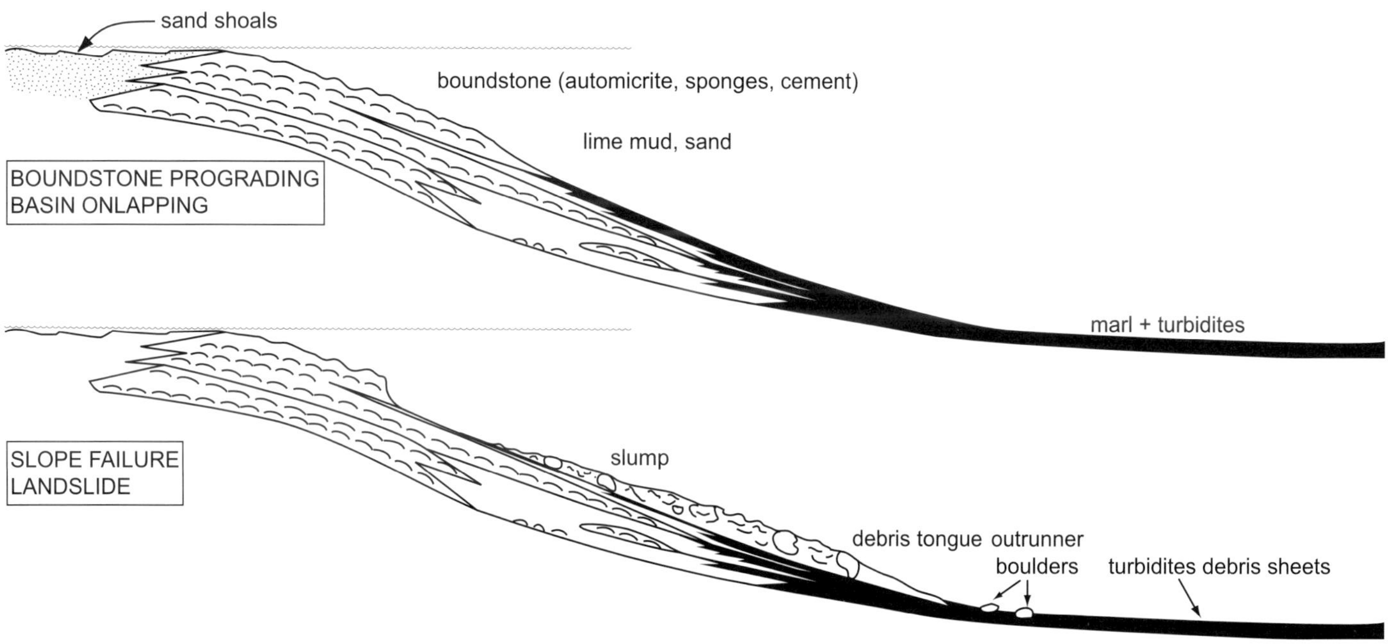

Fig. 7.19.— Model of slope failure by increased shear stress from growth of rigid automicrite lenses or mounds. Model based on observations on Triassic slopes in the Southern Alps. Upper panel: automicrite boundstone grows and overloads slope. Lower panel: slope failure removes excess load; slumps and debris tongues of megabreccias are emplaced on lower slope and basin floor. After Schlager et al. (1991), modified.

tract and the shelf-margin wedge is small and standard seismic data may not resolve it. Lithologic signatures are less pervasive than with type-1 unconformities because only part of the shelf is exposed and subaerial alteration penetrates less deeply. The type-2 unconformity is evidence for a minor fall of relative sea level. It is crucial that an unconformity is assigned type-2 status only if there is evidence for exposure of the inner shelf. It has become common practice to classify unconformities with only questionable evidence of exposure as type-2 sequence boundaries. I recommend to describe such unconformities as sequence boundaries in general and abstain from further specifying the boundary type (see below).

Type-2 boundaries are not common on rimmed carbonate platforms because the rim tends to build to sea level. Under these circumstances, the platform top has no overall seaward dip and a sea-level fall of only a few meters will expose the platform out to the shelf edge and generate a type-1 unconformity. However, Sarg (1988) pointed out that on certain rimmed platforms, such as the Permian of the Guadalupe Mountains, the outermost platform does have a distinct seaward dip that leaves room for the development of a shelf-margin wedge during a modest fall in sea level. The reason for the "hanging shoulders" of these platforms probably lies in the nature of the rim: it consists of the deeper-water automicrite community and extensive marine cement; this system does not necessarily build to sea level – the topographic crest of the platform may be formed by sand shoals that lie landward of the deeper-water rim.

Type-3 boundaries reflect the effects of two independent processes that operate independently or in conjunction to generate marine unconformities: (1) amplification of oceanic tidal waves by sharp topography and (2) platform flooding or drowning, i.e. demise by submergence below the photic zone. Superposition of both processes can generate marine hiatuses that may exceed 100 My and represent some of the most prominent seimic unconformities on record (Fig. 2.25, 7.20, 7.21). It should be noted that type-3 unconformities qualify as sequence boundaries only if one accepts the original definition of Vail et al. (1977). They are not unconformities (nor sequence boundaries) in the sense of Van Wagoner et al. (1988). To stay in line with Van Wagoner et al.'s definition, J.F. Sarg (written communication) proposed an alternative interpretation of the drowning unconformity in Liuhua (Fig. 7.20). He views the top of the platform as a maximum flooding surface and puts the sequence boundary at the exposure horizon within the platform. This interpretation is feasible but unsatisfactory on two counts: the sequence boundary is seismically nearly invisible and a major unconformity and stratigraphic turning point lies within a sequence.

To understand the origin of type-3 boundaries, we need to briefly discuss two processes: platform flooding or drowning, and current-amplification by sharp topography. The sediment record of drowning is variable but always represents a major change in sediment input and dispersal. The basic pattern is a deepening of the depositional environment to below the photic zone and thus below the production zone of the T factory. The drowning process need not go to completion. Platforms may be flooded and submerged to less-than-optimal but still photic conditions. In these instances, one should speak of "incipient drowning" (Read, 1982) or pronounced flooding. Such flooding events often lead to backstepping and re-orientation of the platform margin. The record of flooding or drowning may show a gradual transition from shoalwater to deepwater deposition but major gaps and abrupt changes are more common. A common reason for the punctuated record is current amplification by sharp topography. The drowned platform reinforces the (generally sluggish) oceanic tides and this may trigger intensive and long-lasting marine erosion. The result are major hiatuses between the drowned platform and its (hemi)pelagic cover or within this cover. Many of these hiatuses exceed 10 My, some exceed 100 My in duration (Fig. 2.25; Schlager, 1999b).

The drastic changes in sediment composition and dispersal that accompany platform drowning commonly produce an unconformity called drowning unconformity by Schlager (1989). Geometrically, this unconformity may resemble a lowstand unconformity because of basin-restricted sediment bodies that onlap the slope of the drowned platform (Figs 7.20, 7.21). In reality, however, the unconformity must form during a rise or highstand of relative sea level because drowning can occur only if the platform top is flooded. Drowned platforms and drowning unconformities are common in the geologic record and some of them have been interpreted as the result of major lowstands. This may explain some discrepancies between the sea-level curve from sequence stratigraphy and curves derived by other techniques (Fig. 7.22). Raised rim and empty bucket are common features of drowning unconformities and provide good criteria for their recognition.

Drowning may be a drawn-out affair whereby the productive platform surface shrinks gradually or in discrete steps. The last step in this process often shows the most productive part of the former rim disintegrating into a chain of patches. These patches are mound-shaped because they are deeply submerged and no longer planed by waves (Fig. 7.20, 7.21, 7.23).

Drowning events often appear more prominently in seismic data than exposure events. In fact, seismic reflectors and unconformities resulting from drowning are so prominent that they are often picked as sequence boundaries in seismic stratigraphy. This creates a definition problem where drowning is not preceded by exposure (examples in Wendte et al., 1992; Erlich et al.,1990; Moldovanyi et al., 1995; Saller et al., 1993). The type-3 boundary, i.e. an unconformity between a highstand tract and an overlying transgressive tract without intervening exposure but with intensive marine erosion, avoids this problem. It also acknowledges the status of drowning unconformities as stratigraphic turning points of the first order. In areas where the rate of sea-level fall in a eustatic cycle never matches the rate of subsidence,

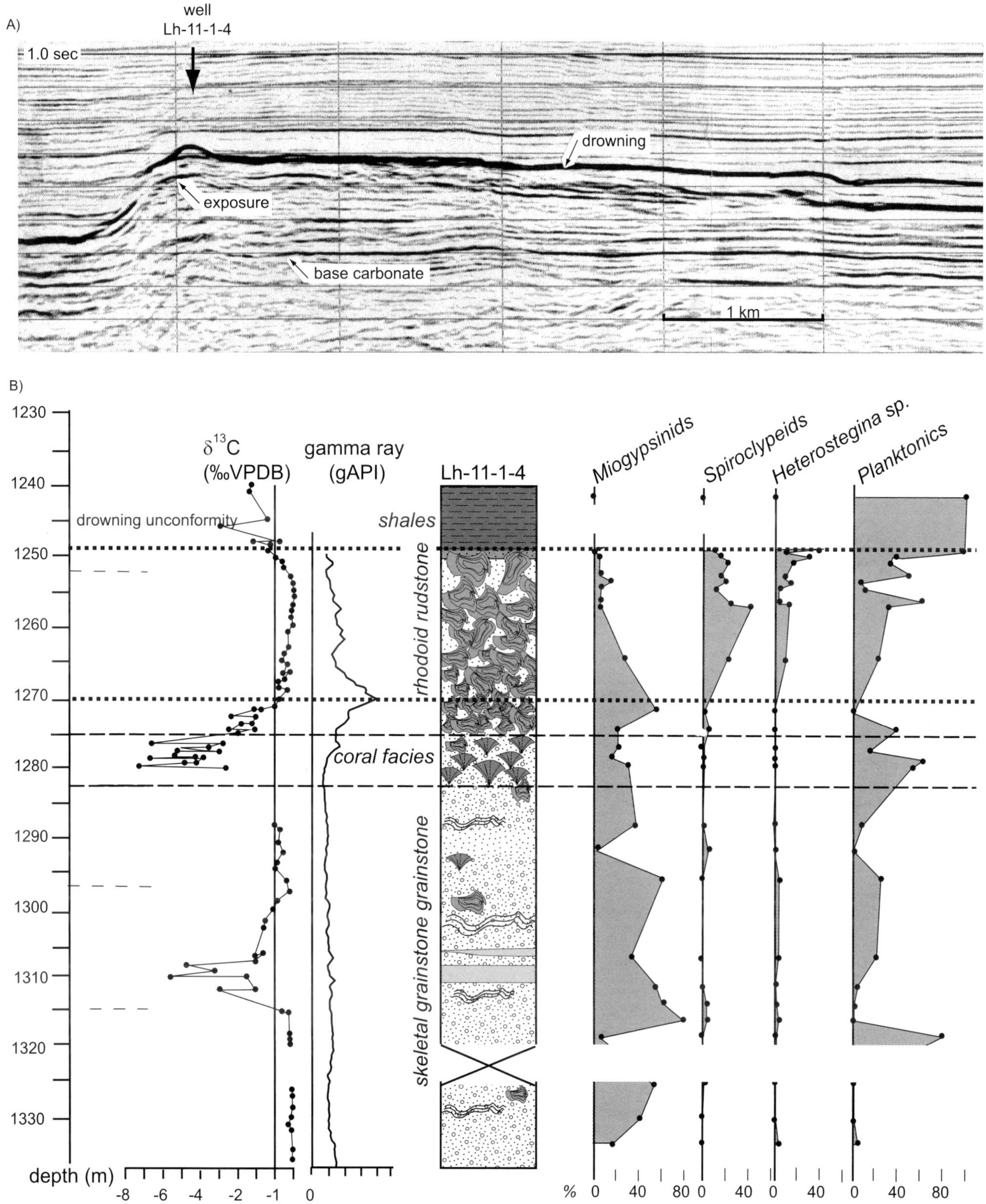

Fig. 7.20.— A) Drowning unconformity (and type-3 sequence boundary) in seismic profile of Miocene Liuhua platform, South China Sea. Note asymmetry of the record: stationary, elevated rim on the left and backsteps on the right . Starved sedimentation and marine erosion pervailed for some time before hemipelagic shale covered the platform. After Erlich et al. (1990). B) Borehole data from Liuhua platform indicate probable exposure at surfaces within the carbonate platform but no evidence of exposure was found at the drowning unconformity. Foraminifera indicate gradual deepening by decrease of Miogypsinids and increase of Spiroclypeids and Heterosteginids as well as plankton. After Erlich et al. (1990) and Sattler (2005), modified.

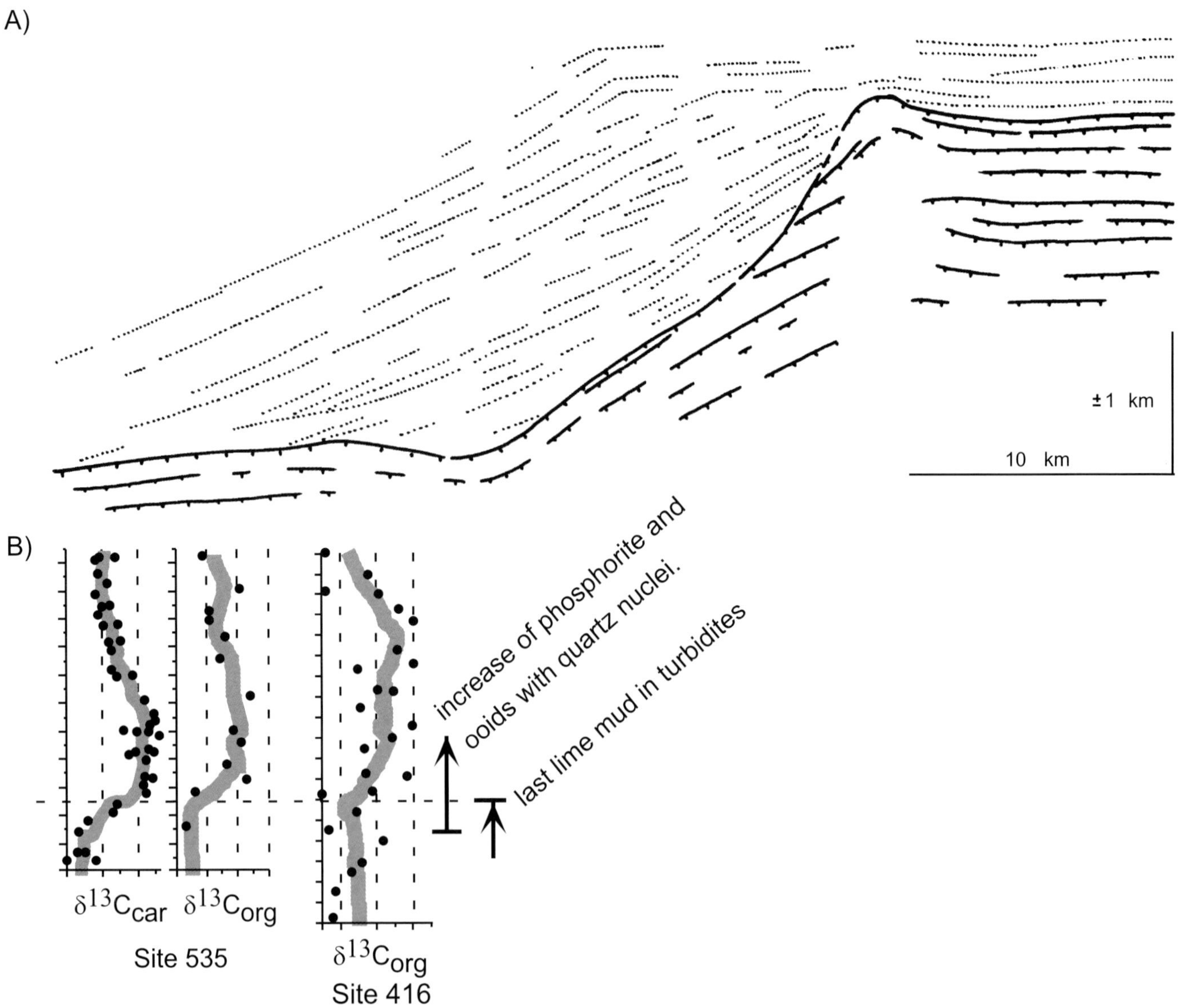

Fig. 7.21.— Jurassic – Cretaceous Mazagan platform off Morocco. Platform (in bold, dented lines) was drowned in the Early Cretaceous (Valanginian) and onlapped by siliciclastics (dotted lines) that prograded from right to left over the platform and also piled up on the seaward side, burying the platform flank from bottom up. This marine onlap is caused by the high declivity of the carbonate slope (about 20°); the geometry resembles that of a lowstand turbidite wedge. B) Deepwater record seaward of the platform belt links platform drowning to paleoceanography. Platform termination is recorded by change in calciturbidite composition of DSDP site 416. Timing of drowning coincides with a shift towards heavier carbon isotopes in carbonates and organic matter, indicating a global anoxic event recorded in site 416 as well as 534 in the western Atlantic. Based on Schlager (1980; 1989) and Wortmann and Weissert (2001).

exposure cannot occur and type-3 unconformities may be the only record of this eustatic cycle at a particular location (Fig. 7.24).

Drowning events or pronounced flooding (=incipient downing) events have been used as sequence boundaries in boreholes and outcrops of carbonates. The characteristic pattern consists of a highstand tract overlain by a transgressive tract without an exposure surface in between (e.g. Bosellini et al., 1999; Van Buchem et al., 2000).

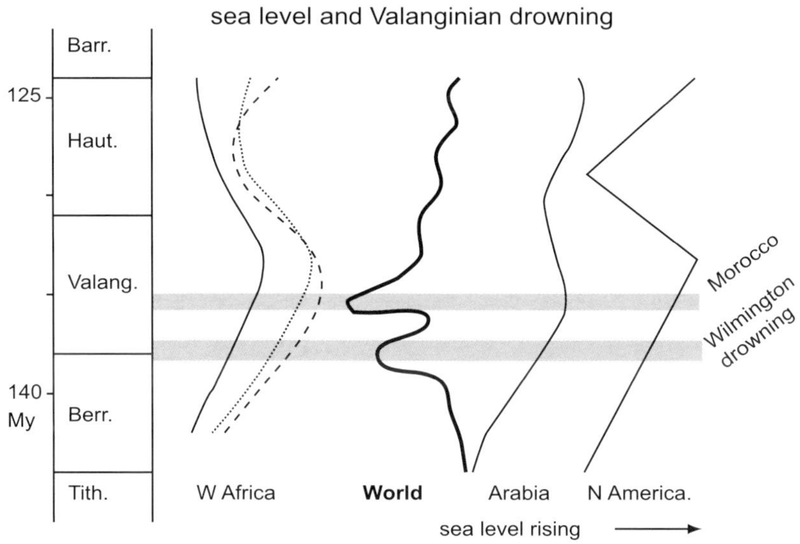

Fig. 7.22.— Early Cretaceous sea-level curves from various parts of the world are rather similar, suggesting strong eustatic control. The global sequence-stratigraphic curve (bold) by Haq et al. (1987) is at variance with the other records, probably because widespread Valaginian drowning unconformities were misinterpreted as lowstands. After Schlager (1989).

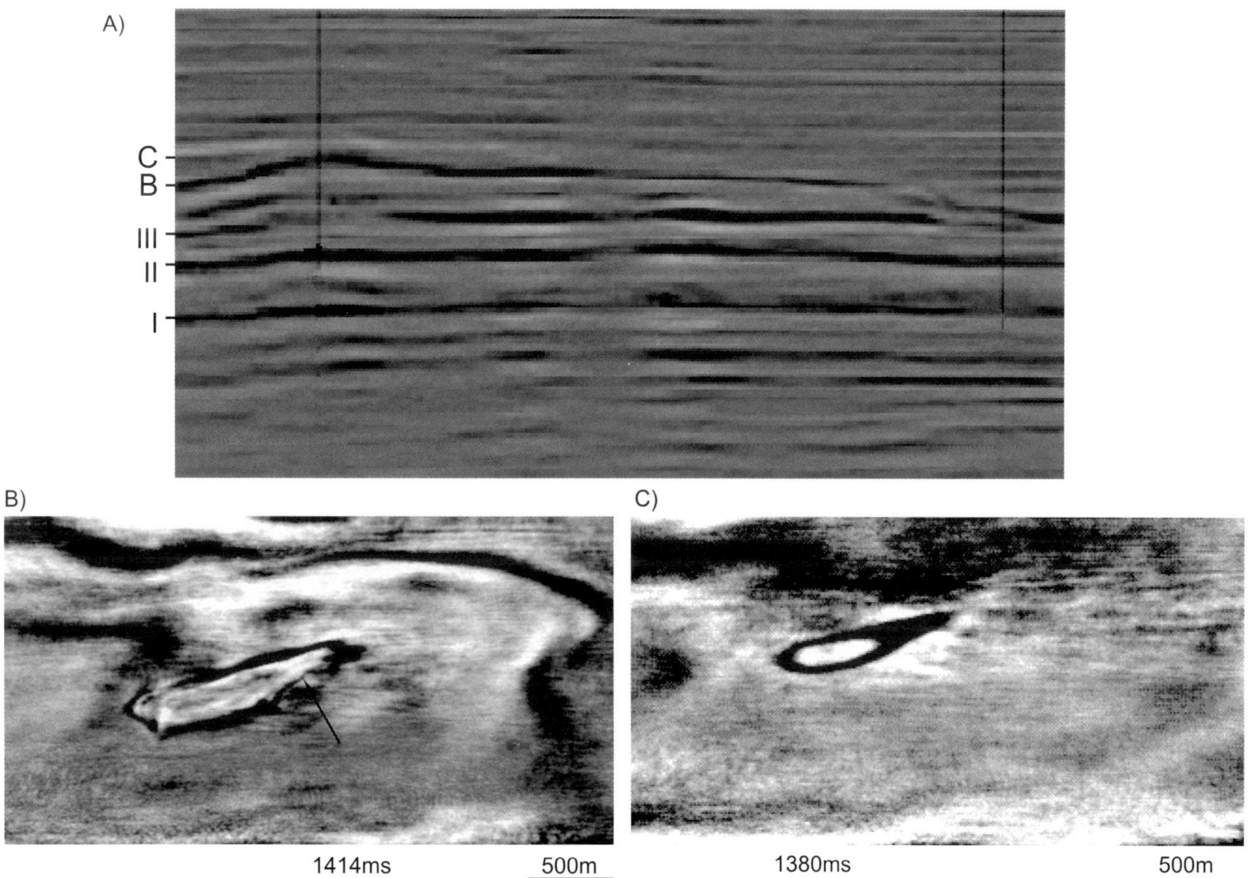

Fig. 7.23.— Backstepping and drowning of a Miocene build-up, South China Sea. A) Depth-migrated section of the build-up and two wells correct the vertical distortions of the more common time sections. Horizon I forms a gentle anticline and marks the initiation of the build-up on gently folded substrate. Horizons II and III are very nearly horizontal and represent the flat-topped platform stage of growth. Above horizon III the build-up backsteps and changes from flat platform to mound, indicating gradual submergence below zone of wave action and drowning. B) Time slice of 3D seismic data, showing build-up during flat-topped platform stage just above horizon III. Arrow indicates slump scar. C) Time slice during final mound stage of growth. Smooth oval outline of build-up (and general absence irregular karst-like morphology) suggest slow submergence and death by drowning. After Zampetti et al. (2004), modified.

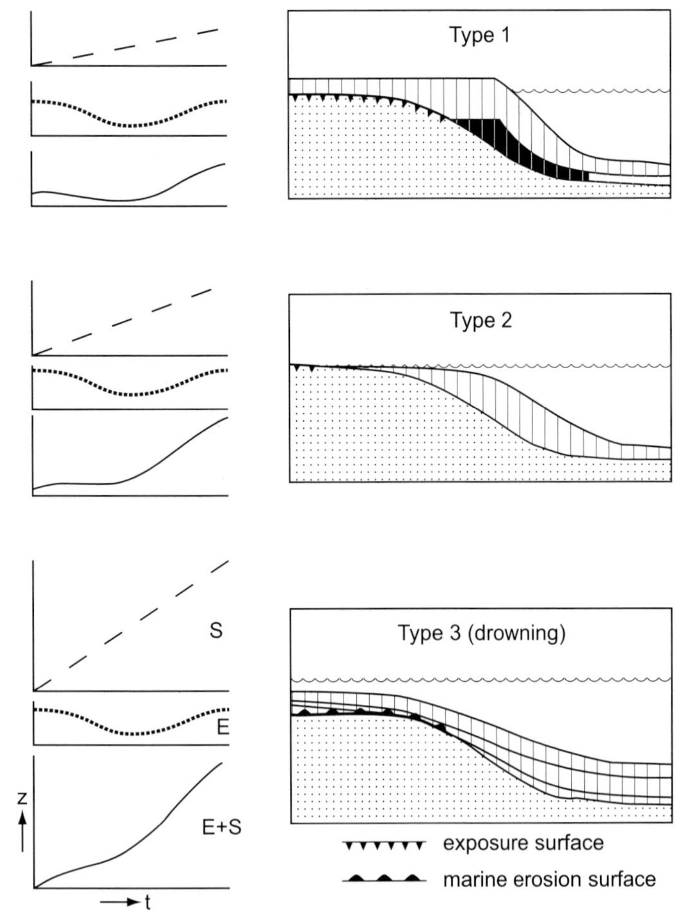

Fig. 7.24.— Conceptual model of the interplay of eustasy and subsidence. The same eustatic cycle creates different types of sequence boundaries depending on the rate of subsidence. Top panel: type-1 boundary forms if the rate of eustatic fall distinctly exceeds the rate of subsidence. Middle panel: type-2 boundary forms if the rate of eustatic fall is approximately equal to the rate of subsidence. Bottom panel: No exposure occurs if the rate of eustatic fall is less than the rate of subsidence; however, the system may be drowned or deeply flooded during the subsequent rise when eustasy and rapid subsidence are in phase; this produces a type-3 boundary. After Schlager (1999b).

Transgressive surface and maximum flooding surface

Both surfaces (or intervals) are common on tropical reefs or platforms albeit with certain differences to the siliciclastic standard.

Transgressive surface. It is highly variable, depending on the extent of the lowstand tract. Where the lowstand tract is narrow and dominated by reef growth rather than accumulation of loose sediment, the transgressive surface may have many meters of depositional relief as it runs up and down over isolated reef bodies. The early Holocene shows examples of this type.

Maximum flooding surface. On tropical platforms and reefs, a level of maximum flooding usually can be defined as the point where the facies trend changes from deepening to shoaling. Sediments of the T factory are rather depth-sensitive such that deepening and shoaling trends can be established with fair certainty. Geometric identification of the mfs as a level of downlap of highstand clinoforms is difficult or impossible because the platform tops are virtually horizontal, bedding near-parallel and lapouts scarce. As a consequence, maximum flooding levels commonly are indicated as intervals rather than surfaces. Techniques for dealing with this "sequence stratigraphy of gradual change" are discussed at the end of this chapter.

Pseudo-unconformities

In chapter 6, it was pointed out that seismic unconformities and outcrop unconformities do not always match. It seems that discrepancies between outcrop record and seismic record are more frequent in carbonates, particularly tropical carbonates, than in siliciclastics. The following characteristics of carbonates may be responsible for the difference.

- Carbonates can build steeper slopes because the shear strength of muddy deepwater carbonates is higher than their siliciclastic counterparts and because the sediments often lithify at or near the sea floor. Thus, the total range of depositional slope angles in carbonates usually is wider than in siliciclastics.
- Carbonate slope angles may change rapidly laterally and with time during the accumulation process. This is so because there are many local sediment sources, the production includes *in-situ* framework and a wide range of grains sizes, and rapid lithification impedes lateral transport that usually reduces variations in slope angle of siliciclastics.
- Changes from the carbonate factory to surrounding siliciclastic domains may be particularly rapid because there are many local carbonate production sites and limited lateral transport because of rapid cementation.

Seismic models of outcrops are one way to directly compare outcrop observation and seismic image. The potential of outcrops in carbonate sequence stratigraphy was elegantly illustrated by Bosellini (1984, 1988) and soon followed by studies presenting actual seismic models of large outcrops in carbonates (e.g. Middleton, 1987; Rudolph et al., 1989; Schlager et al., 1991; Biddle et al., 1992; Stafleu, 1994; Bracco Gartner and Schlager, 1999; Anselmetti et al., 1997; Kenter et al., 2001). It is probably significant that the majority of these models show one or several lapout patterns in the seismic model that were not present in the outcrop nor in the impedance model that formed the basis of the seismic model.

Figs 7.25, 7.26, 7.27, 7.28, 7.29, 7.30 present the basic types and examples of these false lapouts, termed pseudo-unconformities by Schlager et al. (1991). The pseudo-unconformities include pseudo-toplap, pseudo-onlap and

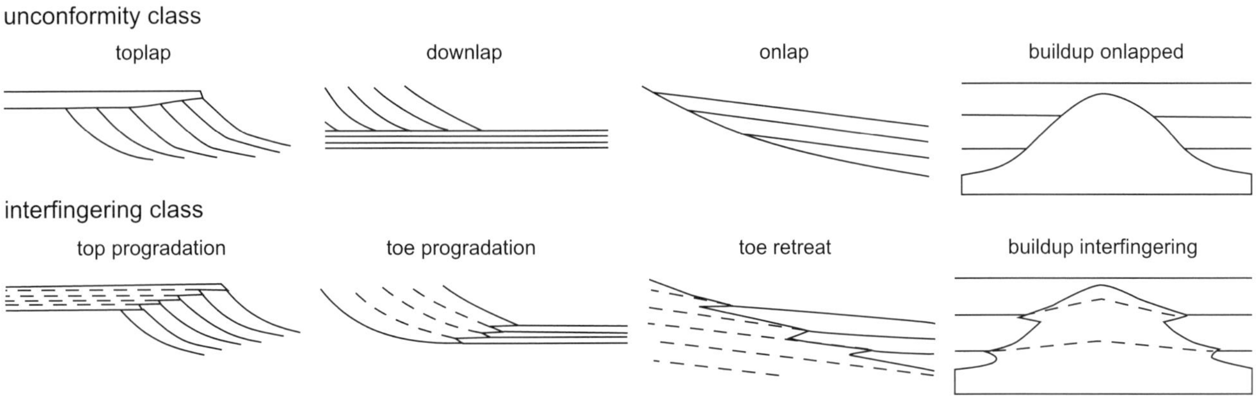

Fig. 7.25.— Depositional geometries of reefs and platforms where genuine unconformities and pseudo-unconformities caused by facies interfingering may be difficult to separate.

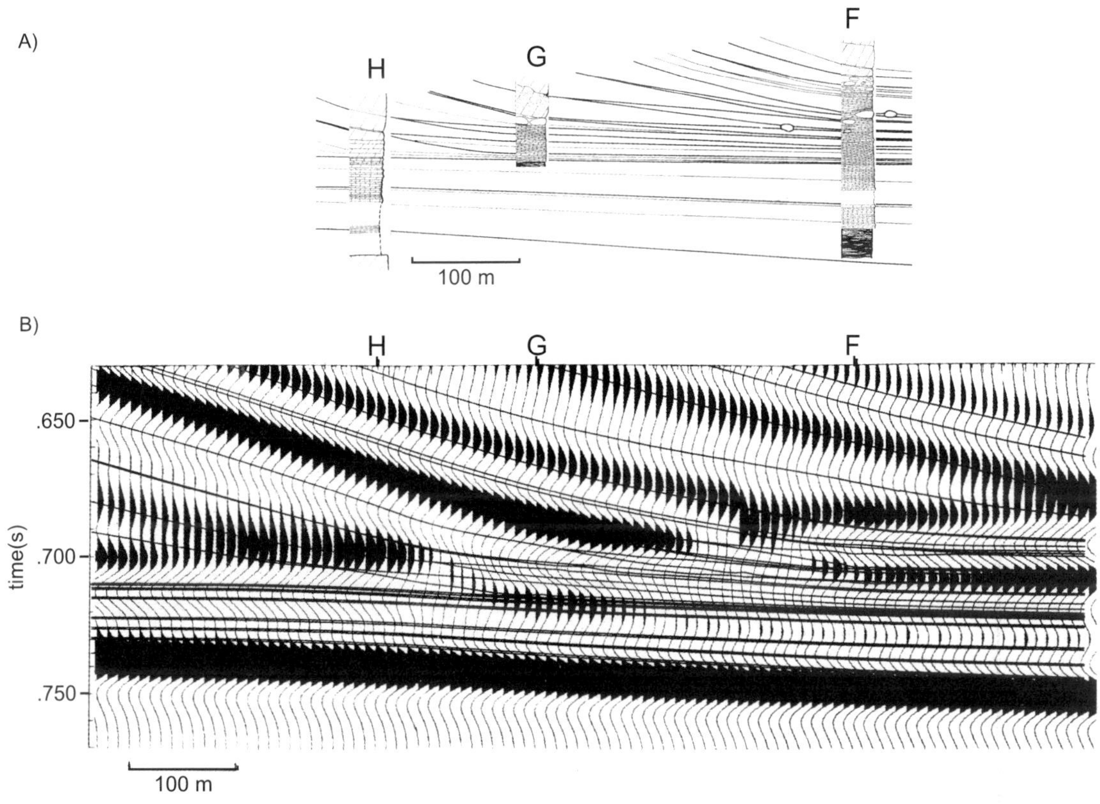

Fig. 7.26.— Triassic slope-to-basin transition of Picco di Vallandro (Dürrenstein) in the Southern Alps . After Rudolph et al. (1989). A) Interfingering of carbonate slope deposits (upper left) with marly basin sediments (lower right). B) Vertical incidence seismic model at 25 Hz shows bedding-parallel reflections within the slope and basin domains but also produces a reflection oblique to bedding - a pseudo-downlap reflection that follows the climbing toe-of-slope. This base-of-slope reflection is discontinuous and shifts upward in the middle third of the diagram but the continuous segments in the left third and the right third of the diagram still demonstrably cross time lines.

pseudo-downlap. In most instances, the true nature of the interfingering pattern is revealed if frequency, and thus resolution, are increased (Fig. 6.10, Fig. 7.28). However a several-fold increase in frequency may be required to solve the problem. This is easy to accomplish in a model but difficult or impossible with real data. Bracco Gartner and Schlager (1999) found that the use of a seismic attribute, instantaneous phase, allowed one to differentiate between onlap and interfingering at significantly lower frequencies than in standard reflectivity displays.

A well-constrained submarine example of pseudo-downlap was discovered in the Neogene off eastern Australia (Fig. 7.29; Isern, Anselmetti, Blum et al., 2002). A very similar configuration off western Australia, not tested by the drill, was interpreted as a genuine unconformity by Cathro et al. (2003).

At present, we do not know what percentage of seismic unconformities are pseudo-unconformities that correspond to facies changes or drastic thickness variations in the rock column. The percentage may be significant, though, and the problem should not be brushed aside.

Pseudo-unconformities occur in at least 50 % of the dozen outcrop models constructed by our group or compiled from the literature between 1989 and 1995 (Schlager, 1996). The principal cause of the common occurrence of pseudo-unconformities in carbonates may be drastic facies changes and rapid thickness variations at the periphery of carbonate factories. Fig. 7.25 summarizes the most common situations where these false lapouts have been observed.

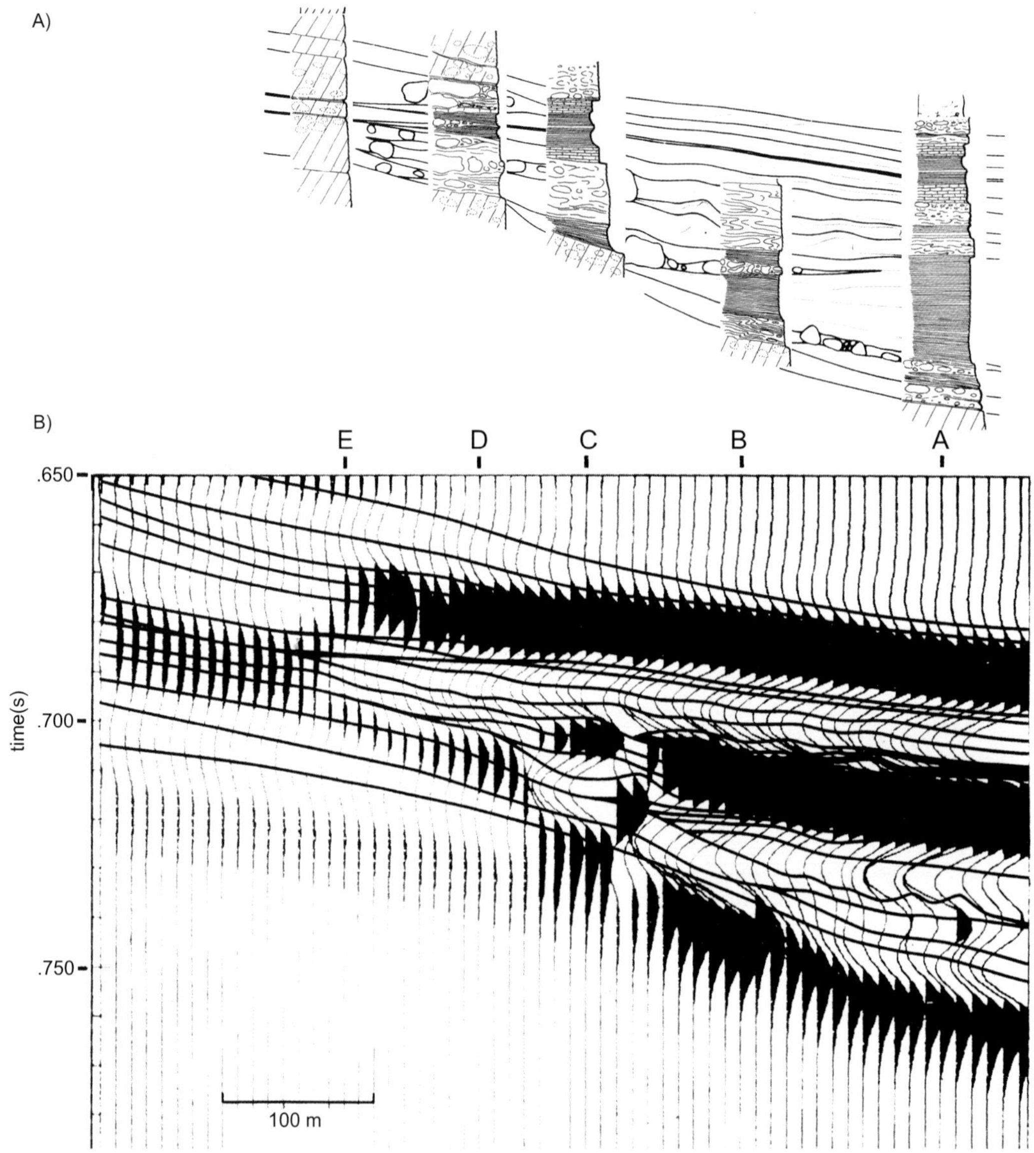

Fig. 7.27.— Details of interfingering at the toe of a retreating platform slope at Picco di Vallandro (Southern Alps, Italy). Compiled after Rudolph et al. (1989) and Bracco-Gartner and Schlager, (1999). A) Interfingering of slope carbonates and argillaceous basin sediments. Slumps and debris flows deposited tongues of carbonate material that were emplaced at a time when accumulation in the basin was faster than slope sedimentation such that basin sediments encroached on the lower slope. B) Vertical-incidence seismic model at 25 Hz shows bedding-parallel reflections in the slope and basin domains, respectively. In addition, a nearly continuous reflection runs oblique to bedding and creates the impression of basin sediments onlapping the slope.

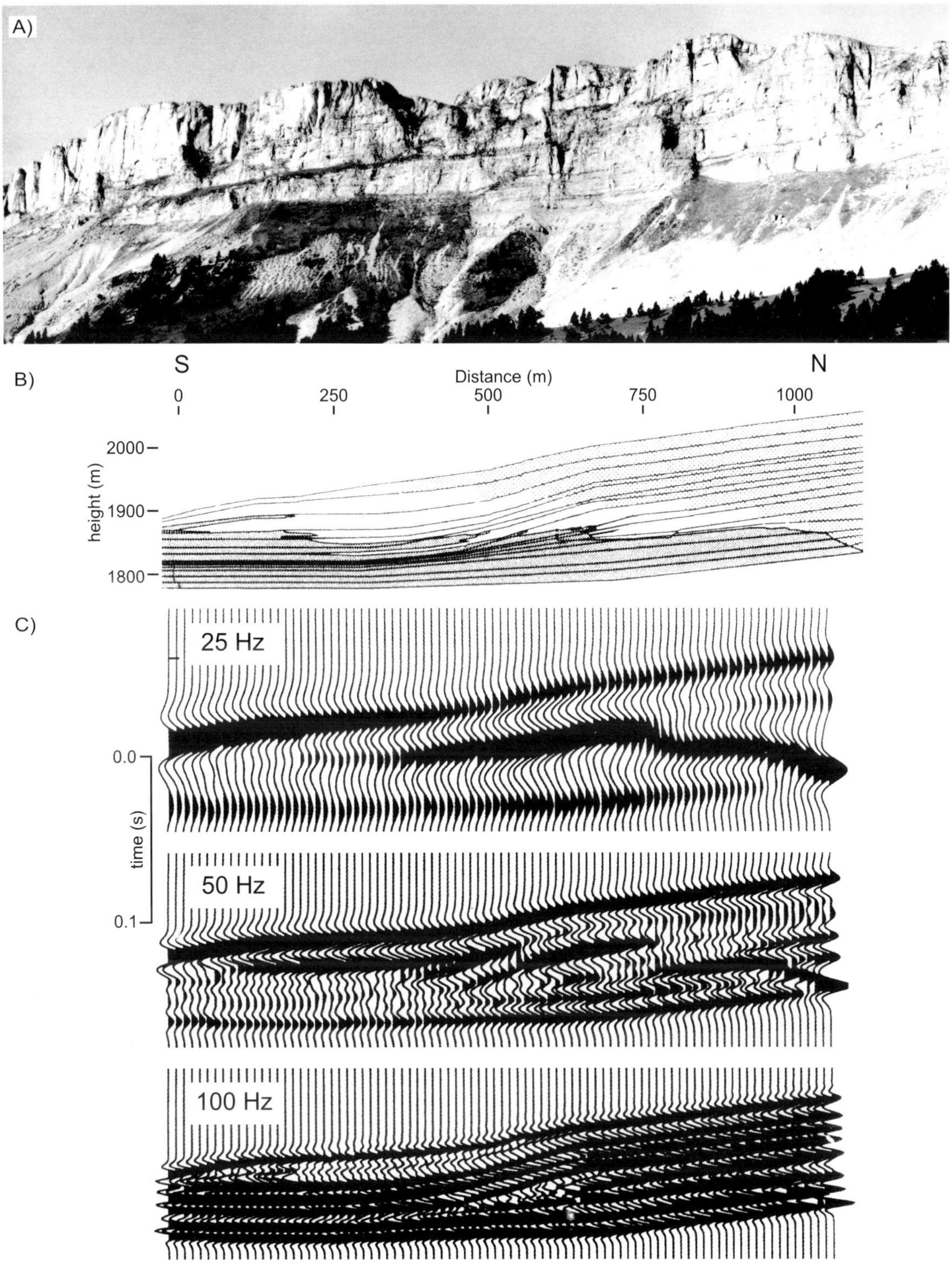

Fig. 7.28.— Sigmoidal clinoforms of a prograding Cretaceous carbonate platform in the Vercors Mountains, Western Alps (France). After Stafleu et al. (1994), modified. A) Outcrop photograph. Clean platform lime grainstones in upper right become gradually muddy and argillaceous as they extend down the clinoform. At the toe-of-slope, the detrital limestones interfinger with argillaceous lime mudstones and marls (largely covered by scree). B) Lithologic model used to construct the seismic model. Lines represent bedding surfaces that bound lithostratigraphic units and cross three different facies: clean lime grainstones – dotted, packstones – wackestones – blank, argillaceous lime mudstones – gray. As different impedance values were assigned to lithostratigraphic units, the bedding surfaces acted as potential reflectors. C) Vertical-incidence seismic models of B) at three different frequencies. 100 Hz model shows the lithostratigraphic patterns in essentially correct form. 25 Hz and 50 Hz models show false lapouts owing to incomplete resolution. 50 Hz model is particularly misleading because the local flattening of clinoforms creates the illusion of a lowstand systems tract onlapping a highstand slope.

SILICICLASTICS AND EVAPORITES IN CARBONATE SEQUENCES

In chapter 4 it was concluded that siliciclastics and evaporites may occur in all carbonate facies belts but the probability varies. The situation with regard to sequences and systems tracts is similar.

Siliciclastics may be found in all systems tracts but they are most common in the basinal parts of lowstand tracts and most exceptional in the shelf margins of highstand tracts. This dichotomy leads to the phenomenon of "reciprocal sedimentation" in mixed systems such as the Permian of the Guadalupe Mts. (Wilson, 1975; Sarg et al., 1999). Reciprocal sedimentation implies that the zone of maximum sedi-

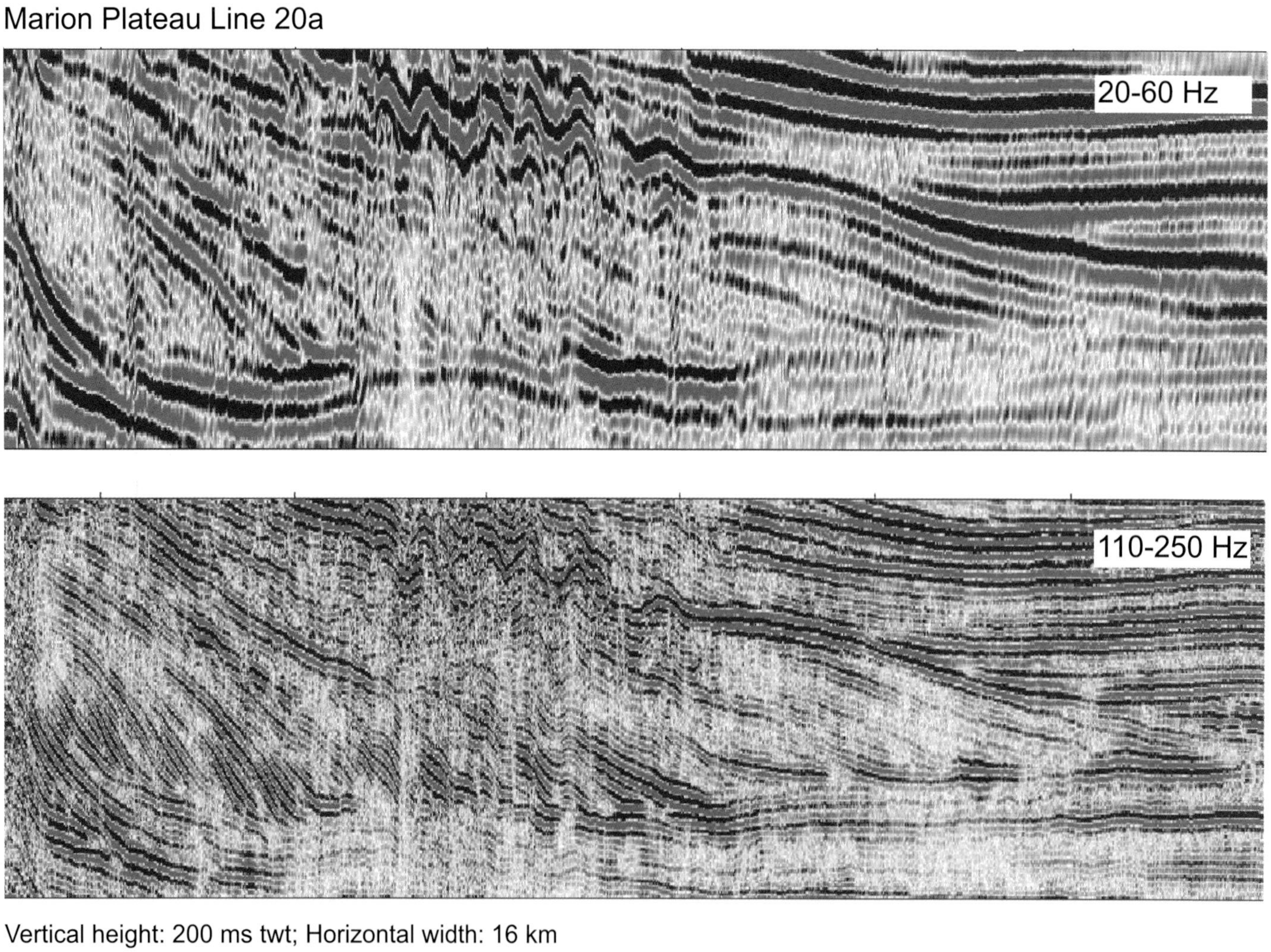

Fig. 7.29.— Downlap at the toe-of-slope of a Miocene platform, Marion Plateau, SW Pacific (Isern, Anselmetti, Blum et al., 2002). Prograding clinoforms are clean calcarenites, flat basin sediments are slightly argillaceous calcarenites. The slight difference in lithology plus the drastic change in dip suffice to produce a pseudo-unconformity (downlap) at 20 – 60 Hz frequency. At 110–250 Hz, the reflections at the toe-of-slope can be seen to bend rather than terminate at several places. Lines courtesy of A. Isern and F. Anselmetti.)

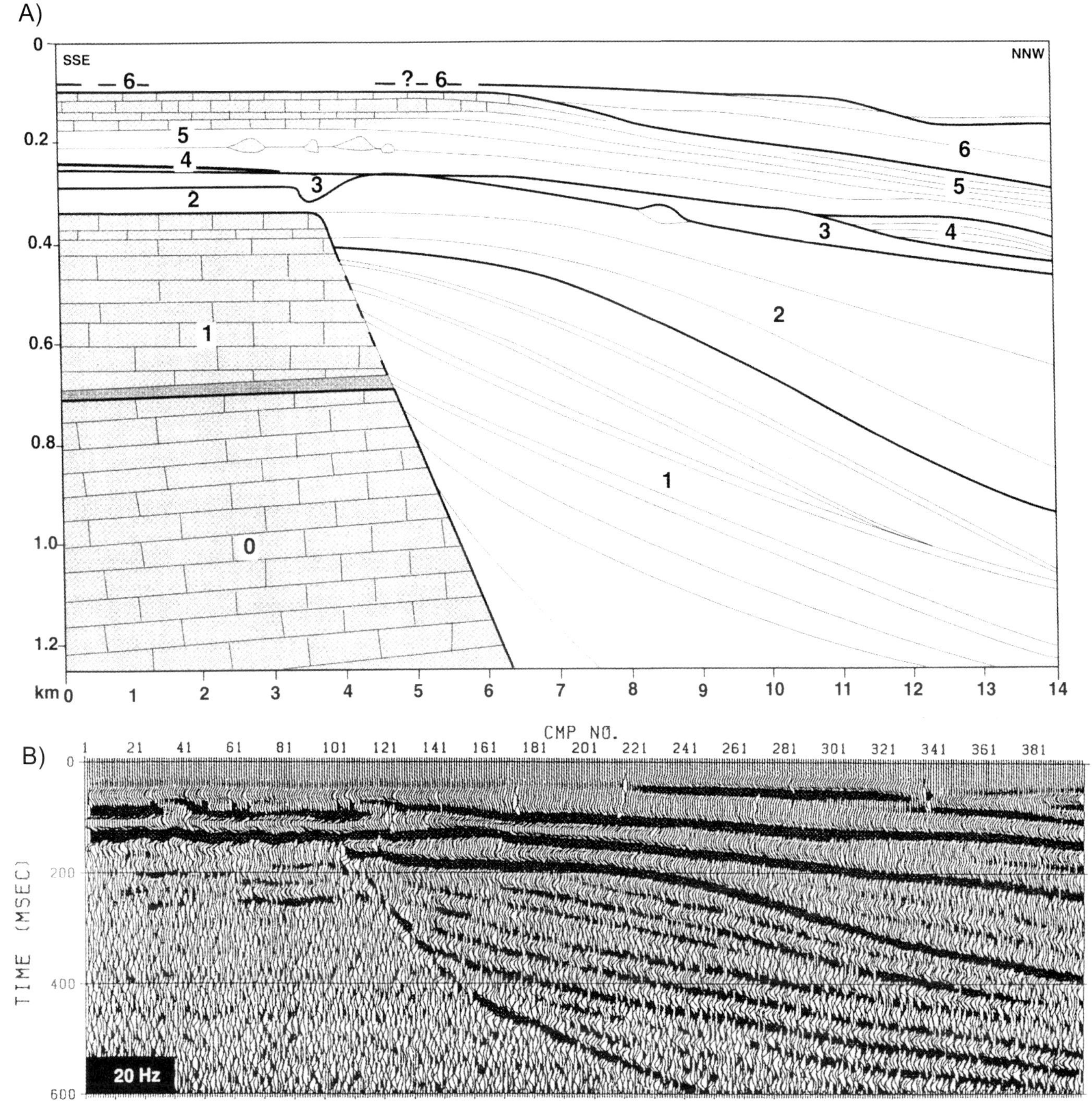

Fig. 7.30.— Stratigraphic anatomy and seismic outcrop model of Maiella mountain, Apennines, Italy, after Anselmetti et al. (1997) . A) Stratigraphic model derived from outcrop observations. Lines represent bedding horizons associated with impedance changes.
B) Synthetic seismic section matches overall anatomy of A rather well but shows distinct pseudo-toplap at the boundary of units 1 and 2 between CMP101-201.

mentation alternates between basin center and basin margin. The basin is sediment-starved when the carbonate platform aggrades and progrades most rapidly; on the other hand, thick siliciclastic units in the basin may have no equivalent on the basin margin. Reciprocal sedimentation is a logical consequence of the principles of highstand shedding of carbonates and lowstand shedding of siliciclastics.

In the platform interior parts of highstand and transgressive systems tracts, siliciclastics are somewhat more common than at the platform margin. In the Permian of the Guadalupes, highstand platform cycles are "cleaning upward", i.e. they are quartz-rich at the base and become clean carbonate at the top (Fig. 7.31). The occurrence of evaporites in carbonate sequences has been reviewed by Sarg (2001). They may occur in all systems-tracts but with different probability (Fig 7.32). Thick evaporites may form lowstand deposits when silled basins become nearly isolated and therfore hypersaline during low sea-level positions. In transgressive systems tracts, significant evaporite bodies may form when the lagoonal depression of a former highstand is being flooded. Highstand systems tracts typically contain evaporites in the sabkhas at the landward end of marine deposition. Finally, gypsum/anhydrite deposits may occupy all depositional environments if the water is hypersaline but sea level still high enough to flood the former carbonate shelf. According to the systems tract definition in Fig. 7.3, this would be a highstand tract largely composed of evaporite deposits.

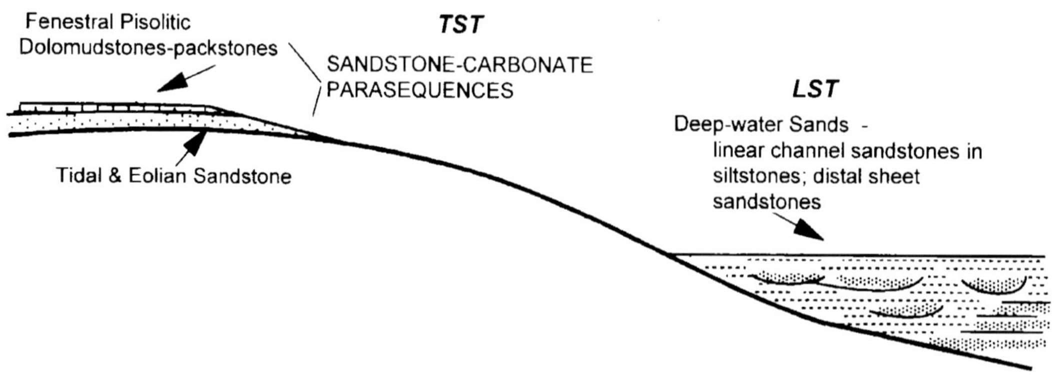

Fig. 7.31.— Distribution of siliciclastics (mainly quartz sand and silt) in the Permian of the Guadalupe Mountains. Major siliciclastic accumulations in the deep basins during lowstands, when the platform is mostly exposed; minor tongues of siliciclastics accumulate between platform carbonates of the transgressive and highstand tracts ("cleaning-upward cycles"). After Sarg et al. (1999), modified.

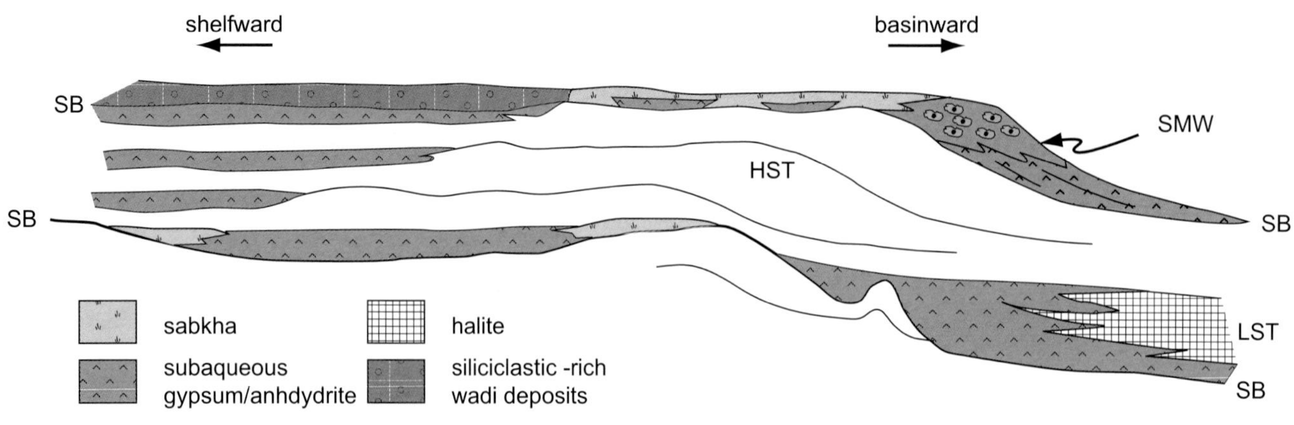

Fig. 7.32.— Overview of evaporite occurrences in carbonate sequences. Evaporite bodies (shaded) may occur in all major facies belts and in all systems tracts. However, they are most common in lowstand and transgressive systems tracts. During basin-wide hypersalinity, continuous anydrite formations may replace carbonates in all settings, including the shelf margin. After Sarg (2001), modified.

TWO NEOGENE CASE STUDIES IN CARBONATE SEQUENCE STRATIGRAPHY

Northwestern Bahamas

The northern part of the Bahama archipelago has become a model and standard for the interpretation of tropical carbonates. Chapters 2 and 3 repeatedly alluded to this role. The area also has become an important calibration point for carbonate sequence stratigraphy by an assembly of research wells drilled along an industry seismic line and its offshore extension by the Ocean Drilling Program. Two platform wells were contributed by the Bahama Drilling Project (Ginsburg, 2001), five wells on the slope and in the basin by the Ocean Drilling Program Leg 166 (Eberli, Swart, Malone et al. , 1997).

Setting. The carbonate platforms of southern Florida and the Bahamas are part of the passive margin of North America, a mature margin that rifted in the Jurassic and whose subsidence rate had slowed to few tens of meters per My in the Plio–Peistocene (McNeill et al., 2001). The carbonate facies indicate tropical platforms, mostly rimmed by reefs or sand shoals. The tropical climate is dominated by easterly-northeasterly trade winds. The sequence-stratigraphic studies concentrated on a margin of Great Bahama Bank that faces leeward with respect to the trades and borders a basin that lies at the juncture of two seaways - the Florida Straits and Santaren Channel. Both seaways are swept by strong northward currents that merge in the study area (Fig. 7.33). They are part of the clockwise subtropical gyre of the North Atlantic. The gyre belongs to the surface circulation of the ocean (Fig. 1.4) but the western boundary currents are very voluminous and therefore fill the Bahamian channels to the bottom. Consequently, sediment accumulations in the study area are shaped by the interplay of carbonate production, leeward transport at the bank top, sea level, and ocean currents that sweep the slopes and basin floors.

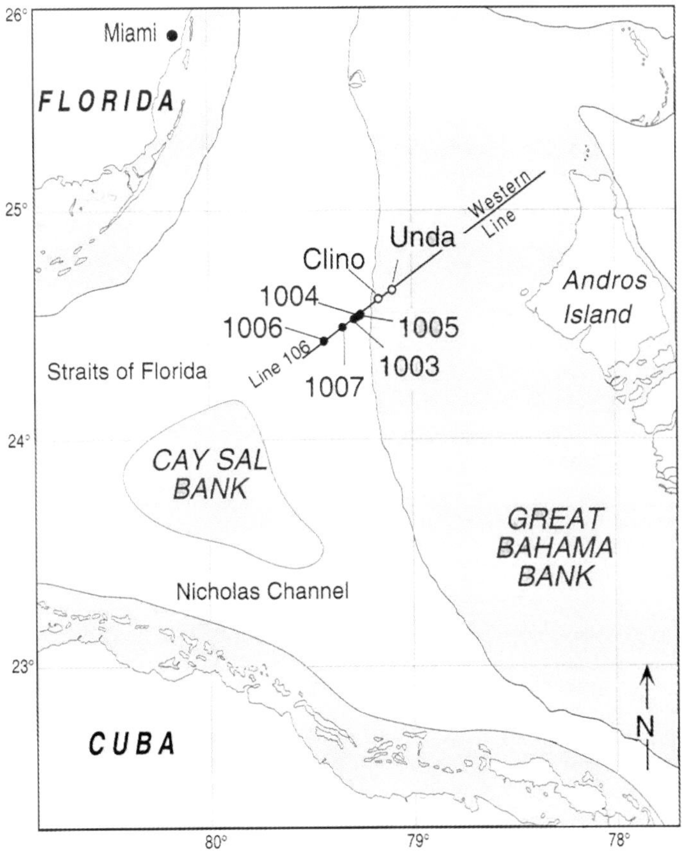

Fig. 7.33.— Location map of western Great Bahama Bank and surroundings showing boreholes and seismic profiles as well as shoal-water carbonate environments (shaded). After Eberli et al. (2002). Copyright 2002, reprinted with permission from Elsevier.

The sequence stratigraphy established by Eberli (2000), Anselmetti et al. (2000), Eberli et al. (2001, 2002) yielded a very consistent pattern. Unconformity-bounded packages were identified in the seismic data and traced from platform to basin (Fig. 7.34). Via check-shots and logs, the boundaries were tied to the seven wells on the transect. Correlation among the wells was accomplished by lithostratigraphic, biostratigraphic and magnetostratigraphic techniques. This exceptionally detailed case study produced several important results.

- Correlations by tracing seismic reflectors and by rock stratigraphy are consistent – the time lines proposed by the various techniques never cross (Fig. 7.35).
- This very satisfactory correlation was achieved by constraining seismic analysis by age dates and lithologic observations from the cores. Pure seismic correlation from platform to basin would face big uncertainties because of repeated merging and splitting of reflections. Currents are the principal cause of this problem – they generate sediment drifts and scours on the basin floor and episodically cut major hiatuses in the slope deposits (Fig. 7.37). On the other hand, currents are probably in phase with sea-level fluctuations such that lowstands coincide with more intensive erosion because of the reduced cross section of the channels (Eberli et al. 2001, p. 258).
- Sequences are bounded by exposure surfaces in the platform domain and by abrupt facies changes or hiatuses marked by hardgrounds in the deep-water domain. Million-year long hiatuses and pervasive marine cementation or authigenic mineralization of hardgrounds almost certainly reflect current activity. The periplatform setting normally is very well supplied with sediment and the long intervals of non-deposition suggest an external cause.
- Lowstand wedges are clearly recognizable but they are small and few have distinct platform margins (Fig. 7.34).
- Highstand shedding dominates this leeward slope. Most progradation occurs during highstands where aggradation is small (Eberli et al. 2001, p. 260). In this

situation bank-top production is high but accommodation small such that most excess sediment is shed down the leeward slope. Slope sediment is mud-rich and low in sandy turbidites. Most turbidity currents probably bypass the slope and deposit their material on the basin floor, similar to the situation in other Bahamian basins (Schlager and Chermak, 1979; Mullins et al. 1984; Harwood and Towers, 1988).

➤ Diagenesis seems to accentuate sequence boundaries because exposure and flooding of the platform during sea-level cycles change the composition and amount of sediment on the slope (Fig. 7.38). During highstands, sediment flux from platform to slope is high and the slope sediment therefore contains large amounts of metastable aragonite. During lowstands, sediment flux is reduced and the slope sediment is low in arago-

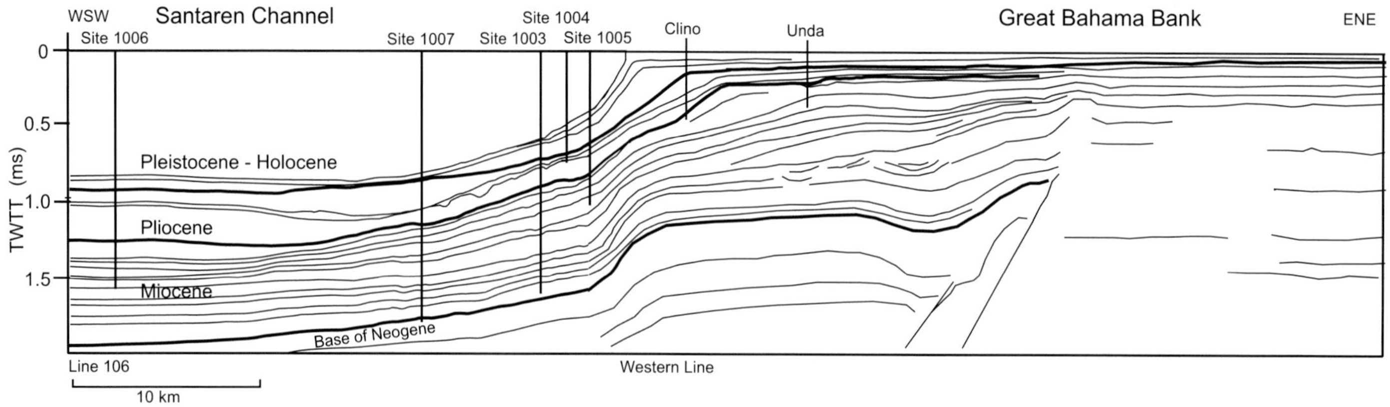

Fig. 7.34.— Interpretive line drawing of seismic profile of southwestern Great Bahama Bank showing platform-to-basin correlation of 17 Neogene sequence boundaries. After Eberli et al. (2002), modified.

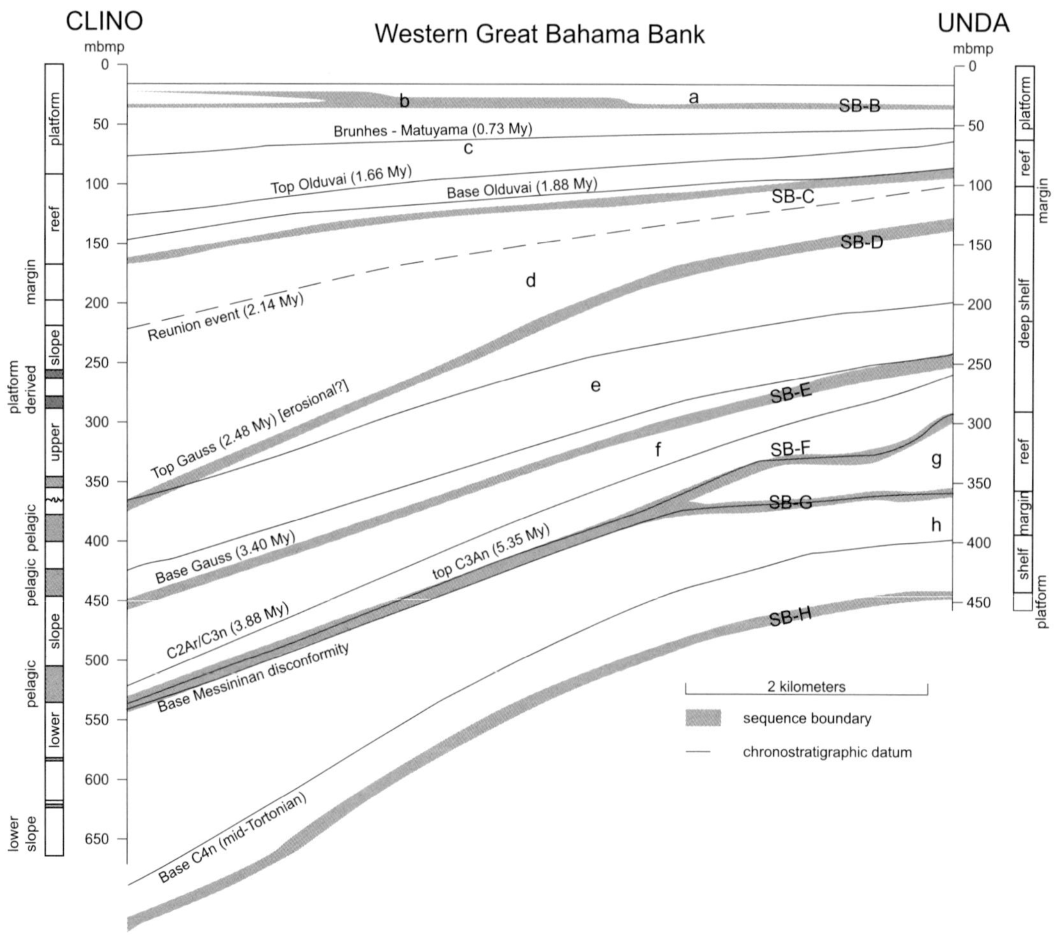

Fig. 7.35.— Time lines of magnetic reversals vs. sequence boundaries. Time lines and sequence boundaries diverge and converge but never cross. After Eberli et al. (2001), modifieds.

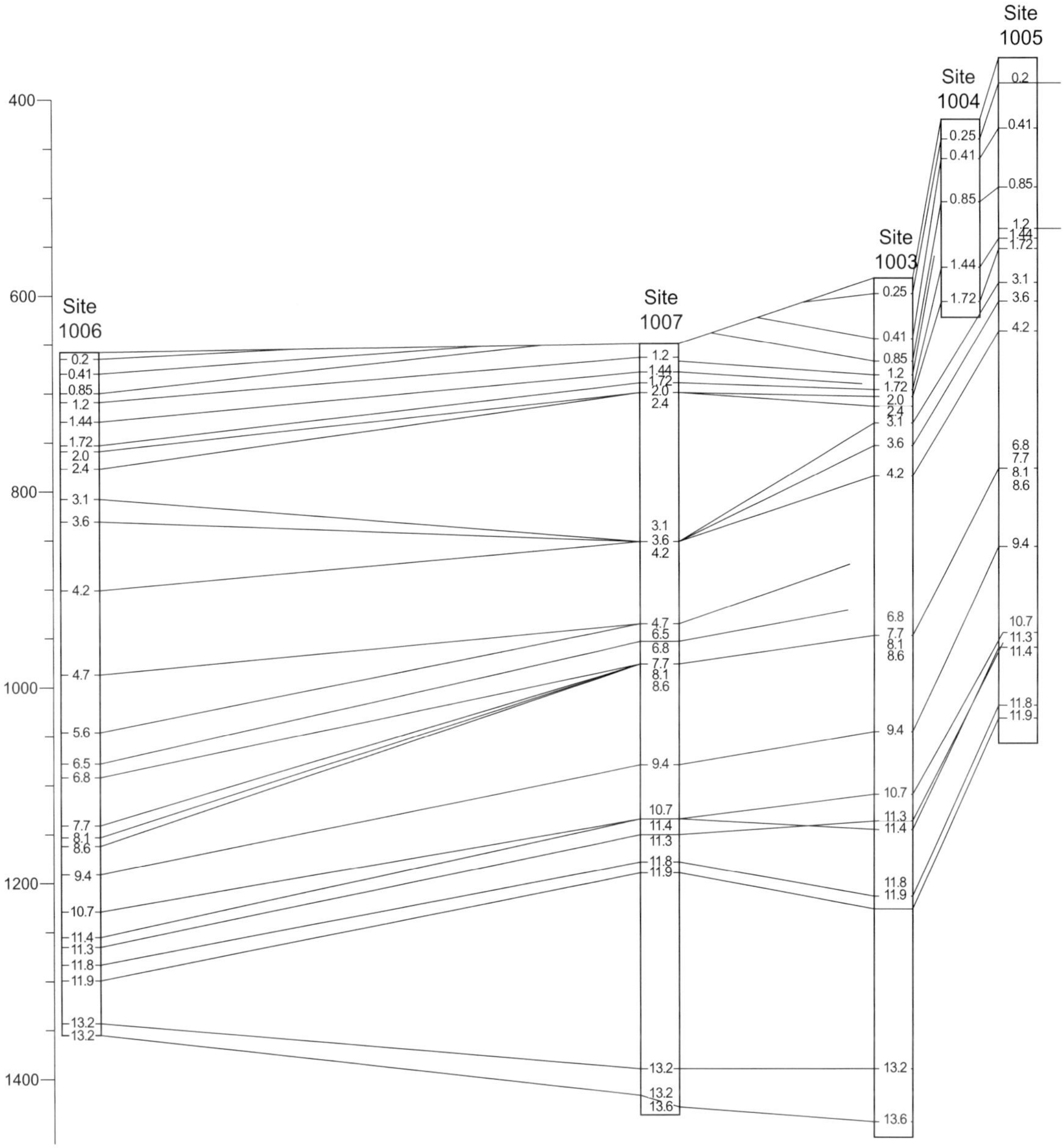

Fig. 7.36.— Time lines correlated between ODP sites shown in Fig. 7.34. Age dating solely by tracing of seismic reflections yields large uncertainties because time lines merge and diverge as sediment drifts thicken and thin. Consistent correlation in Fig. 7.34 was achieved by adding ground truth from the boreholes to the seismic data. After Eberli, Swart, Malone et al. (1997), modified.

nite. Because of the different composition, highstand deposits on the slope are better cemented than the lowstand deposits. The result is a distinct impedance contrast at the sequence boundary with a hard, well-cemented highstand layer overlain by a weakly cemented lowstand layer.

An unsettled question about sequence boundaries is shown in Fig. 7.39. Whereas the record in the Florida Straits suggests that hiatuses form during lowstands of sea level, the record of two boreholes on the Blake Plateau is rather erratic. The borings are only 10 km apart yet their hiatuses differ significantly in age and duration and the correlation with seismic sequence boundaries is not obvious (Austin, Schlager et al. 1986). One possible explanation is that current erosion is in step with sea level if the Gulf Stream is narrowly channeled, such as in the Florida Straits; on the other hand, erosion is irregular if the Gulf Stream is free to meander, such as on the Blake Plateau. Seismic data (Pinet and Popenoe, 1985) indicate rather patchy erosion patterns on Blake Plateau.

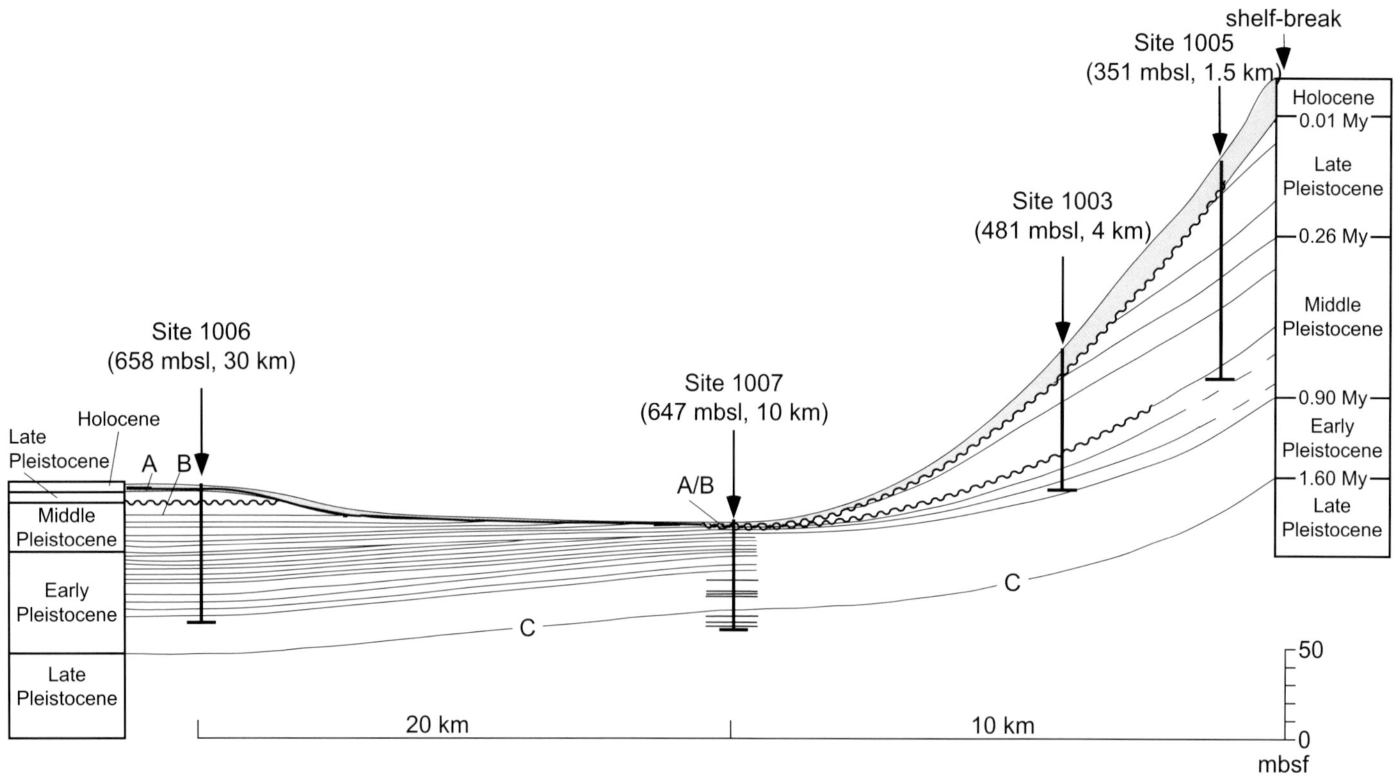

Fig. 7.37.— Erosional truncation of the Bahamian platform slope in the Florida Straits. Erosion is probably caused by steepening of the slope as it grows higher (see Fig. 3.12) and by current scouring from the Gulf Stream). After Rendle and Reijmer (2002), modified.

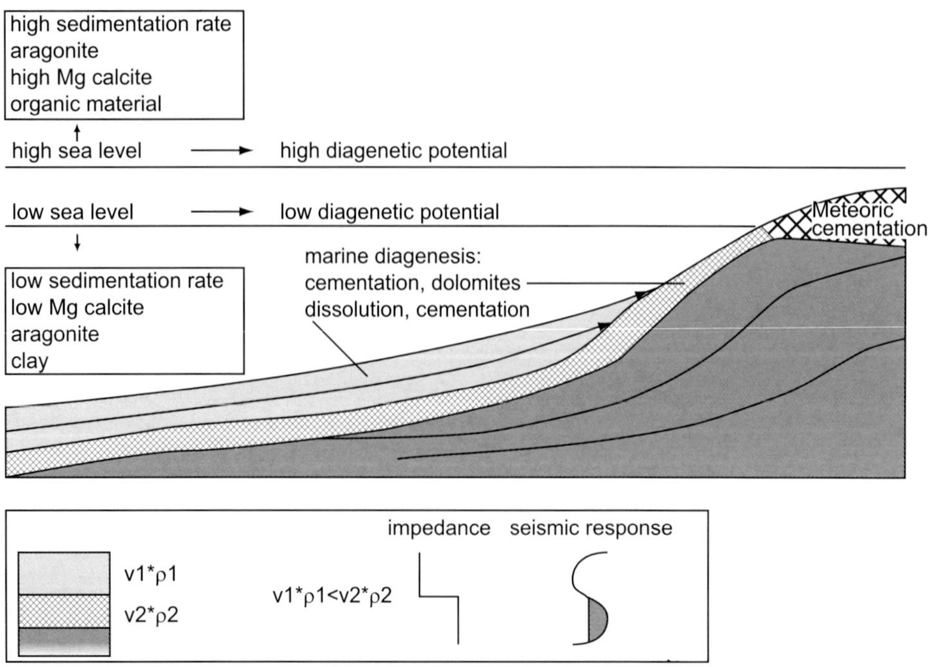

Fig. 7.38.— Model of creating laterally continuous impedance contrast at the boundary of highstand and lowstand deposits by selective marine cementation. Cementation preferentially affects highstand deposits because they are rich in metastable aragonite. After Eberli et al. (2002), modified.

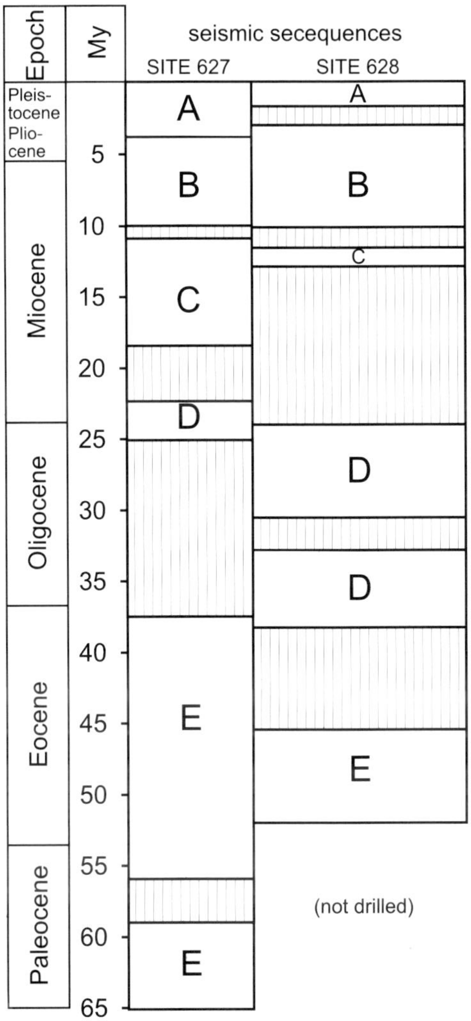

Fig. 7.39.— Hiatuses in the Cenozoic sediments of the Blake Plateau. Boreholes are only 10 km apart, yet the history of sedimentation and erosion differs considerably. Both sites are located on the rise seaward of the slope of Little Bahama Bank. Sediments reflect the interplay pelagic settling, input from the platform by distal sediment gravity flows, and reworking by bottom currents. Gravity flows are at their distal end, thus weak and hardly erosive. The hiatuses, therefore, are probably caused by contour currents. After Austin, Schlager et al. (1986), modified.

Late Miocene of Mallorca

Late Miocene carbonates of the Mediterranean island of Mallorca, documented in exemplary fashion by Pomar (1991, 1993), Pomar et al. (1996) and Pomar and Ward (1999) complement the Bahama case study in several important aspects. Rather than being buried under a modern carbonate platform, the rocks on Mallorca are exposed in many kilometer-long sea cliffs and have been drilled in hundreds of water wells, thus offering a truly exceptional opportunity to examine sequence anatomy at the scale of meters to tens of meters. Detail at this scale has been described from the Bahamian wells but was not resolved seismically or in well-to-well correlations.

Setting. The Miocene limestones, dolomites and marls belong to the sediment fill of intra-montane molasse basins of the Betic Cordillera, an orogenic belt of the Alpine chains of southern Europe. Unlike the Bahamas, the Betic molasse basins experienced long periods of post-depositional uplift and terrestrial erosion. The carbonates on Mallorca are Late Miocene (Tortonian-Messinian) in age; Pomar et al. (1996, p. 197) estimate that the phase of deposition lasted about 2 million years. The facies clearly show the characteristic attributes of the T factory: abundant hermatypic corals and green algae, mud-rich lagoons and fast-growing coral reefs at the platform margin. The platforms grew on a shelf of less than 100 m water depth that subsided slowly during the interval in question. Eustatic sea level fluctuated, but the long-term trend was flat or falling (e.g. Abreu et al., 1998). Thus, the rate of accommodation creation was rather low. The rate of carbonate growth, on the other hand, was persistently high. Consequently, the studied platform prograded 20 km during the examined interval.

The combination of excellent outcrops and careful observation yielded a number of important insights.

- Depositional rhythms appear in a wide range of scales. Individual depositional units, called "sigmoids" (Pomar, 1991; Pomar et al., 1996; Pomar and Ward, 1999), are bundled into sigmoid sets, sets into cosets, and cosets into megasets. The entire formation is considered equivalent to the major portion of a third-order sequence in the standard model. The basic depositional pattern in this hierarchy of cycles remains the same. Pomar and Ward (1995, p. 96) observe that the sigmoid "is not a parasequence ... but ... a small depositional sequence" and "sets, cosets and megasets of sigmoids ... also show characteristics of depositional sequences ...".
- The platform is characterized by strong progradation and very minor aggradation (Fig. 7.40, 7.41). Backstepping of the margin or deepening-upward intervals have not been observed. The succession is an extreme example of a supply-dominated system, one where the rate of carbonate production nearly always exceeded the rate of accommodation creation by a fair margin (see Fig. 7.4). The sequence-stratigraphic characteristics listed below are a direct consequence of this fundamental setup.
- Bounding surfaces are erosional with frequent evidence of terrestrial exposure. Flooding events are extremely subtle; they normally consist of open-shelf limestones onlapping tongues of reef debris in the clinoforms. Tongues of open-shelf sediment extending to the crest of the margin reef are exceedingly rare.
- Transgressive systems tracts as defined in Figs 7.3 and 7.4 are not developed. The record consists of highstand tracts and lowstand tracts. However, there are two patterns that indicate that the rate of accommodation creation by relative sea-level rise sometimes exceeded the rate of carbonate production: (1) The reef rim at the platform margin occasionally rose above the lagoon

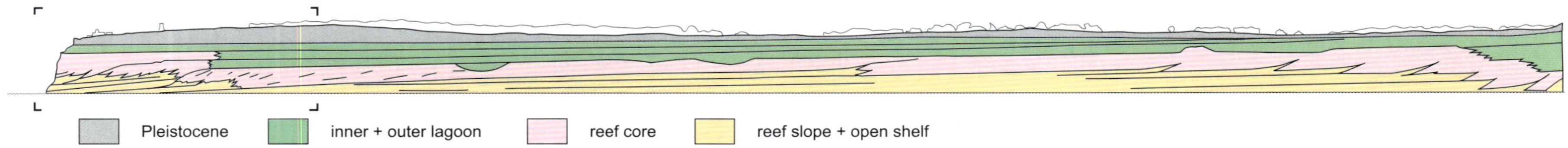

Fig. 7.40.— Overview of 2 km of cliff outcrop. Progradation is virtually continuous but the rate of aggradation varies. Long intervals of progradation with zero or negative aggradation (downstepping) are separated by short intervals with significant aggradation. After Pomar (2000), modified.

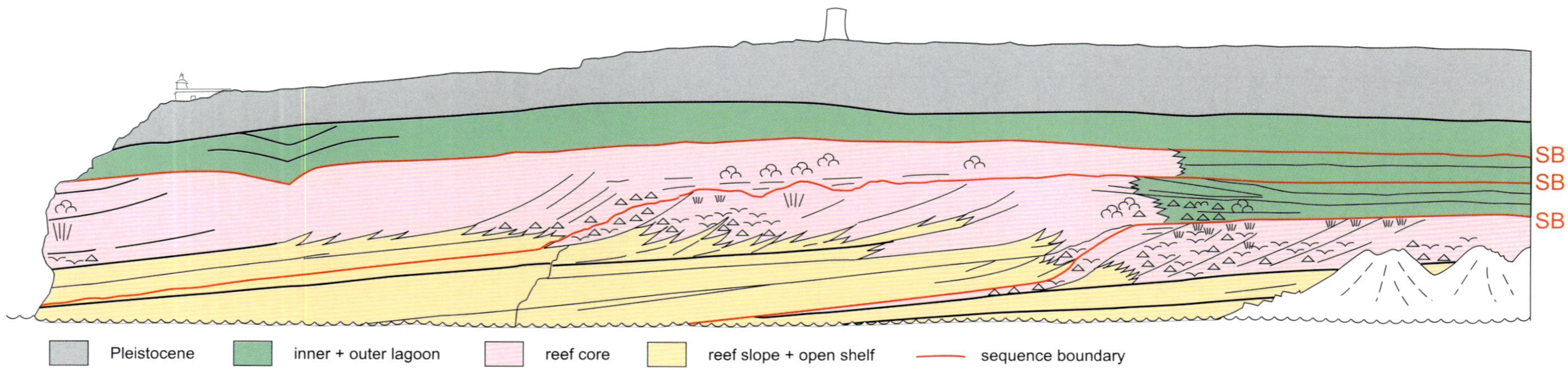

Fig. 7.41.— Field sketch of bedding patterns, facies, and subaerial erosion surfaces in Miocene limestones at Cap Blanc, Mallorca, an example of a supply-dominated system. After Pomar (2000), modified. Progradation is very rapid and interrupted only by erosional exposure surfaces; there is no backstepping. Flooding after exposure leaves no measurable sediment accumulation; the type-1 sequence boundary morphologically coincides with the transgressive surface and the maximum-flooding surface. Bedding geometry and coral morphology indicates that the sigmoids of the unit at lower right are truncated at the top but the coral zones do not step down in the direction of progradation. The same holds for the middle unit of the center. Thus, it remains open if the truncated sigmoids represent offlapping systems tracts in the sense of Pomar (2000) or highstand tracts whose top was eroded during the lowstand. Note raised margin and landward dipping back-reef apron in right part of middle unit.

floor and developed gently landward dipping clinoforms (Fig. 7.41). (2) Seaward of the margin, the open-shelf sediment frequently onlapped and partly buried tongues of reef talus, indicating a temporary halt of progradation.

- The dominance of sigmoidal margin profiles and lack of steep lowstand cliffs indicate that mechanical reworking was very important and the debris-to-framework ratio in the material was relatively high.
- The fine depth zonation on the clinoforms (Figs 7.42, 7.43, Pomar, 1991) allows one to at least crudely estimate the amount of lowstand erosion from the depth level of truncation on the clinoforms. The estimated volumes are quite considerable, again suggesting a system with major mechanical erosion and efficient seaward transport of debris.
- The high resolution of the observations triggered a discussion of the classification of systems tracts. Pomar and Ward (1995) subdivided the highstand tract into the aggrading systems tract, and the high stillstand systems tract. The lowstand tract was split into an offlapping systems tract and a low stillstand systems tract. This classification is useful as a subdivision of the standard categories. It should be noted that the aggrading and high stillstand tracts both satisfy the definition of the highstand tract of Figs 7.3 and 7.4. Similarly, the offlapping tract and the low stillstand tract both meet the criteria of the lowstand tract. The offlapping tract as proposed requires one to distinguish between a systems tract deposited while sea level was falling (the offlapping tract) and a prograding tract whose top part was merely shaved off during the subsequent lowstand. The latter simply represents an eroded highstand tract. Even the detailed depth zonation by corals often remains ambiguous in this respect (e.g. Fig. 7.41).

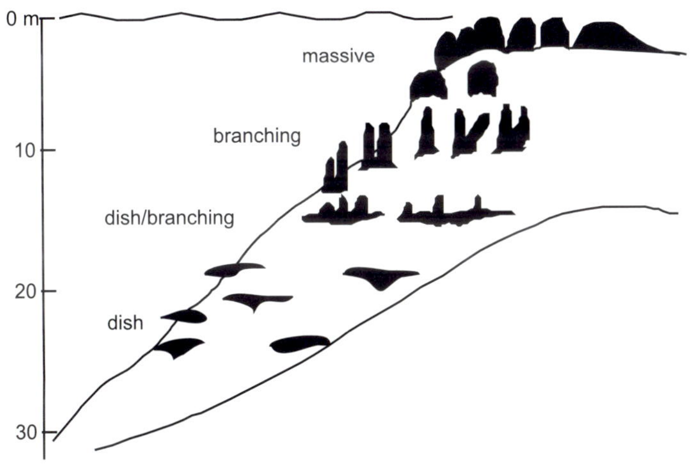

Fig. 7.42.— Depth zonation on fore-reef clinoforms using coral morphology. See Fig. 2.7 for recent analogue. Based on Pomar (1991).

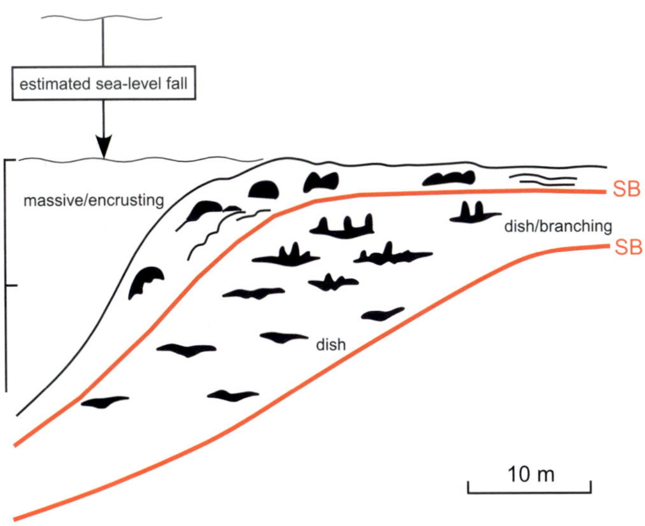

Fig. 7.43.— Field evidence for relative sea-level fall at boundary between two sigmoid units. Zone of branching corals, indicating 10 – 20 m water depth, is overlain at the same depth by massive and encrusting corals indicating 0 – 10 m depth in the younger sigmoid. After Pomar (1991), modified.

THE SEQUENCE STRATIGRAPHY OF GRADUAL CHANGE

The title of this section sounds like an explicit contradiction in terms. The definition of stratigraphic sequences as conformable successions bounded by unconformities clearly implies that breaks in the record are necessary to define sequences. Major breaks, on the other hand, are incompatible with a stratigraphy of gradual change.

The standard sequence model (chapter 6) has abrupt breaks with lapout at the sequence boundary and, less pronounced, at the transgressive surface and the maximum flooding surface. Consequently, a stratigraphic succession consisting of standard sequenes shows highly asymmetric cycles punctuated by pronounced breaks at the sequence boundary. Numerous case studies have documented these punctuations and the asymmetric anatomy of sequences conform the standard model (e.g. Van Wagoner et al., 1990 and Anderson and Fillon, 2004 for siliciclastics; Eberli et al. 2002, and Pomar et al. 1996, for carbonates). However, sequence-stratigraphic case studies also showed that the asymmetric, strongly punctuated standard model has its limitations and that there is a fair number of successions with gradual transitions from transgressive to regressive trends and vice versa. These cycles tend to be symmetrical and the boundaries gradational.

The best known examples of symmetrical cycles with gradational boundaries are the sequences of the 1st and 2nd order (p. 96–98). The 1st order sequences have been recognized from the beginning as nearly symmetrical, transgressive-regressive successions with gradational boundaries, linked to a smooth eustatic sea-level cycle (Vail et al. 1977; Haq et al., 1987). More recently, Duval et al. (1998) depicted the 2nd order sequences, too, as nearly symmetrical units. But the phenomenon of symmetrical sequences with gradational boundaries does not end there. There are also 3rd and 4th order sequences that show almost symmetrical alternations of transgressive-regressive (or deepening-shoaling) trends (examples in Embry, 1993). Carbonate sedimentologists described examples of this phenomenon from flat-topped platforms (Read, 1989; Goldhammer et al. 1990; Montanez and Osleger, 1993). A classic example is the Triassic Latemar platform in the Southern Alps (Goldhammer et al. 1990; 1993). We will use the Latemar to illustrate both the phenomenon and the sequence-stratigraphic approach towards it.

The Latemar Mountain (Fig. 7.44) was a rimmed platform with a flat top virtually at sea level and a composition that

Fig. 7.44.— Triassic Latemar platform. Bedding is virtually horizontal, smooth and devoid of significant angular unconformities. Goldhammer et al. (1990) describe nearly 700 m of measured section in these deposits that form the basis for a sequence stratigraphy of gradual change.

is at the transition between the T and M type. The margin, an unstratified mass of automicrite and marine cement with some sponges and corals, is less than 50 m wide and grades seaward into a 25-35° slope. The evenly bedded rocks of the platform interior lack major breaks but clear subdivisions emerged from mapping facies and stacking patterns of the basic bedding units, called cycles by Goldhammer et al. (1990; 1993). Figs 7.45 and 7.46 display the major results. The lower part of the 670-m section is shoaling upward as indicated by the increase of supratidal and terrestrial facies. The upper part shows the reverse trend. No major break or unconformity has been observed on the platform, nor was there clear evidence of lowstand system tracts on the slope. As major breaks and geometrically differentiated systems tracts were lacking, Goldhammer et al. (1990; 1993) relied on proxy indicators instead. They put a sequence boundary in the interval of most frequent exposure – a unit of stacked tepee horizons and meteoric diagenesis. The underlying, shoaling-upward interval was interpreted as a highstand tract, the deepening interval above as a transgressive tract. The interval of thick subtidal beds with sparse and thin supratidal layers at the base of the section was interpreted as another transgressive systems tract. In line with the definitions used here, Goldhammer et al., 1993 defined a lowstand systems tract as a unit whose marine deposits did not extend to the platform top.

An important tool for the sequence stratigraphy of gradual change is the "Fischer plot" (Fig. 7.46A). It plots cumulative deviation from mean cycle thickness versus cycle number as a qualitative measure of time; in addition, the graph plots subsidence for each cycle, assumed to be constant throughout the section (Fischer, 1964; Read and Goldhammer, 1988; Sadler et al., 1993). Fig. 7.46B shows the Latemar succession of Goldhammer et al. (1993). The sequence boundary interval lies at the center, in a succession of very thin cycles. The maximum flooding levels are again intervals rather than distinct surfaces and lie in the rising limbs of the curve. Their positions are difficult to pick because they lie near the lower and upper boundary of the plot. If one were to replace the Fischer plot by a smooth curve, the sequence boundary would lie at the inflection point of the falling limb of the curve, the maximum flooding level at the inflection point of the rising limb.

Goldhammer et al. (1993) have taken the study of the Latemar succession a step further by creating a computer simulation of the observed stratigraphy. For this simulation they assumed that the bedding rhythm was dictated by the Earth's orbital perturbations with individual beds corresponding to the precession cycle with an approximate duration of 20 ky, and bundles of 4–5 beds corresponding to the Earth's short eccentricity cycle. The long-term trend in cycle facies and stacking patterns can be adequately reproduced by combining the orbital pulses with a eustatic sea-level wave of about 40 m amplitude and 3 My period (Fig. 7.47). With the above assumptions, the measured section would represent a little less than 3 My, the sequence boundary in the middle coincides with an interval of maximum exposure produced by a long-term eustatic fall. The maximum flooding levels lie in the intervals of maximum submergence near the base and the top of the section. The assumptions in the model on the length of the time interval are at variance with recent findings on biostratigraphy and radiometric dating (Brack et al. 1996; Zühlke et al., 2003, vs. Preto et al., 2001). However, the basic approach towards a stratigraphy of gradual change remains valid regardless of the outcome of the chronostratigraphic debate.

From the Latemar case study emerges a clear strategy for the sequence stratigraphy of gradual change on carbonate platforms.

- Use bed facies and bed stacking patterns to reveal deepening and shoaling trends generally considered synonymous to "transgressive/regressive" trends in sequence stratigraphy.
- Identify turning points in these trends and interpret changes from transgressive to regressive as maximum flooding levels, changes from regressive to transgressive as sequence boundaries.

Similar approaches have been used in many subsequent studies. Recently, it has become popular to mark shoaling and deepening trendsby triangles on the vertical section. This is excellent for quick orientation but often leads to very schematic descriptions. I suggest to at least display quantitative observations along with the triangles (e.g. D'Argenio et al. (1999) and specifically identify intervals that lack clear trends (e.g. Immenhauser et al., 2004).

A consequence of defining sequences where significant breaks are not obvious is that sequence boundary and maximum flooding surface are no longer surfaces but intervals (Goldhammer et al., 1990; Schlager, 1992; Montanez and Osleger, 1993, p. 322). Only stratigraphy with very high resolution or firm evidence for the presence or absence of significant lowstand wedges can determine if the gradual trends are punctuated by major gaps. Where large gaps and stratigraphic turning points are lacking, correlation of sequence boundaries and systems tracts becomes much more arbitrary and sequence-stratigraphic analysis converges with the tried and true sedimentologic practice of delineating shoaling and deepening, coarsening and fining, thickening and thinning trends in a section. In most instances, these trends simply reflect the gradual change in the balance of the rate of accommodation creation and the rate of sediment supply. Embry (1993) made a strong case for the power of this T-R ("transgressive-regressive") stratigraphy. However, if transgressive and regressive intervals smoothly grade into one another, it becomes very difficult to distinguish between trends caused by relative sea-level fluctuations and trends caused by changes in the rate of sediment supply. Thus, the search for major breaks and lowstand systems tracts remains crucial for the reconstruction of sea-level history.

The above discussion leads to the following conclusions on the sequence stratigraphy of gradual change:

- Sequence stratigraphy of gradual change sometimes is the only way to identify sequences in a responsi-

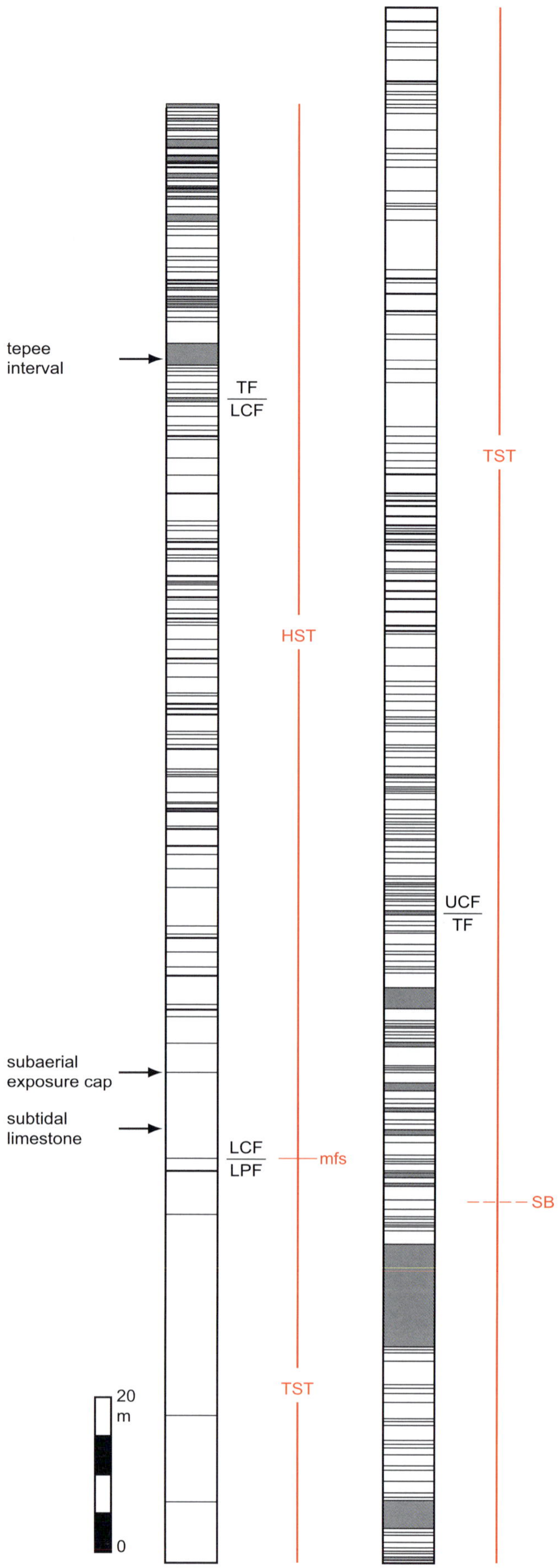

Fig. 7.45.— Sequence stratigraphy of gradual change, demonstrated on the Latemar. Columns show over 400 m of rhythmic platform deposits with bottom at lower left, top at upper right. Black lines and shading = terrestrial and supratidal deposits, blank = subtidal deposits. LPF – largely subtidal cycles; LCF – lower subtidal-supratidal cycles; TF – tepee cycles, dominated by supratidal deposits; UCF – upper subtidal-supratidal cycles. Abundance of supratidal/terrestrial facies first increases, reaches a maximum in the middle of the section, then decreases again. Shoaling and deepening trends are used to identify systems tracts indicated in red. TST = transgressive systems tract, HST = highstand systems tract, SB = sequence boundary, mfs = maximum flooding surface. After Goldhammer et al. (1993), modified.

ble way, i.e. without "forcing" sequence boundaries through conformable stacks of strata.

- Trends of shoaling and deepening (synoymous to transgressive and regressive in sequence terminology), are the essential building blocks for systems tracts under these circumstances.
- Criteria used for delineating shoaling/deepening trends should be clearly indicated. In the sediment record, water depth needs to be estimated via proxies such as fair-weather wave base, depositional base level, base of the euphotic zone etc. Unlike hydrostatic pressure, these proxies relate to depth in a relative way and the trends may not always correspond.
- Flat-topped carbonate platforms often require the gradual-sequence approach because the lateral variation in accommodation is almost nil and lapout patterns therefore extremely subtle and scarce. Bedding is essentially parallel throughout.
- Wherever possible, the platform record should be correlated with the deeper-water record beyond the platform margin. Margins and slopes need to be searched for lowstand systems tracts because this is the most direct way of determining if major breaks are hidden in the record of gradual change on the platform top.

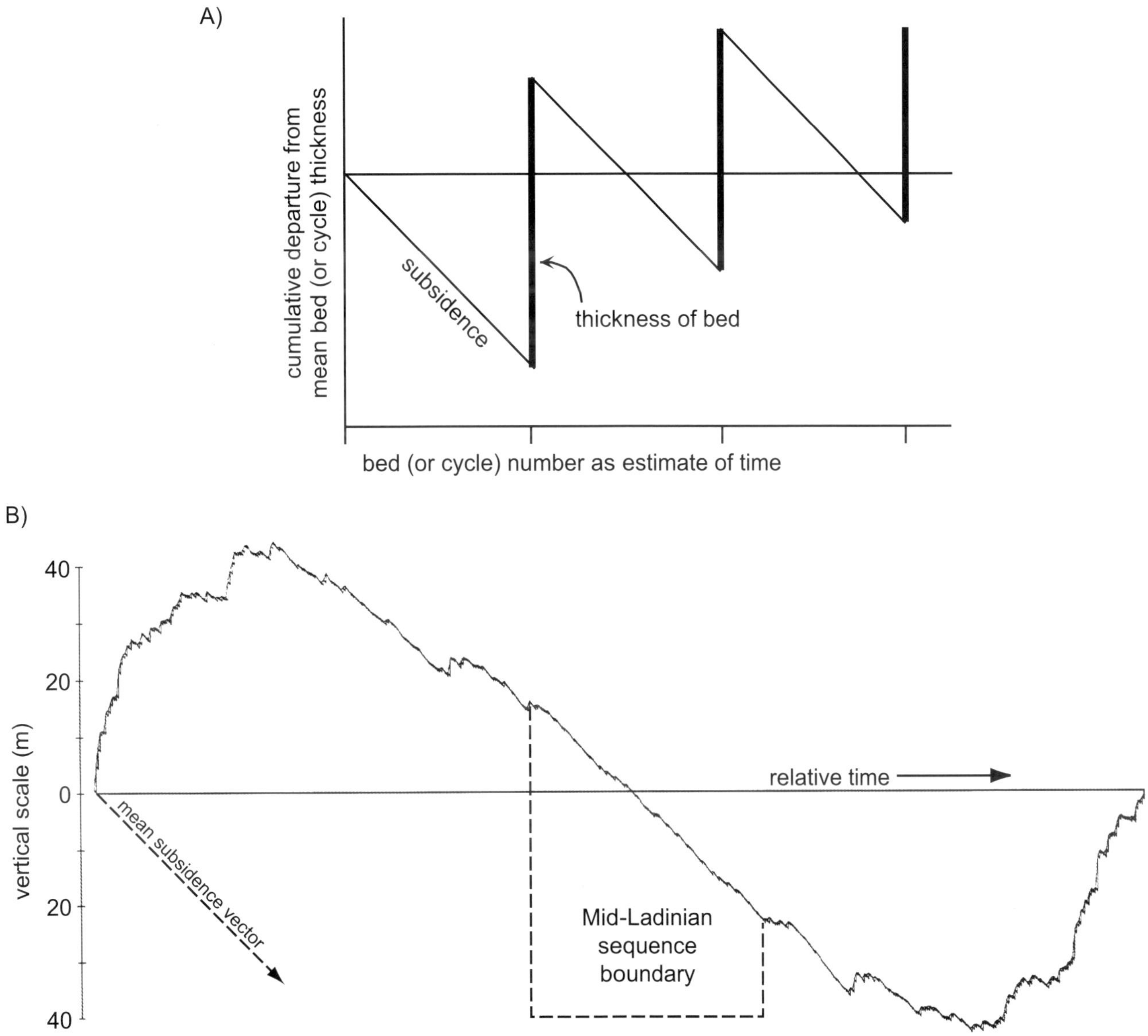

Fig. 7.46.— Fischer-plot technique applied to the Latemar. A) Principles of the Fischer plot: it displays cumulative departure from mean bed (or cycle) thickness on the vertical against cycle number (as a qualitative measure of time) on the horizontal. Subsidence is assumed to be constant and each cycle is shifted downward by the appropriate amount of subsidence. B) Fischer plot of the Latemar section. Rising limbs represent intervals of thicker-than-average beds, falling limbs thinner-than-average beds (or cycles). Sequence boundary was put in the interval of thinnest beds at the center; maximum-flooding levels would lie in the rapidly rising portions near the bottom (left) and top (right) of the section. In contrast to classical sequence stratigraphy, sequence boundary and maximum flooding levels are intervals rather than discrete surfaces. After Goldhammer et al. (1993), modified.

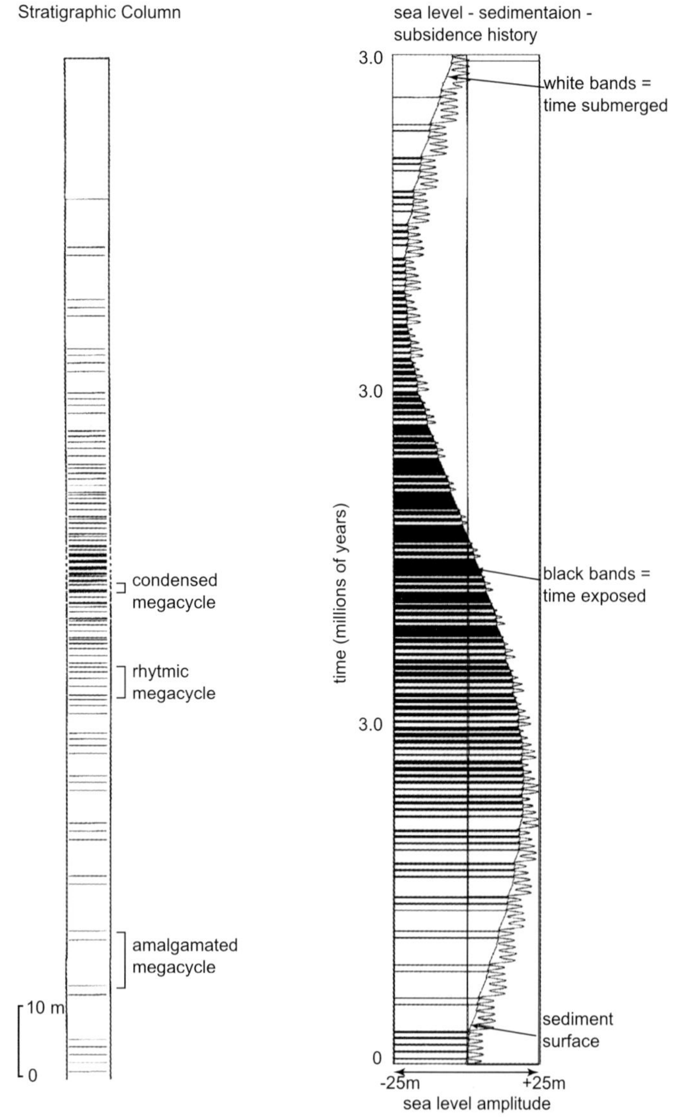

Fig. 7.47.— Latemar section of Fig. 7.45 modelled as the result of superposition of sea-level cycles of periods of 20 ky, 100 ky and 3 My. Sequence boundary lies in the zone of maximum exposure (black) in the middle of the section and is caused by a long-term fall of sea level. The trends observed in Fig. 7.45 are satisfactorily reproduced even though the absolute chronology remains debatable. After Goldhammer et al. (1990).

HIGH-RESOLUTION SEQUENCE STRATIGRAPHY OF CARBONATES

Originally, sequence stratigraphy was a tool for the seismic interpreter and therefore limited by the resolution of the seismic data. Application of sequence stratigraphy to outcrops, cores and wireline logs immediately demonstrated that unconformity-bounded units also exist at much finer scales than those resolved by seismic data. The concepts of sequence orders as well as the fractal model of sequences (chapter 6) are attempts to apply sequence stratigraphy to a wide range of scales in time and space.

In shoal-water carbonates, particularly of the T factory, numerous studies have meticulously documented facies and the nature of bedding surfaces, sometimes with a resolution of centimeters (Eggenhoff et al., 1999; Strasser et al., 1999; D'Argenio et al., 1999; Homewood and Eberli, 2000; Preto et al., 2001; Van Buchem et al., 2002; Della Porta 2003, p. 212-222; Immenhauser et al., 2004) . The recurring pattern is an alternation of deepening and shoaling trends, dissected by surfaces indicating breaks in sedimentation. As thinner and thinner beds are logged and correlated, the question arises: What does it all mean? What is global or regional signal, what is local "noise"in this fine web of beds, layers and surfaces?

There are no hard and fast rules for interpreting the fine detail of shoal-water carbonate sections in terms of depositional environments and stratigraphic sequences. However, several rules of thumb have emerged from the studies of the past decades.

- Bounding surfaces. It is very important to distinguish between: (1) exposure surfaces where marine deposition was temporarily interrupted by terrestrial conditions with at least incipient soil formation, and (2) flooding surfaces where a shoaling-upward trend peaked in the shallow-marine, intertidal or supratidal environment, and the environment subsequently changed to deeper or more open-marine conditions. The distinction is critical because flooding surfaces may be generated by internal feedback in the depositional system whereas exposure surfaces require a drastic seaward shift of the shoreline that is practically impossible without a fall of relative sea level (chapter 6). Moreover, incipient soils usually take 1 ky or more to develop (p. 98-99); this yields a minimum duration of recognizable exposure surfaces.
- Estimate the depositional relief that existed within the individual facies belts. Use direct observation of channels, reefs, sand bars etc., augmented by data from analogous modern settings. During sea-level falls, the horizontal sea surface intersects this morphology in various ways, creating lens-shaped sediment bodies and merging and diverging bounding surfaces.
- Estimate local variations in sedimentation rates from the anatomy of subtle oblique bedding surfaces that are truncated by master bounding surfaces. Also obtain the local range of sedimentation rates in comparable modern settings. Both data sets together provide guidelines on how quickly shoaling trends may change laterally into deepening trends and vice versa. (See Holocene systems tracts in the Bahamas; p. 110f).
- Trace beds and bounding surfaces laterally as far as possible (e.g. Immenhauser et al. 2004). Statistically speaking, global or regional signals must exceed local signals in lateral extent.

Fig. 7.48 depicts a platform section recorded in great detail and subsequently used for regional correlation and time series analysis. The description is based on continuous drill cores and outcrops. The rocks were classified by facies, ranked in six categories from open marine to highly restricted, with two categories added for emergent conditions at the landward end.

The choice of parameters for recording change in the stratigraphic section is arguably the most critical step in high-resolution sequence stratigraphy. Ideally, the parameters should relate to properties that change gradually or in small steps over a wide range. They should be rather insensitive in space to small, accidental features and in the time domain to short, strong, but accidental events such as storms, sudden gravitational collapse etc. On very flat platforms, the degree of restriction and emergence is an attractive choice. Ramps on continental margins may be open marine right to the shoreline. In such settings, wave energy as a proxy for water depth may be an appropriate choice. A third possibility are relative light levels as indicated by

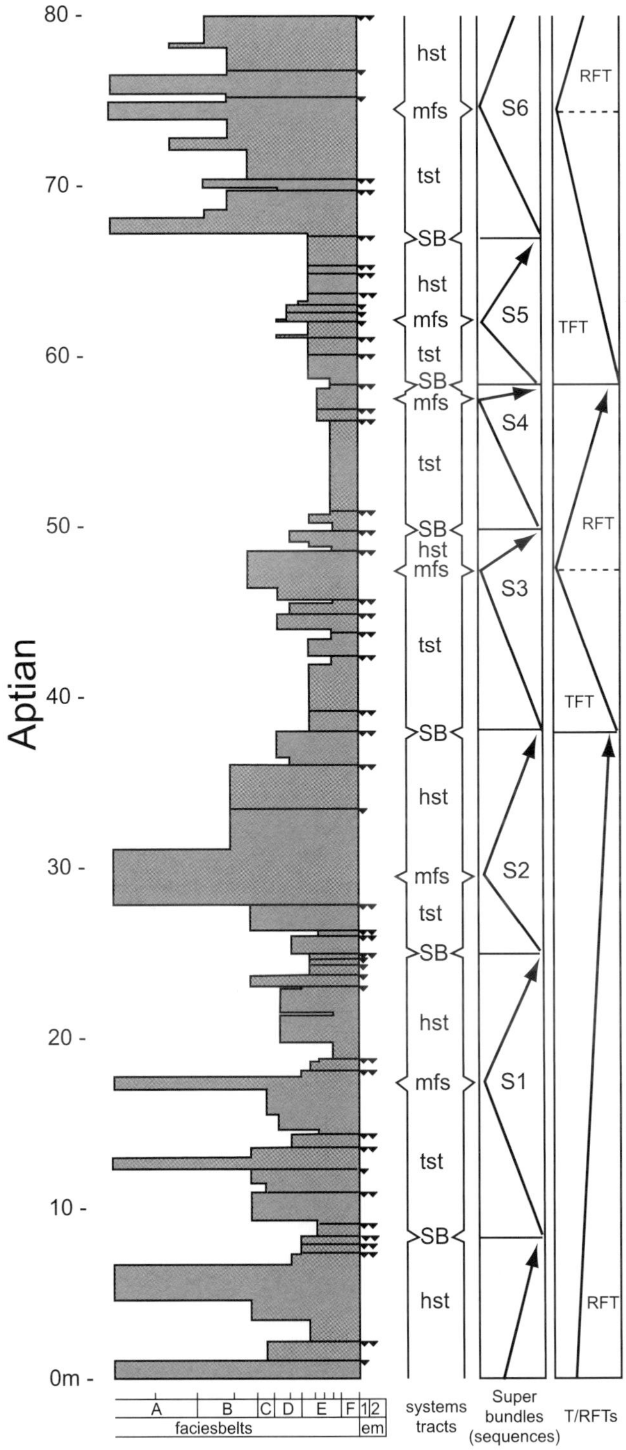

Fig. 7.48.— Bedding record of an Early-Cretaceous (Aptian) carbonate platform in the Apennines, Italy. Profiles were measured with 2-centimeter resolution, only large patterns are shown. Left: Age and thickness scale in meters. Shaded: Lithologic column showing facies varying from fully open-marine on the left (facies belt A) to restricted supratidal on the right (facies F). Sedimentation is interrupted by two types of exposure surfaces: type 1 – supratidal, possibly terrestrial; type 2 – demonstrably terrestrial. Central column shows sequence classification. Arrows on the right: shoaling and deepening trends of two different orders. After D'Argenio et al. (1999), modified.

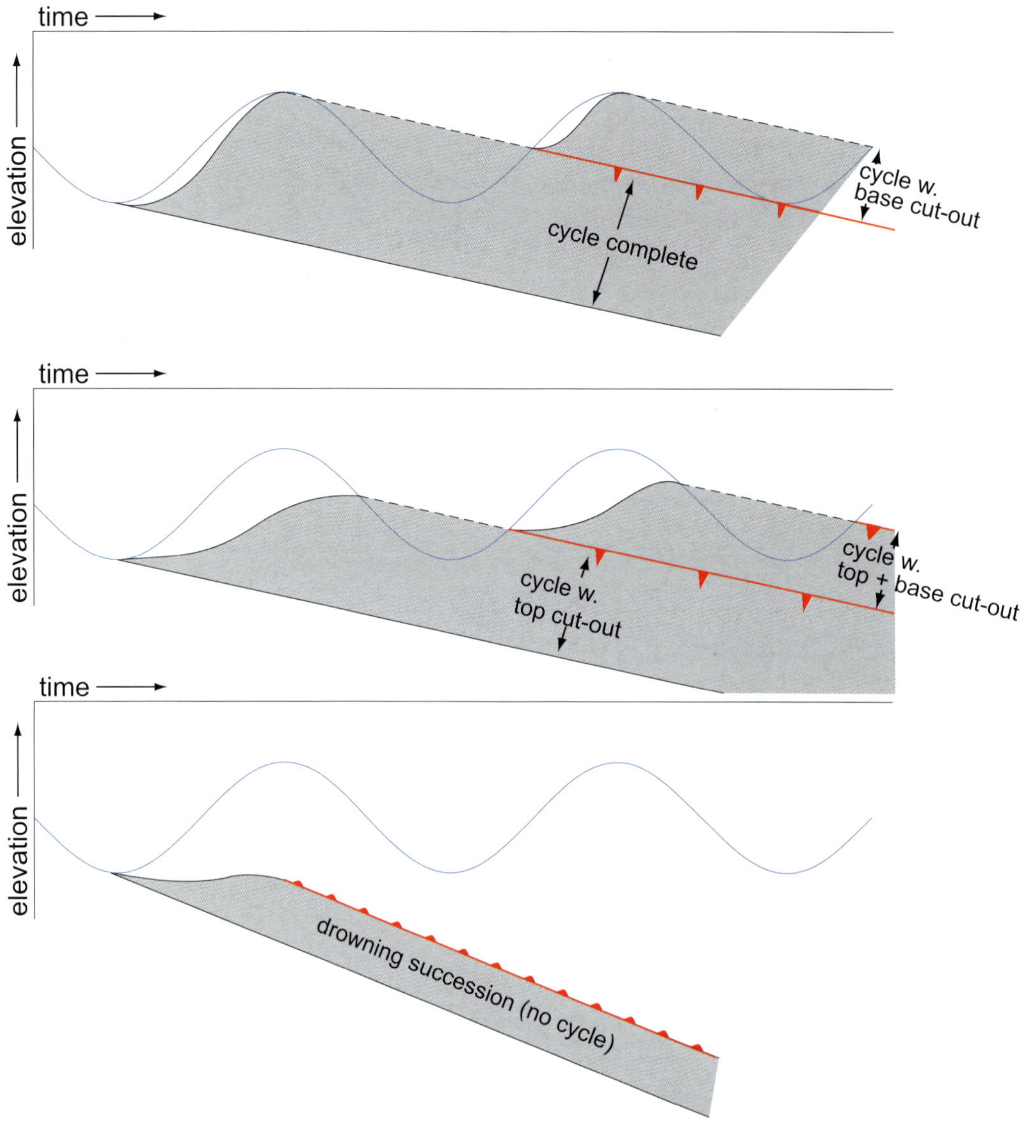

Fig. 7.49.— Cartoons of the sea-level records on very shallow carbonate platforms. The interplay of eustasy, subsidence and sediment supply produces various kinds of truncated cycles where part of sea-level history is lost in hiatuses. Blue – sea level, red – hiatuses. Based on Soreghan and Dickinson (1994), Strasser et al. (1999), Tipper (2000).

sessile benthos such as corals (Figs 2.7,7.42). Grain size as principal parameter for estimating depth or distance from shore is less straightforward in carbonates than in siliciclastics because carbonate grains vary enormously in density and shape; moreover, their abundance varies because of localized production.

The art in facies ranking lies in the choice of parameters and in the right balance between lumping and splitting. Descriptions based on oversplit categories drown in meaningless flicker; descriptions based on broadly lumped categories lose useful information.

Sea-level signals are particularly important in sequence stratigraphy. Fig. 7.49 summarizes the record one can expect on flat-topped platforms. The interaction of eustasy, subsidence and sediment supply creates a record that is replete with hiatuses. Usually, only a fraction of the total sea-level fluctuation is recorded in the sediment accumulations (see Soreghan and Dickinson, 1994; Hillgärtner and Strasser, 2003, for detailed analysis). If longer sea-level waves are superimposed on the short oscillations, the platforms are likely to miss large parts of the record (Fig. 5.5).

CHAPTER 8

Sequence stratigraphy of C and M factories

INTRODUCTION

The discussion of sequence stratigraphy started in chapter 6 by examinig the standard model and its observational support based largely on siliciclastics. Chaper 7 introduced carbonate sequence stratigraphy with a detailed look at the T factory - the best known carbonate system and the most productive one. The present chapter deals with the sequence stratigraphy of the C and M factories. It does so by highlighting differences to the T factory. This is not to say, however, that the principles of marine carbonate production outlined in chapters 2 and 7 are no longer valid simply because they are not discussed at length here. In the C and M factories, just as in the T factory, sediment is largely produced by organic activity within the depositional setting. Environmental factors therefore strongly influence the sediment type and the rate of production. Consequently, environmental change is a major competitor of sea-level change in shaping the sequence record.

C FACTORY

Overview

The hallmark of C-factory sequences are coasts formed by cliffs or sandy beaches, consistently seaward-dipping shelves, and sigmoidal shelf breaks that bend down to relatively gentle slopes. Reefs, if present, have low-lying crests, and are widely scattered over the outer parts of the shelf and the upper slope. The shelf breaks of the C factory generally lack wave-breaking rims. Therefore it is not surprising that there are no detailed reports of empty buckets either. With regard to sequence stratigraphy, one can say that the deposits of the C-factory behave similarly to siliciclastics in the marine environment and their response to exposure is somewhere between that of siliciclastics and tropical carbonates.

These characteristics and their causes will be discussed in more detail below. The following attributes of the C factory are crucial in a sequence-stratigraphic context.

- The output of the C factory consists almost exclusively of skeletal particles ranging in size from coarse silt to pebble. Mud, if present, stems from abrasion and biodegradation of the coarse particles and from occasional input by the planktic factory of the open ocean.
- The depth window of production is much wider than that of the T factory. Like the T and M factories, the C factory cannot produce above sea level; only clastic accumulations of marine material may occur in the form of eolian dunes. The lower limit of C-factory production usually is set by influx of terrigenous or planktic fine material. Below a certain depth, skeletal production simply drowns in mud. However, on current-swept sea floors the C factory may produce even in the deep bathyal and abyssal environment.
- Lithification within the depositional environment usually is scarce or absent. The reason is the scarcity of soluble aragonite in the sediment and the low carbonate saturation of the sea water. Consequently, sea-floor lithification distinctly increases in warm temperate environments where the C factory gradually passes into the T factory.
- The ability to build protective rims is weak. Reef-building communities exist but they are either entirely independent of light, or include only a small fraction of phototrophic organisms, mostly red algae. As a consequence, the reef comunities of the C factory rarely build to sea level. A structure that rises slightly above the adjacent sediment on the shelf or upper slope will do. As they remain below the zone of permanent wave action, the tops of the reefs are not planed off. The typical geometry is that of a convex mound. Internal lithification is scarce and the structures are easily reworked. It seems that the cool-water reefs are less long-lived than tropical reefs. Reef patches start and die more frequently. The result are swarms of reefs scattered in loose sediment.

Sequence facies and bounding surface

The facies succession from coast to basin of the C factory has been presented in chapter 4. The facies belts typically develop on a ramp - a smooth surface that dips seaward with angles of less than 1.5 °. The ramp facies model applies to shelves of continents and islands as well as epeiric seas. This is advantageous because the C factory, too, may develop in all these settings.

On the shelves and upper slopes, standard sequence-stratigraphic techniques can be applied but the differences to T carbonates described in chapter 7 remain significant. The scarcity of syndepositional cementation, together with the lack of protecting rims, leads to wholesale reworking of shelf sediment when sea level falls and the high-energy belt shifts downward.

Reworking notwithstanding, the internal anatomy of cool-water carbonate deposits often records shoaling or deepening trends that aid in recognizing systems tracts. Shoaling cycles bounded by exposure or flooding surfaces are particularly common (Fig. 8.1; James, 1997; Knoerich and Mutti, 2003). Changes in water depth can be gleaned from organisms (e.g. Betzler 1997; James et al. 2001; Brachert et al., 2003), and from hydrodynamic structures. Such structures are more common than in tropical carbonates because the sediment consists mainly of sand and gravel and is less burrowed than tropical carbonates. Finally, mud content is a fairly reliable, relative depth indicator. This is so because there are no wave-breaking rims that can protect shallow lagoons or muddy tidal flats as in tropical carbonate settings. Mud, therefore, accumulates in deeper water and the content of mud, carbonate or terrigenous, has been shown to consistently increase with increasing water depth on cool-water carbonate shelves and slopes (Collins, 1988; Henrich et al., 1997; Gillespie and Nelson, 1997; James et al., 1999; James et al., 2001). The "mudline" concept of Stanley et al. (1983) can easily be transferred from siliciclastics to cool-water carbonates.

Recognition of sequence boundaries and discrimination between exposure and flooding surfaces usually is more difficult than in tropical carbonates because lithification is slower and reworking more severe.

The preservation of exposure surfaces in C carbonates is highly variable and depends on the rate of cementation. Marine cementation is slow and areally limited. Based on observations on Tertiary limestones of Australia and New Zealand, Nelson and James (2000) argued that cool-water marine cementation occurs only in specific circumstances (Fig. 8.2). Surlyk (1997) made similar observations in the Cretaceous of NW Europe. Where an exposure surface is superimposed on a submarine hardground, it is more likely to be preserved (e.g. Nelson and James, 2000, p. 616).

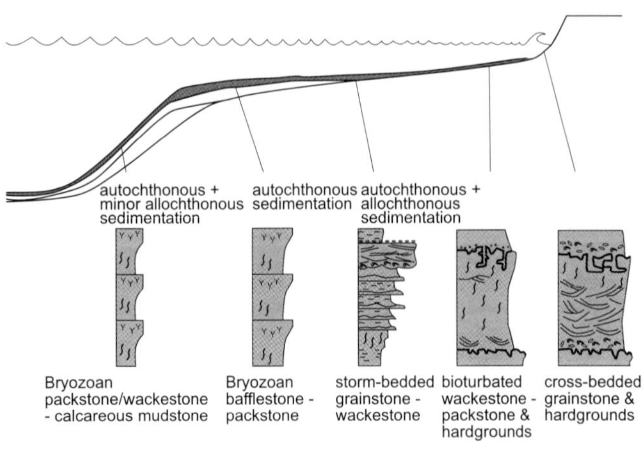

Fig. 8.1.— Shoaling and deepening trends in C carbonates. After James (1997), modified.

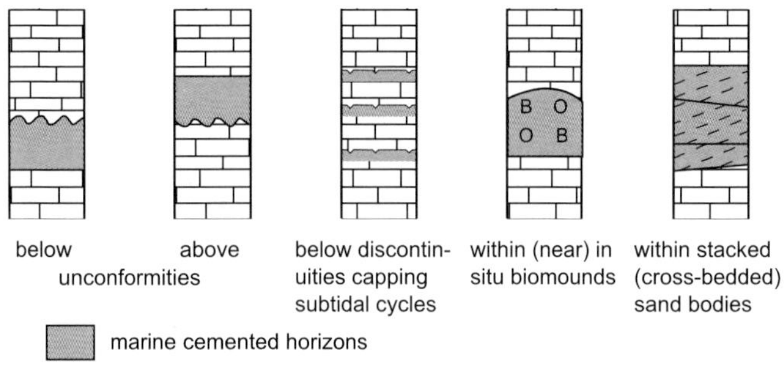

Fig. 8.2.— Favorable settings for submarine lithification in cool-water carbonates. After Nelson and James (2000), modified.

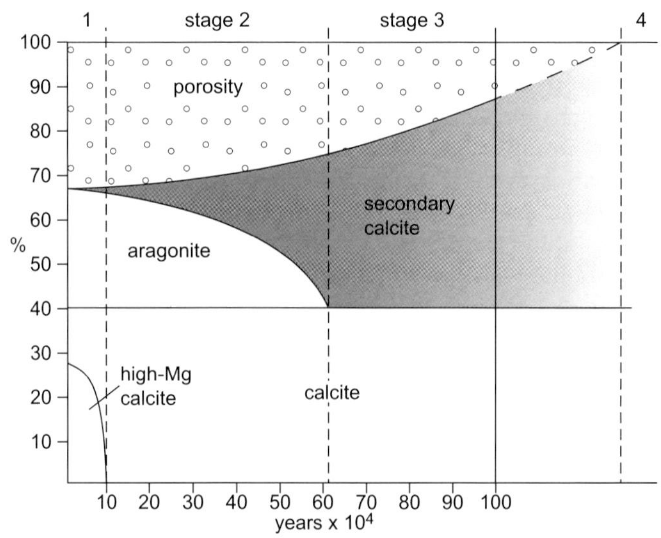

Fig. 8.3.— Meteoric diagenesis in C carbonates, Pleistocene-Holocene, Warrnambool, SE Australia ; based on Reeckmann and Gill (1981). Diagenetic pathways resemble those of tropical carbonates: aragonite dissolves and re-precipitates as blocky calcite cement, whereby lateral transport of dissolved carbonate may be important; rock shown here has received extra material to close all pore space. Magnesian calcite converts to calcite in-situ by dissolution-precipitation reactions.

Meteoric cementation primarily depends on the aragonite content of the sediment and this may vary between 0 – 60% (Fig. 2.15; Nelson 1988; James 1997). Aragonite selectively dissolves during early diagenesis and thus provides ions for precipitation of calcite cement. Magnesian calcite, on the other hand, has been found to expel its magnesium by in-situ recrystallization and thus stabilize to calcite without wholesale dissolution (Reeckmann and Gill, 1981; James and Bone, 1989; Knoerich and Mutti, 2003). The rates of meteoric cementation of C carbonates are not well known but a few case studies offer important constraints. Fig. 8.3

summarizes work by Reeckmann and Gill (1981) on Quaternary calcarenites in southern Australia. After 10 ky, only few percent of calcite cement had formed and the sediment was well cemented after 60 ky of exposure. In southern Australia, too, James and Bone (1989) showed that calcarenites virtually devoid of primary aragonite are still loose or friable after 10 My of exposure, whereas an overlying unit with 20-30% of primary aragonite content is now well cemented and the aragonitic shells have been replaced by molds.

A transgressive surface or transgressive interval is frequently well developed on modern cool-water shelves. Beautiful examples of this phenomenon occur in southern Australia (Fig. 8.4; James, 1997): The shelves are hundreds of kilometers wide and have shelf breaks far below 100 m. The large swell generated in the roaring fourties of the southern ocean moves sand at depths of 100 m and more. Carbonate production extends from the shoreline to the upper slope but the middle part of this C factory is what James et al. (2001) called a "shaved shelf" - a zone covered with relict Pleistocene material but virtually devoid of Holocene carbonate. This zone of Holocene erosion and reworking separates a coastal zone of deposition from an outer-shelf-and-slope zone of deposition. James (1997) notes furthermore that at least part of this shaved shelf was also bare during the Pleistocene sea-level falls.

Geometry of systems tracts

C-factory sequences resemble siliciclastic sequences with rounded shelf breaks and gentle slopes. There are two important exceptions to this general similarity: C-factory systems tracts lack point sources of sediment input from rivers and they have the ability to build seismically recognizable reefs, albeit not at sea level but at greater depth. Morphologically, the C-factory reefs have convex tops because they are not planed off by wave action. This difference to the flat, wave-breaking reefs of the T factory can be resolved seismically. Unfortunately, tropical systems that slowly drown and subside below the zone of wave action also develop convex tops (Figs 7.21, 7.23).

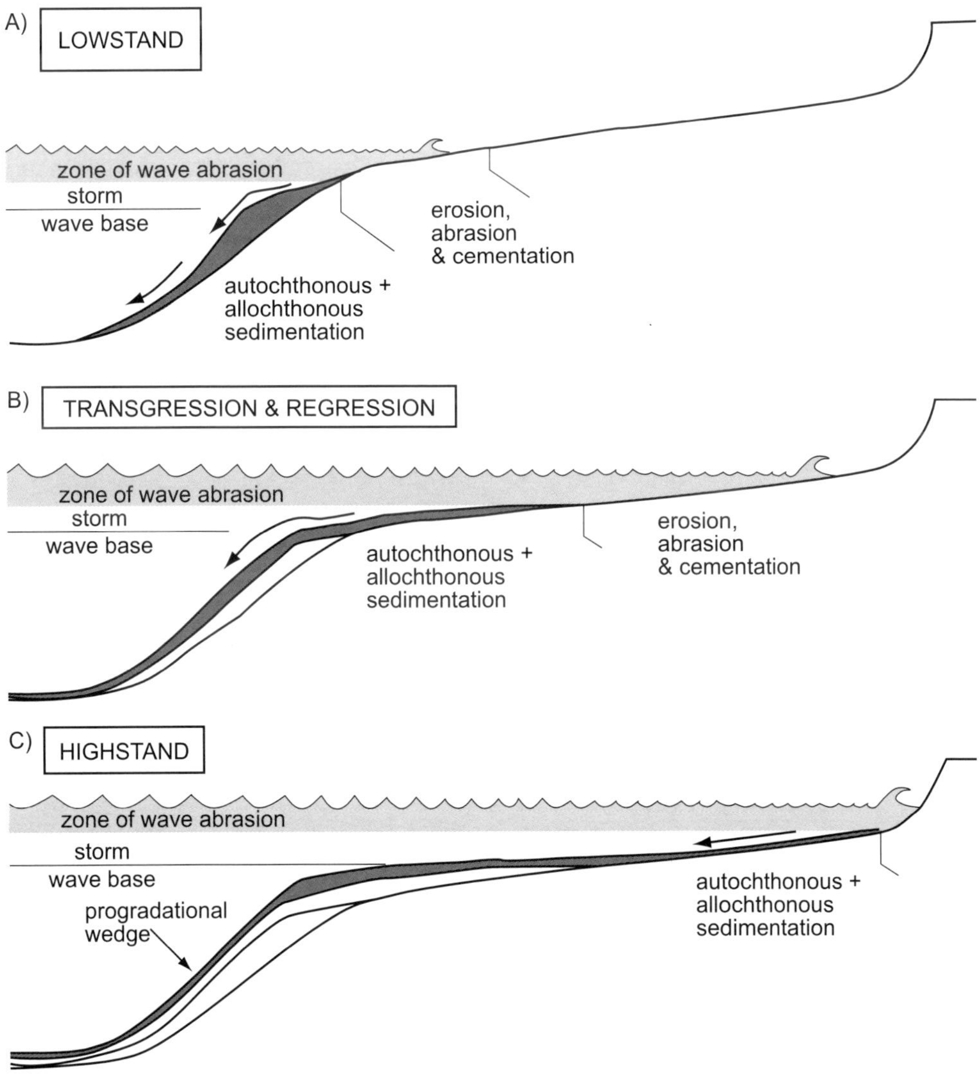

Fig. 8.4.— Transgression and regression on Australian shelves of the C factory. Slow marine lithification leads to intensive reworking of sediment, particularly on ocean-facing shelves where waves that move sand at 100 m depth are common. The result is a broad zone of non-deposition that is being reworked and abraded during both transgressions and regressions. After James (1997), modified.

Sediment reworking is fast and effective since cementation is slow in both the terrestrial and the marine environment. As in siliciclastic sequences, significant portions of the highstand systems tract may be shaved off during sea-level fall such that the highstand tract has to be reconstructed from the preserved parts of the prograding clinoforms(Feary and James, 1998; Saxena and Betzler, 2003; for siliciclastic analogues see Anderson et al., 2004).

One of the best examples of sequence stratigraphy in C carbonates is the Neogene – Quaternary Eucla shelf off southern Australia (Fig. 8.5; Feary and James, 1995; 1998; James, 1997). Unconformity-bounded sequences are well defined in the high-sedimentation area of the outer shelf and upper slope. Bounding surfaces converge and partly merge on the middle shelf where reworking and non-deposition are most intensive. A second high-sedimentation belt is in the coastal zone.

Highstand shedding or lowstand shedding?

The standard model of sequence stratigraphy postulates that the influx of sediment to the slope and basin is high during lowstands of sea level when shelf accommodation is small and rivers discharge their load on the outer shelf. We have seen that the sequences of the T factory approximately swing in antiphase to this model (p. 116f). Tropical platforms export most sediment during highstands of sea level when the production area is large, and waves and tides can move the sediment efficiently to the margin. Export during lowstands is small because the production area is reduced and rapid cementation of exposed sediment precludes lowstand flushing, i.e. extensive mechanical erosion of sediment from the preceding highstand.

The effects that cause highstand shedding in the T factory are small or absent in the C factory as the following analysis shows.

- The depth window of production is 5 – 10 times wider than that of the T factory. Consequently, the width of the production window exceeds the amplitude of nearly all eustatic fluctuations in the sequence-stratigraphically relevant time domain of 10^3 – 10^6 years.
- The production area directly affected by sea-level cycles is a steadily seaward-dipping shelf with a round shelf break. Thus, the difference in production area between highstands and lowstand is much smaller. (It is nil on a perfectly planar epeiric ramp).
- Cementation in the meteoric and the marine environment is slow because seawater is less supersaturated with calcium carbonate and there is less metastable aragonite to dissolve.

Based on these considerations one expects the C factory to differ from the T factory and closely obey the standard model.

The question of highstand or lowstand shedding in cool-water carbonates has not been studied as extensively as in tropical carbonates. Nelson et al. (1982) confirmed lowstand-shedding for the most recent glacial cycle on a New-Zealand shelf. These authors found that sediment export from the shelf was high during the last glacial lowstand and has virtually ceased in the Holocene highstand. On the Eucla shelf of southern Australia, Saxena and Betzler (2003) found that lowstand sedimentation had a distinct maximum on the upper slope whereas sedimentation of transgressive and highstand tracts was more uniform; they also showed that lowstand erosion largely removed the shallow (most characteristic) portion of the transgressive tract and highstand tract. The cumulative result of several cycles is a thick shelf-margin wedge that thins both downslope and shoreward (Fig. 8.6).

Sequence stratigraphy of the C factory in deep-water environments

One of the most profound differences between the T factory on the one hand and the C and M factory on the other is the depth window of production. The window of the T factory is extremely narrow and restricted to the sunlit part of the ocean. The other two factories can produce in a much wider depth range. The window of the C factory certainly extends to abyssal depths: production is virtually independent of water depth and temperature. It may commence wherever currents sweep the sea bed, remove the fines and provide sufficient nutrients for benthic growth.

The depth-indepence of production has profound implications for sequence stratigraphy. The siliciclastic standard model rightly proceeds on the premise that any sediment accumulations in the deep sea must somehow be connected to a terrestrial source, and that the succession of environments from shore to abyss represent one system. This premise also holds for the T factory, albeit with the modification that the source is in the shallowest part of the sea.

For the C factory, the concept of a connected system linking the shore with the deep sea need not be valid at all. There are linked environments, to be sure. Canyon-fan systems exist even though well-documented examples are rare. One example is the Tertiary of the Gippsland Basin where major canyons were filled with redeposited C-carbonates from the shelf (Wallace et al., 2002). However, there are other areas where the C factory produces large volumes of carbonate material at bathyal and abyssal depths without any connection to shoal-water. One example is the northern North Atlantic (De Mol et al., 2002). Part of this deep production accumulates in reefs or mound-shaped buildups, another part is transported by deep-sea currents and remodeled into sediment drifts, often mixed with terrigeneous and pelagic sediment.

Stratigraphic sequences can be recognised in these environments if one defines them as conformably layered packages bounded by unconformities as proposed by Vail et al. (1977, p. 55). The link of sediment anatomy with sea level is more subdued than on the shelf but the stratigraphic control may be excellent if sufficient plankton is preserved. In the Quaternary, for instance, pelagic sediments intercalated in cool-water carbonates may allow one to correlate them with eustatic highstands and lowstands (e.g. Passlow, 1997).

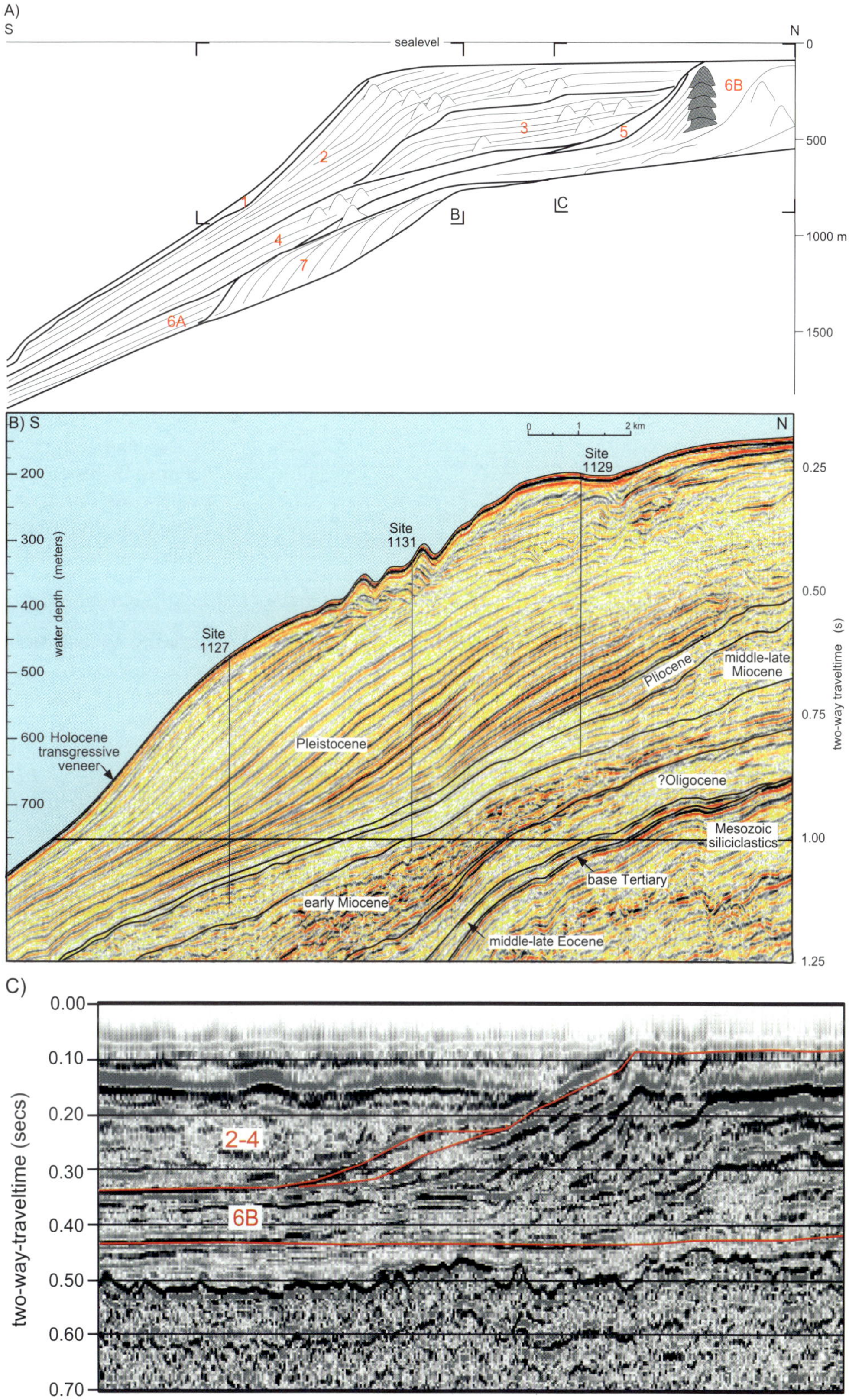

Fig. 8.5.— Sequence stratigraphy of the South Australian margin in the Cenozoic. Compiled from Feary and James (1998), James et al. (2000). A) Overview of sequences and their internal geometry. Note difference between C-factory sequences 1 – 5 and 7, and the probable T-factory sequence 6. B) Seismic image of sequences 1 and 2 (Quaternary). Note rounded shelf break and bryozoan mounds on the upper slope. C) Sequence 6B, interpreted as rimmed, tropical platform, onlapped by sequence 5 (debris of reef escarpment) and cool-water carbonates of sequences 2 – 4. Note difference between the interpreted rimmed T platform and the rounded, deep shelf margin of cool-water carbonates in Fig. B.

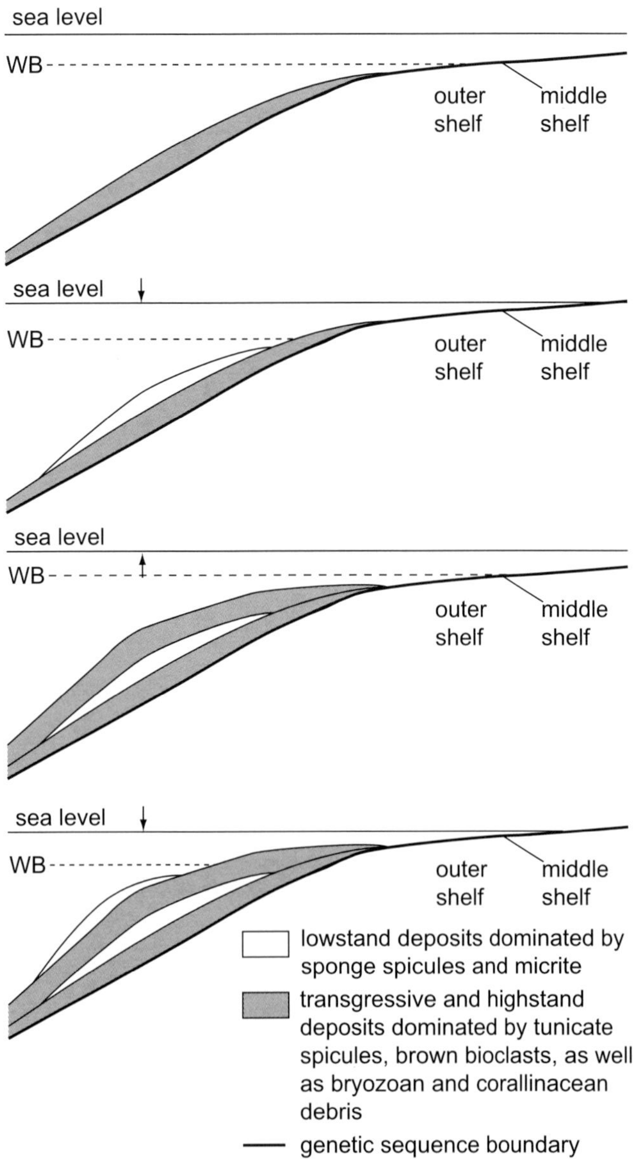

Fig. 8.6.— Model of systems tracts in Quaternary glacial-interglacial cycles based on ODP drill holes and seismic data, Eucla shelf, southern Australia. After Saxena and Betzler (2003), modified. Lowstand tract (unshaded) forms a lens on the upper slope. Transgressive and highstand tracts extend with fairly uniform thickness across upper slope and outer shelf but the part on the outer shelf is eroded during lowstands. The result is a prograding sediment wedge on the upper slope and a zone of reduced sedimentation or erosion on the shelf.

M FACTORY

Overview

The characteristic accumulations of the M factory are either swarms of mounds that may coalesce to a network of ridges, or rimmed platforms that geometrically resemble T platforms.

Sediment anatomy of the M factory is directly related to the production process and its environmental controls, similar to the relationship observed in the T and C factories. For sequence stratigraphy, the following characteristics of the M factory are particularly important.

- The primary material ranges from clay-size lime mud to pebble-size peloids or oncoids and also includes hard framework that may extend over thousands of cubic meters. Some of this framework may be reworked to boulder-size clasts in the depositional environment. Overall, the size spectrum (not the grain types) resembles that of the T factory.
- Depth window of production extends from the shoreline to bathyal, perhaps abyssal depths. However, competition by the T factory may block the shallow part of the window in the Phanerozoic.
- Lithification in the depositional environment is as rapid as in the T factory and probably even more pervasive.
- Where the M factory builds platforms, it develops rims at the shelf edge. In most instances, these structures seem to be less robust than those of the T factory. However, some Precambrian rims have been truly elevated such that the platform developed an empty bucket (Adams et al., 2004).

Sequence anatomy and bounding surfaces

Where the M factory builds shoalwater platforms, the facies belts of these M platforms fit Wilson's standard model (see chapter 4). The crucial element is a platform-margin belt consisting of a rigid framework of automicrite and marine cement (Keim and Schlager, 2001; Stephens and Sumner, 2003). This margin may crest in tens of meters of water or build to sea level. Shallow rims shed much excess sediment onto the platform and may develop debris aprons analogous to the backreef aprons of coral reefs (e.g. Stephens and Sumner, 2003, p. 1285).

Platform-interior deposits generally are flat-bedded, often showing shoaling cycles bounded by flooding or exposure surfaces, just as in T carbonates. Small stromatolite mounds (e.g. Chow and George, 2004) can be viewed as microbial analogues of patch reefs on modern tropical platforms.

The slopes seaward of the rim are shaped by the interplay of gravity-driven downslope transport and in-situ production of automicrite and cement. Where in-situ production dominates, slope angles may exceed 50° and debris aprons at the toe-of-slope are small (Lees and Miller, 1995, p. 199). Where gravity-flows dominate, slope angles peak at 35 – 40 ° - the angles of repose of poorly sorted sand and rubble, and debris aprons are larger (Keim and Schlager, 2001).

The wide depth window of production allows the M factory to develop autonomous production centers in deep water - the mud mounds. In the Phanerozoic, M platforms are rare and mud mounds are the most characteristic products of the M factory. They form groups and roughly contour-parallel belts of mounds where the individual structures may be tens to hundreds of meters thick and thus recognizable in seismic data.

Mound flanks are often steep but debris aprons are small or absent. The scarcity of debris and the fact that flank declivity commonly exceeds the angle of repose of sand and rubble (Fig. 3.13), indicate that the flanks are part of the stiff framework of the mounds rather than a talus cone that covers the framework. The toe-of-slope progrades as the mound increases in height. The vast Devonian exposures in the Sahara show that chains of mounds may coalesce to kilometer-long ridges (Fig. 8.7, Wendt et al. 1993; Wendt and Kaufmann, 1998). Spectacular networks of mud-mound chains were revealed by 3D seismic data of the Permo-Carboniferous in the Barents Sea (Fig. 8.8, 8.9; Elvebakk et al. 2002). The buildups grew during 35 My, extend over hundreds of km^2 and reach thicknesses of 350 – 1,200 m. Geometrically similar, albeit smaller, patterns develop in coral reefs of the T factory where isolated patch reefs have been observed to expand and coalesce to networks with polygonal depressions in between.

Convex mounds that turn flat during growth are of special interest to the sequence stratigrapher (with the proviso that the flat surface is not the result of coincidental coalescence of several mounds). Flattend mounds are likely to have grown into the zone of wave action - waves, after all, are the most effective planing tool in marine sedimentation. For some mounds, growth into shoal-water can be demonstrated. Fig. 8.10 is an example, contributed by Calvet and Tucker (1995). The Triassic mounds turned into mini-platforms as they grew higher; simultaneously, the facies changed from a framework of automicrite, sessile microbes and cement to skeletal sands with coral patches. The change to a wave-swept environment goes hand in hand with increased lateral progradation, probably because of increased sediment supply to the slope.

Bounding surfaces of depositional units develop in similar fashion as in the T factory. Marine cementation is rapid and pervasive. Consequently, flooding events frequently are accompanied by hardgrounds. This also applies to M platforms where sedimentation fills all accommodation and ends in supratidal flats. Hard crusts were common on these flats and are likely to turn into hardgrounds or layers of lithoclasts during the subsequent transgression.

Patterns and rates of karst formation on M factory carbonates have not been studied in detail. The fact that the fraction of metastable carbonates is similar to the T factory suggests that the rate of karst formation is similar, too. Terrestrial cementation rates may be lower because the aragonite content of M carbonates seems to be somewhat lower (Fig. 2.15). From the Devonian M platform of the Canning Basin, Playford (2002) reports karst with karren and caves that

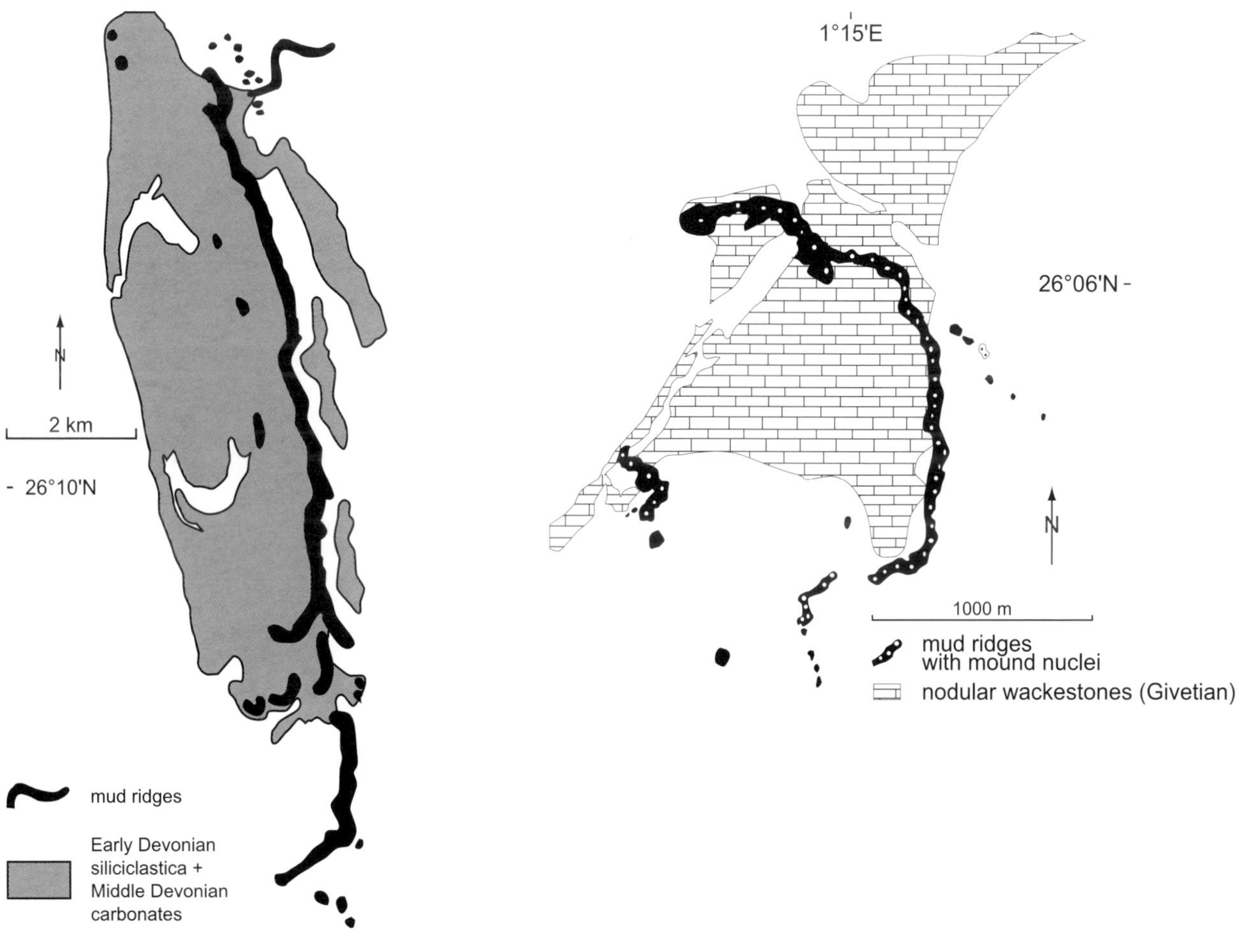

Fig. 8.7.— Two examples of mounds that coalesced to kilometer-long ridges in the Devonian of southern Algeria. Compiled from Wendt et al. (1993) and Wendt and Kaufmann (1998).

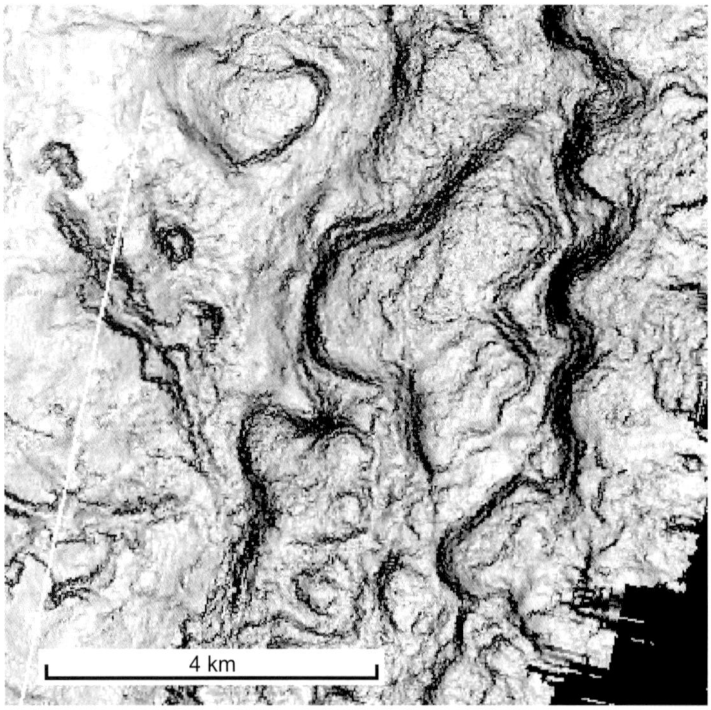

Fig. 8.8.— Networks of mound chains mapped by 3D seismics in the Carboniferous-Permian of the Barents Sea. Ridges are hundreds of meters high. After Elvebakk et al. (2002). Copyright 2002, reprinted with permission from Elsevier.

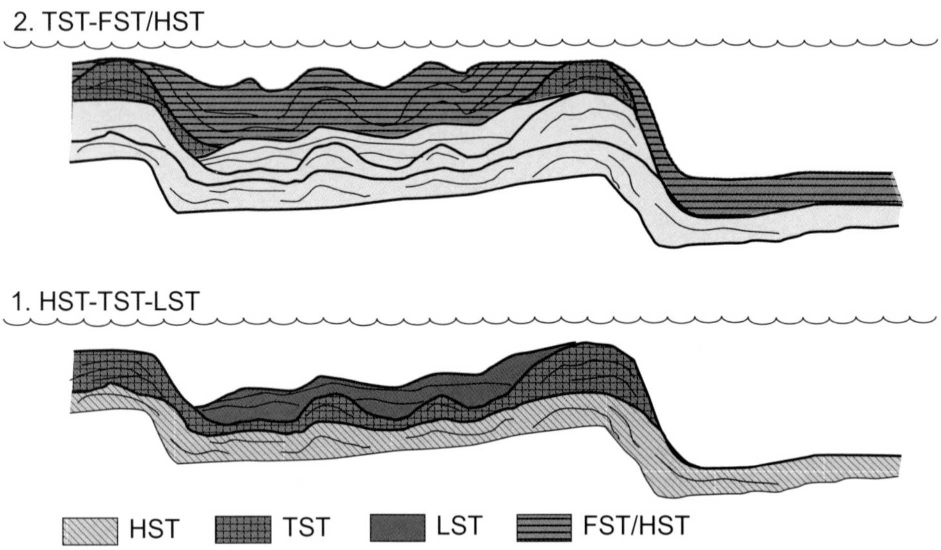

Fig. 8.9.— Idealized cross section in the polygonal mound pattern of Fig. 8.8. Systems tracts can be defined geometrically. Lowstand tracts are confined to the spaces between the mound crests; transgressive tracts show general retrogradation and local drowning of mounds; mounds of basal highstand aggrade and prograde. Mounds of falling-stage systems tract (topmost unit) also prograde but their tops are truncated by erosion surfaces. Note that there is no clear seaward-landward differentiation in the network of ridges. Systems tracts need to be defined in relation to the highest mounds. After D. Hunt, Norsk Hydro, written communication.

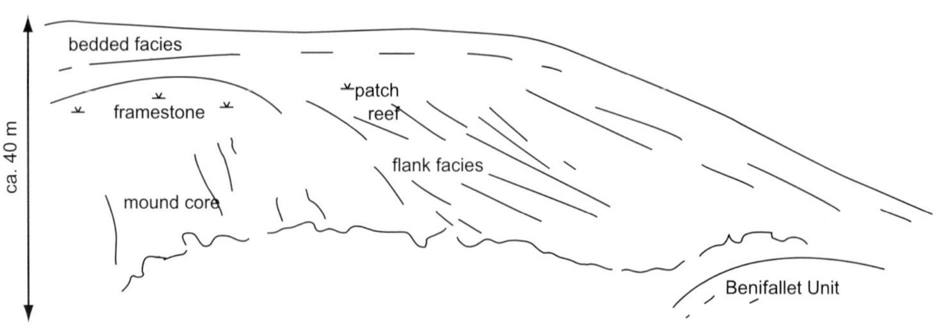

Fig. 8.10.— Convex mud-mound that built into shallow water and became flat during growth. Concomitant facies change from automicrite to skeletal sands and coral patch reefs indicates that the mound reached the sunlit and wave-agitated environment. Note that significant lateral progradation commenced after the change to skeletal facies. After Calvet and Tucker (1995), modified.

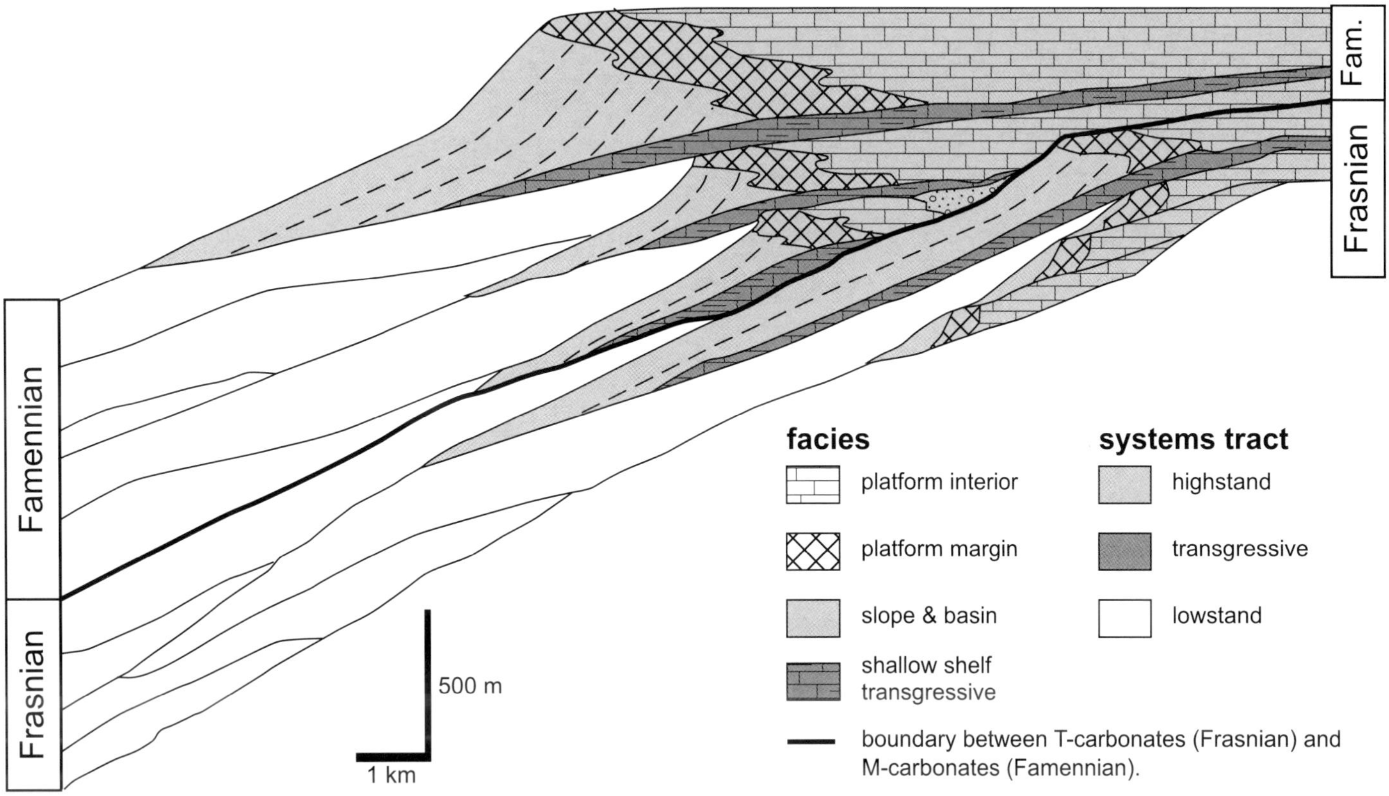

Fig. 8.11.— Sequence model of Devonian carbonate platform of the Canning Basin, NW Australia, based on wells and seismic data. Frasnian rocks belong to the T factory, Famennian rocks to the M factory. Note that the sequence geometry is virtually identical. M platforms and T platforms are practically indistinguishable at this large scale. After Kennard et al. (1992), modified.

formed during an intra-Famennian event and was sealed by biostratigraphically only marginally younger shale.

Systems tracts on M platforms can be expected to resemble those of the T platforms. The Late Devonian of the Canning Basin supports this view (Fig. 8.11; Kennard et al., 1992). No significant difference has been observed between systems tracts of the T and M factory.

Further support for the notion that the systems tracts of the T and M factories are very similar may be derived from recent studies in Late-Proterozoic carbonates (Adams et al., 2004).

CHAPTER 9

Looking back, moving forward

Single-author books inevitably present a biased view of things. In this last section, I will add insult to injury by exposing some very personal thoughts on the future of carbonate sedimentology and sedimentary geology in general. These thoughts are guided by basic patterns and lines of reasoning that have emerged in the expose on the preceding pages.

With regard to marine carbonates, this book – I hope – has illustrated their intricate connection with the chemical conditions and the biota of the ocean. The mode of precipitation, notably the degree of biotic influence, determines not only the texture and composition of carbonate sediments but to a significant extent also the anatomy of the accumulations and their distribution in space and time. The precipitation modes, therefore, provide a basis for distinguishing the different production systems, or factories, introduced in chapter 2.

The eminent role of chemical and biotic factors distinguishes carbonate systems from siliciclastic ones. In siliciclastics, source and sink fundamentally depend on rates of uplift and subsidence as well as topographic gradients – i.e. factors largely controlled by tectonics. The siliclastic accumulations themselves faithfully record hydrodynamic conditions and sediment supply but they are rather insensitive to many other environmental factors. Hydrodynamic factors being equal, a wave-dominated delta will look very much the same at the equator and at 60^0 latitude, in normal marine, hypersaline and fresh-water settings. The respective differences in carbonate accumulations are profound and chapters 2 – 4 summarize important trends.

The different environmental sensitivity of siliciclastics and carbonates must not distract from the fact that the two sediment families have a common basis in the hydrodynamics of sedimentation and erosion. Therefore, many rules apply to both families and this leads to effective cross-fertilization. In chapter 4, for instance, the siliciclastic shore-to-shelf model was quite easily applied to the carbonate ramp. The same holds for turbidites, debris flows, slumps and many other phenomena.

Sequence stratigraphy, discussed in chapters 6 – 8, also constitutes a fundamental concept that applies to siliciclastics and carbonates as well as sulfate evaporites, volcaniclastics and other sediments not examined here. Some modifications are required if one applies the siliciclastic standard model to other sediment families but usually the necessary modifications can be deduced from first principles of sedimentology. For instance, the adaptations of the standard model required for carbonates are a direct consequence of the way carbonates are produced, and altered by early diagenesis, as illustrated in chapters 7 and 8.

The concept of sequence stratigraphy includes both the technique of recognizing unconformity-bounded units as well as the identification of systems tracts. The differentiation of shoal-water accumulations into prograding, retrograding and downstepping units is - in my opinion - a fundamental property of the sediment record. It must be noted, however, that only the downstepping units are reliable indicators of relative sea-level change. Changes from progradation to retrogradation and vice versa are ambiguous. On a subsiding substrate, progradation and retrogradation depend on the balance between the rate of accommodation creation and the rate of sediment supply. Thus, a large change in sediment supply may override the rate of accommodation creation caused by sea-level change (chapter 6, 7).

Where do we go from here? The future of sedimentary geology as a dynamic discipline may depend on how successful we are in discovering general principles behind the formation of sedimentary rocks and their stratigraphic successions.

Sequences and systems tracts illustrate one such principle that governs many sediment accumulations - scale invariance in time and space. The principle was recognized early in sedimentology. For instance, Potter (1959) in his succinct summary of the emerging concept of facies models already observed "that scale is not a critical factor". In this book (chapter 6), it was argued that sequences and systems tracts are invariant on length scales of at least 6 orders of magnitude (from $10^{-}1$ to 10^5 m). Time required to produce these scale-invariant sequence patterns may vary from 10^3 y to 10^8 y, i.e. by five orders of magnitude if one includes the slow process of soil formation. For purely geometric aspects of sequences, the scale-invariant domain is significantly larger, probably 10^{-1} y to 10^8 y.

Fig. 9.1 shows an example of scale-invariant sediment geometry that extends beyond the sequence domain: sediment accumulations fed by a point source assume the shape of a delta with dendritic feeder channels. Van Wagoner et al. (2003) explain them as energy-dissipation patterns, a geometric expression of the principles of thermodynamics.

The trio of topset-foreset-bottomset provides another example of scale invariance, particularly if one considers it together with the analogous classification of undaform-

Fig. 9.1.— Valley-delta systems in plan view – invariant sedimentation patterns that reflect the dissipation of kinetic energy. Total observed range of scale invariance is more than 14 orders of magnitude, shown examples cover only a very small part of this range. The important point is: size and shape are uncorrelated in this example. After Van Wagoner et al. (2003), modified.

clinoform-fondoform discussed in chapter 4. Thorne (1995) contributed a quantitative analysis of the unda-clino-fondo trio under the assumption that the entire system consists of loose sediment. He found that the governing equations can be applied in a wide range of scales, certainly including the geologically particularly relevant dimensions shown in Fig. 9.2. The governing principle for the formation of prograding clinoforms is high sediment supply at the top, gravity transport down the slope and decay of transporting power on the flat basin floor. Carbonates are not always deposited as loose sediment. They may include bodies that are hard immediately upon formation such as reefs, automicrite mounds or early-lithified sands. However, even the complex carbonate systems produce scale-invariant anatomies in a considerable range of scales. A well-known example is the repetition of the atoll structure at different scales. There exists a continuous size spectrum that ranges from mini-atolls, 10 m in diameter, to oceanic atolls exceeding 100 km in diameter. The underlying principle in the atoll example seems to be the advantage of the rim position for benthic growth.

Scale invariance is not limited to geometry. The relationship of sedimentation rates to the length of the observation span is invariant under changes of scale over 11 orders of magnitude (p. 102). Wave phenomena are another example of scale invariance. Especially relevant for sedimentology and sequence stratigraphy is the statistical behavior of sea-level fluctuations: the power (= square of the absolute amplitude) of sea-level fluctuations is related to cycle period by a power law (p. 102). It seems that sea-level fluctuations, much like sedimentation rates, have fractal characteristics. The observations on the scaling of sedimentation rates and sea level changes provide support for the fractal model of stratigraphic sequences in chapter 6. It is important to note, however, that the power-law scaling of sea-level changes and sedimentation rates demonstrated to date pertains to first-order trends. There are islands of order in the data and the exponents in the power laws vary – clearly there is much additional information in the data and more work needs to be done.

Scaling laws in sedimentology and stratigraphy are both a challenge and an opportunity. They are challenging because the fundamental concepts lie outside our discipline, are still evolving and are not always mathematically rigorous (see, for instance, the many ways of defining fractals; Falconer, 1990; Hergarten, 2002). On the other hand, scaling laws represent opportunities because they often lead to quantifiable relationships among important variables. Two major benefits may arise from work in this area. First, scaling provides a basis for quantitative prediction of sediment bodies that cannot directly be examined, such as subsurface reservoirs for hydrocarbons or water. Second, scaling laws may reveal fundamental principles governing sedimentation, erosion and formation of the stratigraphic record. Some potential applications of fractals and power-law scaling have been outlined in chapter 6. In biology, the study of scaling laws relating mass of organisms to metabolic rate started in the 1930's (Fig. 9.3). It has led to important insights and triggered considerable discussion on the laws governing life processes (e.g. Calder, 1984; West et al., 2002; West

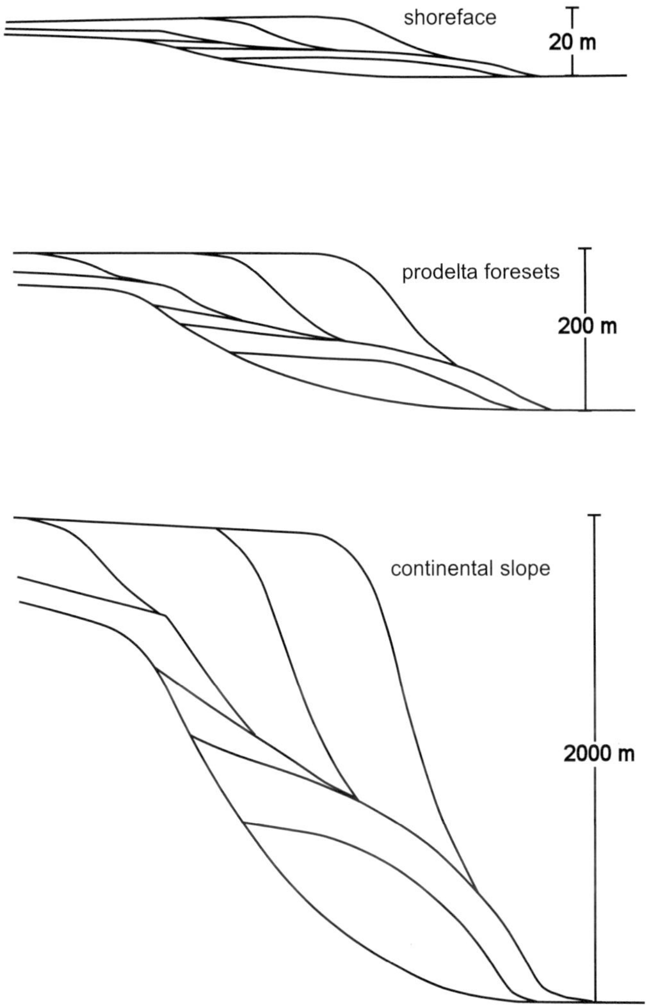

Fig. 9.2.— Prograding clinoforms - another example of scale invariant sediment anatomy, after Thorne (1995), modified. Examples show prograding clinoforms that downlap on older units and differ by three orders of magnitude in height. This range could be extended by at least 2 orders on the fine scale. In the illustration, height was allowed to increase as a qualitative indication of the increasing size. In reality, all spatial dimensions grow at about the same rate.

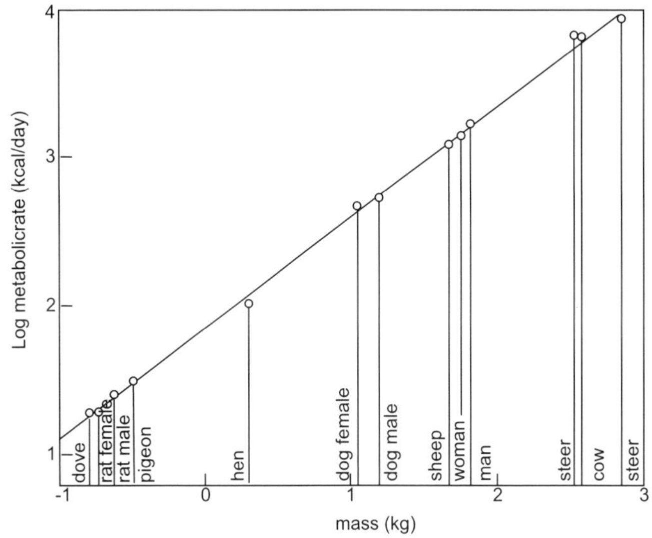

Fig. 9.3.— Relationship of mass and metabolic rate – a classic power law in biology. Display presents data on mammals and birds. Metabolic rate scales with the 3/4 power of mass – a relationship that is now known to extend over 27 orders of magnitude of mass. After West and Brown (2004), modified.

and Brown, 2004).

In sedimentary geology, the study of scaling laws is slowly gaining momentum. The potential of the approach was demonstrated by work on sedimentation rates (Figs 2.19,6.20; Sadler 1981; 1999; Plotnick,1986), depositional patterns (Rankey 2002; Tebbens et al. 2002) and on sea-level fluctuations (Fig. 6.21; Harrison 2002).

We do not yet know how important scaling laws and geometric scale-invariance will turn out to be and to what extent they will improve the prediction of sedimentary systems. I am optimistic because these principles describe properties, albeit statistical ones, that apply to different sediment families, different depositional environments and different geologic times. Thus, insights in one area can be brought to bear on other areas. With this unifying quality the general principles counterbalance the trend towards specialization and strengthen the cohesion of our discipline - just as arches and floors stabilize the columns and pillars of gothic cathedrals and allow them to rise higher. Scientific disciplines, too, need both - vertical pillars and horizontal stabilizers (Fig. 9.4). In sedimentary geology, the drive to push upward into the unknown with the aid of new analytical tools and powerful computers is well developed. This book attempted to illustrate the importance of general principles as cross-connecting stabilizers. They may deserve more attention than they currently receive.

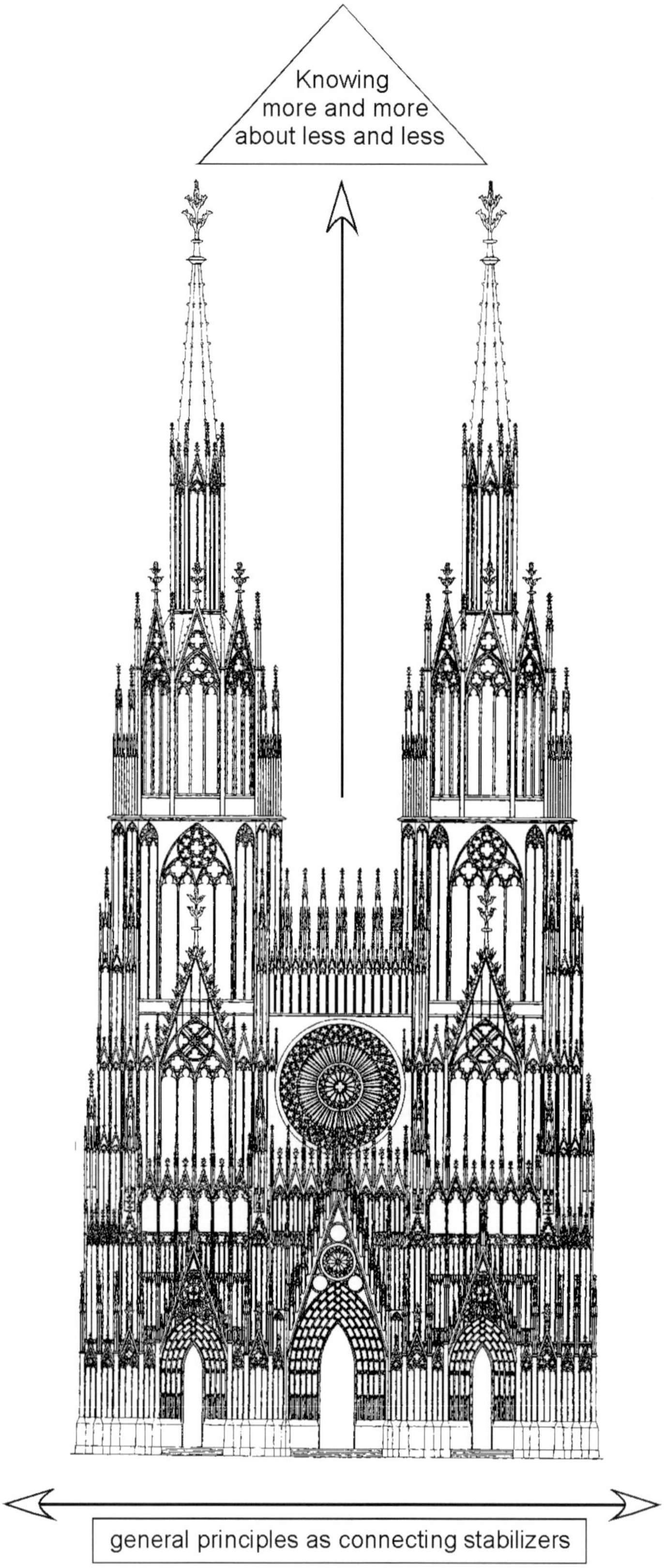

Fig. 9.4.— The gothic cathedral – a model for the internal structure of a geologic discipline? Height and stability of the cathedral derive from the combination of vertical pillars and columns connected by horizontal arches and floors. In science, specialized knowledge in narrow fields can rise to great height if it is stabilized and connected by knowledge of general principles valid across disciplines.

Appendix A

Fractals

The term "fractal" was coined by Mandelbrot (1967) for objects or sets of fractional dimension. For want of a generally accepted definition of fractals, I will characterize them by listing essential properties and examining some classic examples. We will then proceed to methods for determining the fractal dimension and the fractal character of natural phenomena. For more information, the reader is referd to Falconer (1990) for mathematical background, Schroeder (1991) for a discipline-crossing overview of fractals, and to Turcotte (1997) and Hergarten (2002) for overviews directed to earth scientists.

What is a fractal?
A fractal is an object or set of objects that

- is self-similar, i.e. looks the same at all scales,
- has a fine structure,
- has a fractal dimension larger than its topological dimension.

The topological dimension is, for most practical purposes, equal to the Euclidean dimension. Several different formulas are used to calculate the fractal dimension; all of them quantify the fine structure of the fractal, for instance by measuring the degree to which a wiggly curve fills an area, or a series of aligned dots starts to form a one-dimensional line.

For natural scientists, it is important to recognize the difference between abstract fractals and natural fractals. Fig. A.1 shows an example of an abstract fractal conceived in 1904 by the Swedish mathematician H. von Koch. It is created by erecting an equilateral triangle over the middle third of a straight line segment and repeating this operation with smaller and smaller line segments to infinity. The result is a wiggly curve that looks the same at all scales. This means that any part of the curve properly enlarged looks like the whole. The curve is self-similar because of its fine structure and has a fractal dimension between 1 and 2, between the Euclidean dimension of a line and that of an area. We may say that the curve is so wiggly that it starts to fill part of an area. The topological dimension of the curve is one. Topology is also known as "rubber-sheet geometry". If the Koch curve in Fig. A.1 were made of a rubber string then it would be possible to change it by continuous transformation without any cutting or pasting into a straight line or any other smooth curve.

Fig. A.2 shows another mathematical fractal – the Cantor set, also called Cantor bar, conceived by the German mathe-

Fig. A.1.— The Koch curve – a geometric fractal. It is generated by erecting an equilateral triangle over the middle third of a line segment and repeating the operation at smaller and smaller segments to infinity. After Falconer (1990), modified.

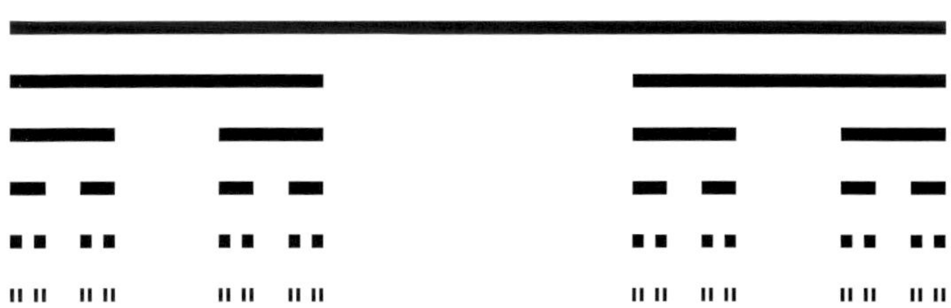

Fig. A.2.— The Cantor bar (or Cantor set) – a mathematical fractal well known in stratigraphy. The initiating step is a line segment. In the generating step, the middle third of the line is removed and this operation is repeated to infinity. After Schroeder (1991), modified.

matician Georg Cantor in the nineteenth century. The fractal is constructed by erasing the middle third of a line segment and again, as with the Koch curve, repeating the initial operation an infinite number of times. The result is a fractal set with a dimension between zero and one, the Euclidean dimensions of a point and a line respectively. The topological dimension of the Cantor set would be zero.

In Fig. A.3 we see a natural object with fractal characteristics – the rugged coastline of Norway formed by the intersection of erosional topography and the sea surface. Like the Koch curve, the coastline is a very wiggly curve with a fine structure in a wide range of scales. We will see below that its fractal dimension is between 1 and 2, again similar to the Koch curve. Two important differences between this natural object and the Koch curve should be noted: First, the coastline is a fractal only in a statistical sense. Magnifying part of the coast to the scale of the entire feature will produce a curve that resembles the entire coast in a statistical sense but does not provide an exact match. Second, the coastline has fractal characteristics in a certain scale range, the fractal trend breaks down at the scale of individual crystals or sediment grains. The terms "infinity" and "at all scales" are inappropriate in a description of natural fractals.

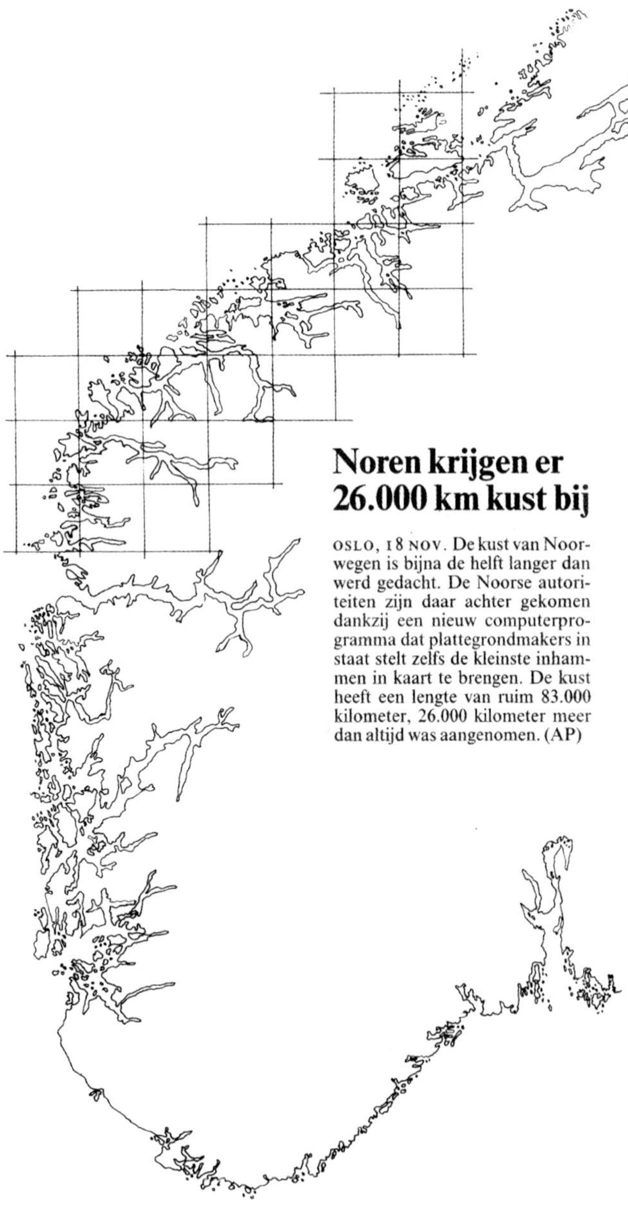

Fig. A.3.— Coastlines were among the first natural phenomena whose fractal nature was recognized. Here we see part of the coast of Norway. The clipping from a Dutch newspaper reports that the Norwegian coast has become 26,000 km longer because it was surveyed in finer detail. Based on Feder (1988).

The Koch curve illustrates an important property of fractal curves: their length depends on the length of the measuring rod. Because of the fine structure of the curve, it gets longer as the length of the rod is reduced and finer wiggles can be taken into account. This crucial relationship is expressed in the formula

$$N = 1/r^D \qquad (1)$$

where N is the number of times one has to put down the measuring rod of length r to step along the curve and D is the fractal dimension, a measure of the wiggliness of the curve. Equation (1) represents a power law that yields a straight line in plots of $logN$ vs. $logr$ with the slope of the regression line equal to D.

Equation (1) yields the fractal dimension by re-arranging terms

$$r^D = 1/N$$

taking logarithms

$$D \cdot logr = -logN \qquad (2)$$

and substituting the values for the first step (the generating step) of the Koch curve in Fig. A.1. If we assume that the line segment at the top has length 1, then the values for the generating step $r = 1/3$, and $N = 4$. This yields

$$D_{koch} = -(log4/log1/3) = log4/log3 \approx 1.26$$

The result satisfies the earlier statement that the fractal dimension of a fractal curve is larger than its topological dimension which is 1 in this case.

Equation (1) can also be used to calculate D for the Cantor bar. If we again assume that the line segment at the top has length 1, then the generating step in Fig. A.2 yields $r = 1/3$ and $N = 2$.

$$D_{cantor} = -(log2/log1/3) = log2/log3 \approx 0.63$$

Again, the fractal dimension of the Cantor bar is larger than its topological dimension, which is 0.

Fractal dimensions of natural curves are usually not determined by stepping along them with measuring rods of different sizes but by box counting. Fig. A.4 illustrates this widely applied technique. Curves with fractal characteristics yield straight-line relationships between N and r with the slope of the regression line equal to the fractal dimension. The box-counting plots of natural fractals show that the domain of the straight line is limited and does not extend to infinity as is the case with mathematical fractals. In most instances, the power-law domain is bracketed for large r's by finite-size effects and for small r's by the resolution of the data. However, there also may be limits that indicate changes in the natural system.

Tests of fractality can also be performed on time series and waves. Fig. A.5 summarizes the approach. The data are first transformed from the time domain to the frequency domain by a Fourier transform. In the frequency domain, the time series appears as a plot of frequency versus power. Power is equal to the absolute value of the square of the amplitude shown in the time plot. The power spectrum in the frequency domain is then expressed as a log-log plot of power vs. frequency. Fractal time series show a straight-line trend, i.e. a power-law relationship between spectral power and frequency. The slope of the trend is approximately equal to a term called spectral index, denoted β. The spectral index is related to the fractal dimension, D, of the geometric fractal discussed above.

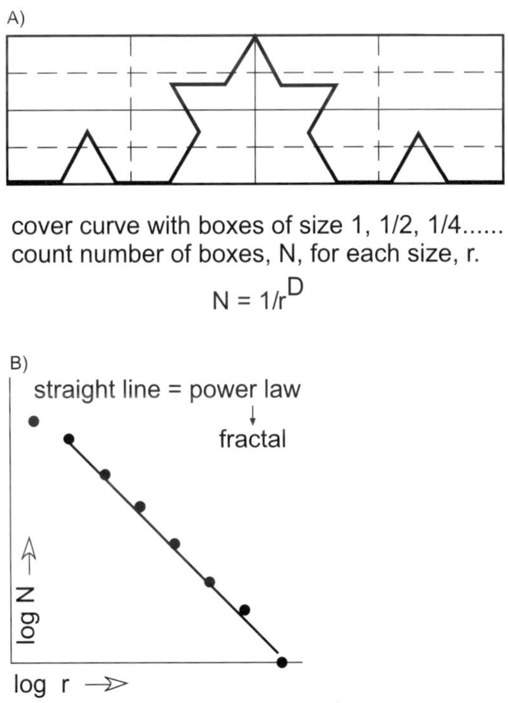

Fig. A.4.— The box-counting technique applied to an early stage of the Koch curve. A) In the first step of box counting, a rectangular box is drawn to snugly cover the feature. Next, the side length of the box is cut in half and the number of boxes needed to cover the feature is counted. The procedure is repeated with smaller and smalle boxes. B) A log-log plot of box size and number of boxes provides a test of the fractal nature of the curve. Data from fractals plot on a straight line because of the power law in equation (1). The slope of the curve is equal to the fractal dimension, D.

Rescaled-range analysis is another method for recognizing fractals in time series. The technique determines the range, R, of the values of a variable, and the (estimated) standard deviation, S, of the values. The crucial graph shows $logR/S$ versus $logN$, with N being the interval in the time series. Again, a straight-line relationship of $logR/S$ vs. $logN$ indicates a power law and fractal characteristics of the time series. Many natural time series have been found to obey the relationship

$$R_N/S_N = (N/2)^{Hu}$$

where Hu is called the Hurst exponent. For a certain range of conditions, the Hurst exponent, too, can be related to the fractal dimension, D.

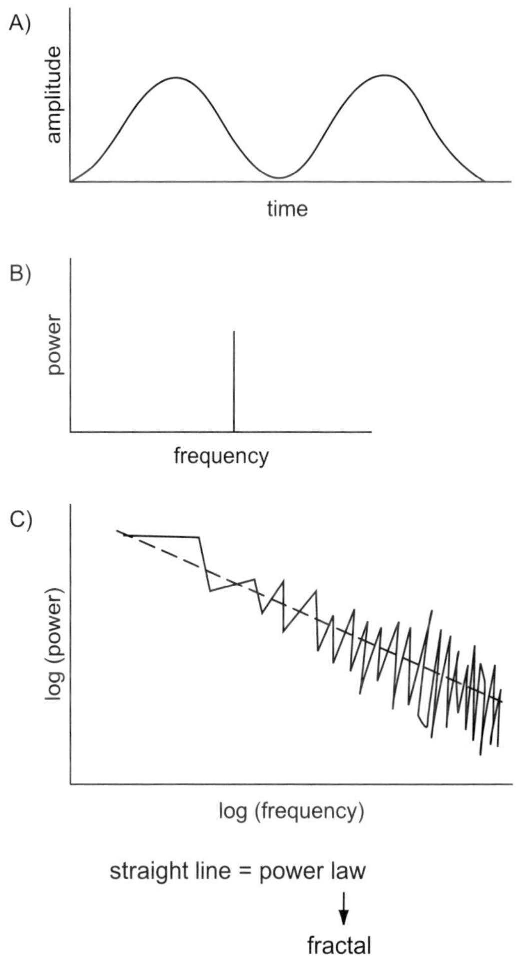

Fig. A.5.— Testing for fractal characteristics in time series or wave phenomena. Data that exist in the time domain A) are Fourier-transformed to the frequency domain B); the periodic sine wave in A) appears then as a single bar. Natural patterns consisting of a range of frequencies yield a broad power spectrum C). Fractal time series (or continuous wave patterns) show a straight-line trend in a bilogarithmic plot of power and frequency. The domain of the power-law relation is limited at low frequencies by limited data and at high frequencies by the resolution of the measured time series.

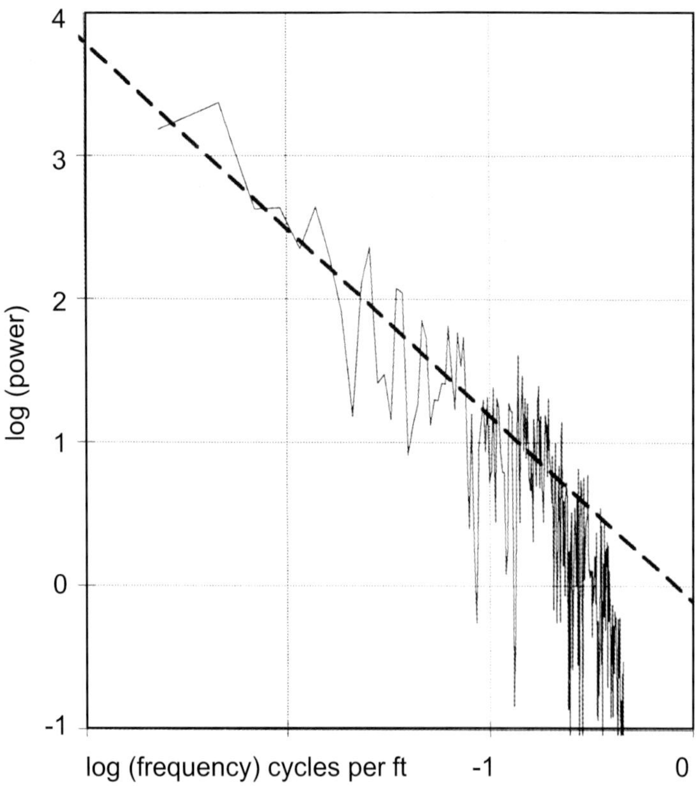

Fig. A.6.— Power spectrum of a gamma ray log. The log measures the natural gamma radiation of the rock that is largely a function of the rock's clay content. The variation of gamma radiation therefore reflects the variation of clay content in the stratigraphic column and usually correlates with depositional rhythms. First-order trend (dashed) indicates power-law relation between power and frequency, suggesting fractal properties of the underlying depositional rhythms. Linear trend breaks down on the right near the limit of resolution of the logging tool. After Tubman and Crane (1995), modified.

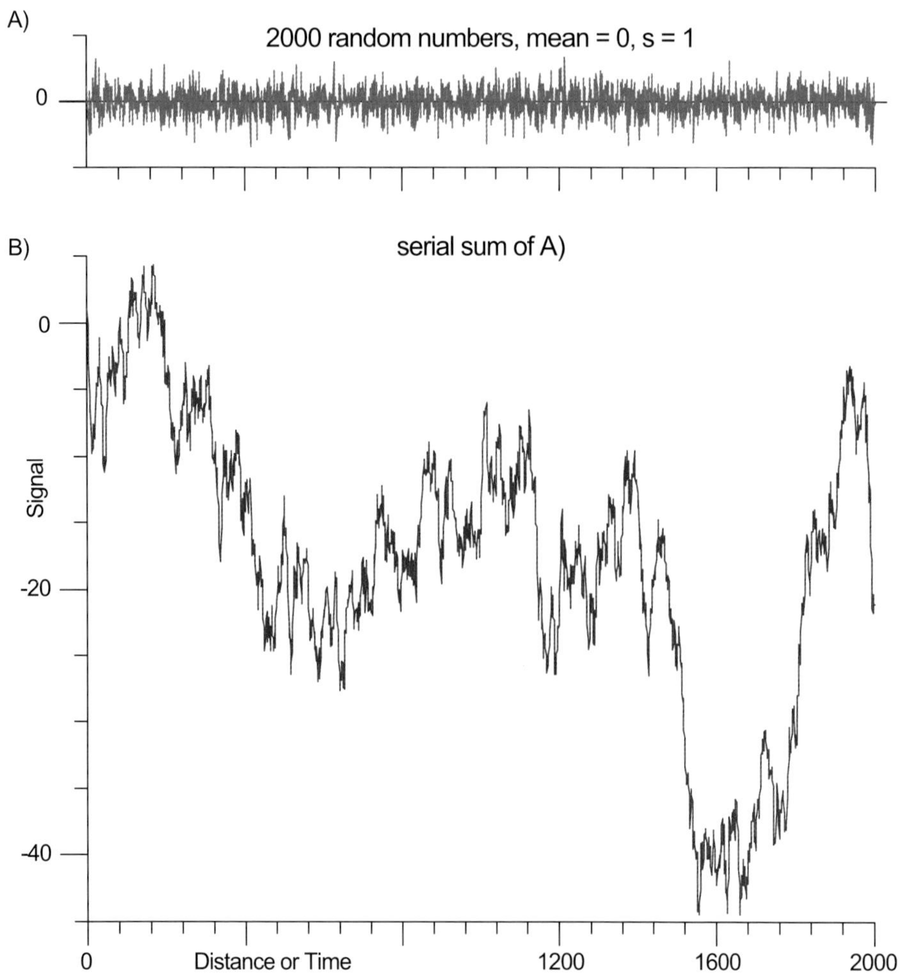

Fig. A.7.— A) Graph of random numbers with the characteristics of white noise. B) Random walk generated by the running sum of the numbers of A). There is a distinct impression of a hierarchy of cycles in B) but the pattern is different for each set of random numbers and reflects no ordering principle. After Harrison (2002).

Appendix B

Introduction to modeling programs STRATA and CARBONATE 3D

In the past three decades, numerical modeling of depositional systems has steadily gained momentum. It has become a versatile tool that forms a bridge between direct observation of natural systems and their theoretical analysis based on principles of physics, chemistry and mathematics. A particularly important role in geology is that numerical models offer a quantitative framework for our free-wheeling thoughts.

In this book, modeling runs have been shown to illustrate the effect of a principle or to compare natural examples with predictions from first principles. All model runs were made with two programs whose structure and theoretical underpinnings are described below. Only the most essential elements of the programs are discussed. Complete descriptions can be found in the quoted publications.

The philosophy of STRATA, CARBONATE 3D and most other sedimentologic modeling programs is to capture long-term trends. Large sediment accumulations were formed by many individual events of sedimentation and erosion. Rather than attempting to model individual events, most programs rely on statistical laws for sediment input and dispersal that provide good approximations of the cumulative effect of the individual events.

PROGRAM STRATA

Overview

STRATA is a 2D modelling program for siliciclastics and carbonates, developed for large-scale basin studies (Flemings and Grotzinger, 1996). Elegant simplicity and concentration on few fundamental principles is a hallmark of the program.

The program produces cross sections. Siliciclastic material is supplied to the system from external sources on the side, carbonate material is produced within the system in designated areas. Distribution of siliciclastics and carbonates in the system, i.e. deposition and erosion, are governed by the diffusion equation discussed below. Subsidence and sea-level changes can be prescribed and the output can be viewed in spatial cross sections as well as time-distance plots ("Wheeler diagrams") with time on the vertical axis.

The program can calculate first-order grain-size distributions based in the decrease of grain size with water depth in a beach-to-shelf transect (Fig. 4.1). In addition, the program models a number of processes and parameters not used here, e.g. lithospheric flexure under sediment load, sediment compaction during burial, physical properties and synthetic seismic traces.

Diffusion equation for sediment dispersal

STRATA assumes that sedimentation and erosion at the Earth's surface are governed by the diffusion equation, one of the fundamental equations in geophysics (e.g. Turcotte, 1997, p.202). Its application to sedimentation and erosion is intuitively plausible and can be justified as follows. If one watches how sediment is transported in rivers or, during sheet floods, over the land surface, one gets the distinct impression that transport is faster on steeper slopes. Measurements confirm this qualitative insight. In first approximation, sediment flux is proportional to slope. In mathematical terms

$$q = -\partial h/\partial x$$

where q is the sediment flux (expressed as sediment volume per unit time), h is the elevation, x the horizontal distance along the x axis, and k is a coefficient of proportionality, called diffusion constant in STRATA. The coefficient is given a negative sign because material moves from high to low values of h. The dimension of the diffusion coefficient in STRATA is m^2/y because the program models in two dimensions. In real-world experiments the coefficient would be m^3/y. The partial derivative of height with respect to horizontal distance, $\partial h/\partial x$, is the tangent of the slope angle, or simply "the slope". If one assumes that no material is dissolved or precipitated during transport, then our system also satisfies the condition that sediment volume is conserved during erosion and deposition. In quantitative terms

$$\partial h/\partial t = -(\partial q/\partial x).$$

This equation states that the rate of change of elevation at a given position is equal to the rate at which material is added to or removed from this position. Combining the two equations yields the diffusion equation

$$\partial h/\partial t = -k \cdot \partial^2 h/\partial x^2.$$

The name "diffusion equation" refers to its original use in calculating the mixing of solutions at the molecular level. In sedimentology and geomorphology, the equation provides good approximations for the cumulative effect of many separate events of sedimentation and erosion. For instance, the transfer of siliciclastic sediment from a terrestrial source across the coastal zone into an ocean basin is well described by the diffusion equation.

The diffusion equation is the centerpiece of the STRATA program. The equation states that the time rate of change in height at a given point is proportional to the second derivative of the height with respect to the horizontal distance. It is written as a partial derivative because the second horizontal dimension, y, is not considered. Fig. B.1 shows a conceptual diagram illustrating the meaning of the first and second derivatives of a topographic profile. The first derivative of height versus horizontal distance, $\partial h/\partial x$, is the slope angle. The second derivative, $\partial^2 h/\partial^2 x$, , is the rate of change in slope angle along the x-axis (Fig. B.1C). Large values of $\partial^2 h/\partial^2 x$ indicate that the surface is strongly curved in the vicinity of this point, low values indicate that it is nearly planar (even though it may be dipping at a high angle). The diffusion equation states that sedimentation (or erosion, i.e. negative sedimention) are most intensive, and morphologic changes most rapid where the Earth's surface is most strongly curved. In a landscape that evolves according to the diffusion equation sharp edges are being eroded and concave depressions filled by the erosion products.

Adjustments to the diffusion equation can be made by changing the diffusion coefficient, k. STRATA allows one to choose separate coefficients for the marine and non-marine environment. Non-marine coefficients typically are 30,000–60,000 m^2/y, marine cofficients 100 – 300 m^2/y. This is plausible. Transport by running water on land is more efficient because it is fast and always goes in the same direction, namely downslope. In the sea, currents are slower and may reverse direction. The program assumes that the transition between the non-marine and marine diffusion domains occurs in the uppermost part of the sea. In this transition zone, the diffusion coefficient changes according to the formula

$$k = k_{marine} + (k_{nonmarine} - k_{marine}) \cdot e^{-\lambda w}$$

where w is water depth and λ a decay constant that determines how quickly the effect of the non-marine diffusion coefficient dies out with increasing water depth.

The diffusion equation is a versatile tool for modeling sedimentation and erosion. However, it also has its shortcomings. Three of them are listed below.

1. The diffusion equation describes only the cumulative effect of many sedimentation events and tells us very little about the processes that are active during the individual event. For instance, the diffusion model describes the cumulative downslope sediment movement by such diverse processes as slumping, turbidity currents and bioturbation under the influence of gravity.

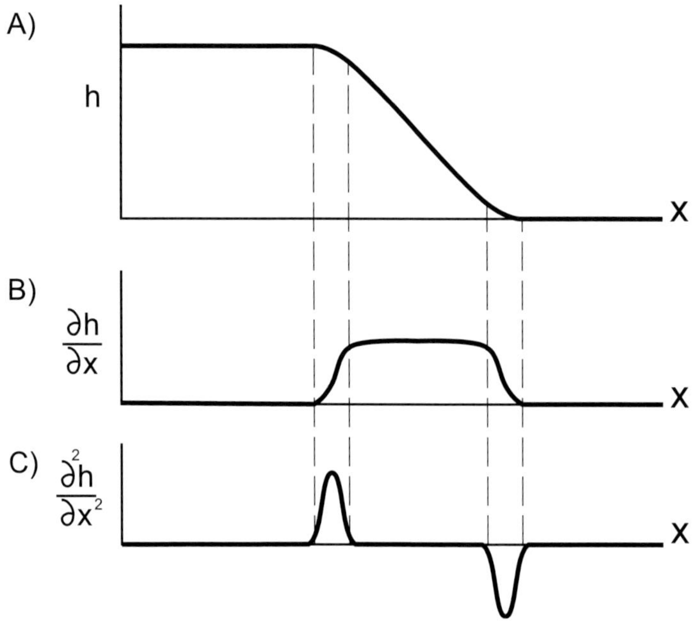

Fig. B.1.— Conceptual diagram showing (A) cross section of topography; (B) the change in slope angle, i.e. the first derivative of topography; and (C) slope curvature, i.e. the second derivative of topography. According to the diffusion equation, maximum erosion occurs at the locations of maximum convex slope curvature, the maxima of the second derivative. Conversely, maximum deposition is predicted at the locations of maximum concave curvature, the minima of the second derivative.

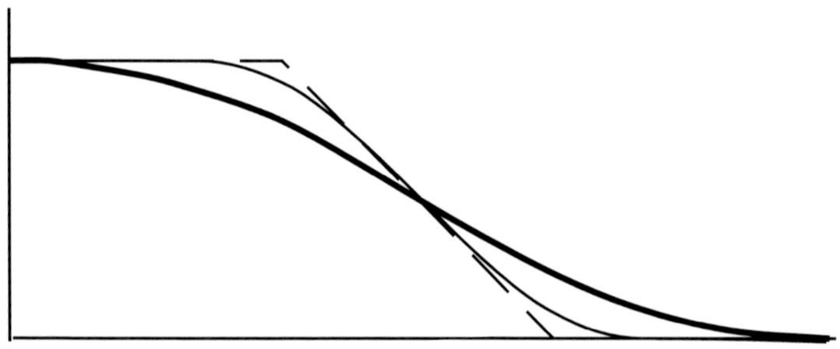

Fig. B.2.— Landscape evolution according to the diffusion equation. Dashed: original profile. Sharp edges are worn down, concave parts filled in.

2. There are certain, common situations that are difficult to model with the diffusion equation. Abyssal plains are a case in point. They represent the flattest parts of the surface of the solid Earth and are essentially composed of turbidites with intercalations of pelagic sediment. Individual turbidite layers have been shown to cover over 50,000 km^2 of the surface of abyssal plains. The turbidity currents depositing such beds are highly efficient transport agents; they are probably propelled by the momentum of the large volumes of water and sediment that are entrained in the flow and are being

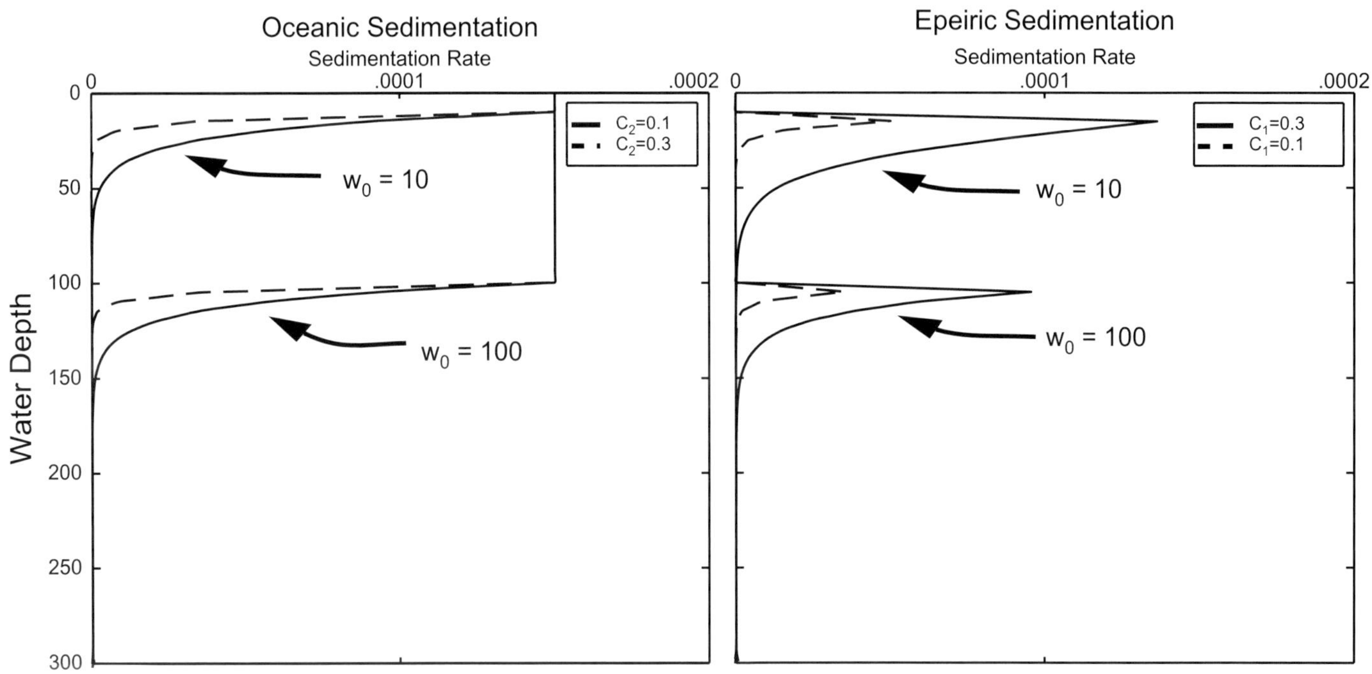

Fig. B.3.— Carbonate production rate vs. depth for epeiric settings (left) and oceanic settings (right). After Flemings and Grotzinger (1996), modified.

accelerated to velocities of over 50 km/h during the descent on the continental slope (Allen, 1985, p.238). Modeling such situations with a diffusion equation requires an abrupt switch to very high diffusion coefficients at a depth where the models assume low values.

3. Another problem area for the diffusion approach are the margins of carbonate platforms. They have sharply convex curvatures and thus should be the site of intensive erosion. Instead, we observe rapid upward growth of reefs or sand shoals in this setting. The reason is that organic construction and intensive abiotic precipitation of carbonate cement override the effects of erosion. In addition, organic binding and marine cementation harden the sediment thus slowing down erosion.

Carbonate production functions

In carbonate environments, the dispersal of material can also be modeled by the diffusion equation. In contrast to siliciclastics, however, the material for sedimentation, is not provided by an eroding hinterland but produced within a designated area by a carbonate production function. STRATA considers essentially tropical systems (T factory in our terminology) and offers two production functions depending on the setting – oceanic and epeiric (Fig. B.3.

In oceanic carbonate settings, the basic control on production is light (see Figs 2.3, 2.4). STRATA approximates the tropical production function by combining a zone of constant maximum production with a zone of exponential production decrease below. The formula is

$$dS/dt = c_1 \cdot exp[-c_2(w - w_0)]$$

where dS/dt is the production of sediment per unit time, c_1 is the maximum production rate (occurring in the uppermost part of the water column) , c_2 is the constant of the exponential decrease, w_0 is the depth to which the maximum production rate extends and w is the water depth at the point under consideration. The formula above holds for the range $w > w_0$; for $w < w_0$, sediment production $dS/dt = c_1$.

In epeiric settings, light is again the dominant control. However, in contrast to oceanic settings, restriction, clastic influx and other nearshore effects are assumed to prevent maximum production close to shore. Consequently, the production maximum lies offshore at some modest water depth. Production rate, dS/dt, is given by

$$dS/dt = c_1 \cdot (w/w_0) \cdot exp[1 - w/w_0]$$

where c_1 and w_0 are the rate and depth of maximum production respectively, and w is again the water depth under consideration.

Oceanic and epeiric production functions may be combined by specifying their respective domains in the model.

CARBONATE 3D

Overview

The program was mainly designed for creating detailed models of outcrops or specific objects in the subsurface, such as hydrocarbon-bearing carbonate rocks and their surroundings. Consequently, CARBONATE 3D differs significantly from STRATA whose main objective is to model first-order trends in the anatomy of sedimentary basin fills.

The program is genuinely three-dimensional. Space is divided into cells and calculations regulate the transfer of mass and energy between each cell and its neighbors.

Carbonate production

The program is very flexible in modeling carbonate production. To start with, three production types are distinguished: shallow, open-marine; shallow, restricted marine; and pelagic production. For each type, an optimal production rate can be specified and then reduced as desirable, depending on local environmental stress.

1) The shallow, open-marine production type has a growth function similar to the T factory (chapter 2). Important stress can be generated by water depth, high rates of sedimentation and high sediment load in the water. Linear increase of any of these factors leads to exponential decrease of production. The exponential decay with depth starts below a zone of optimal production (Fig. B.4). The function is very similar to the open-ocean production function of STRATA. Stress by restriction is modeled as progressive replacement of the open-marine production type by the restricted-marine type as a function of water depth and distance from the shelf break (Fig. B.5).

2) Restricted marine production is perceived as production by predominantly algae that can tolerate restriction but require protection from high water energy. Consequently, water energy is modeled as a stress on this biota. Water energy, in turn, is assumed to decrease with distance from the platform margin. The substitution of open-marine by restricted production and vice versa is shown in Fig. B.5. This approach is rather unusual and the assumed co-variance of restriction and protection in depositional environments is at variance with observations at extant platforms (e.g. Fig. 2.12). From a computational point of view, however, the most important thing is that the program makes it possible to change total production, i.e. the sum of open-marine and restricted production, as a function of restriction.

3) Pelagic production is the material of carbonate plankton that lives in the top few hundred meters of the water column. The program models plankton production as a function of the thickness of the water column (Fig. B.4). Plankton production is zero at sea level (i.e. a water column of zero) and increases exponentially with depth, asymptotically approaching the optimal pelagic production rate. The rationale is that in environments with very shallow sea floors the shallow-water (benthic) production overshadows planktonic production. The balance shifts as benthic growth slows down with increasing depth and decreasing light.

Production rate is calculated separately for each production type. The shallow, open-marine type may serve as an example. Production rate as a function of the horizontal coordinates, $P(x, y)$, is expressed as

$$P(x, y) = M \cdot S(x, y)$$

where M is the optimal production rate and $S(x, y)$ the stress function. Stress is considered a function of the stress factors

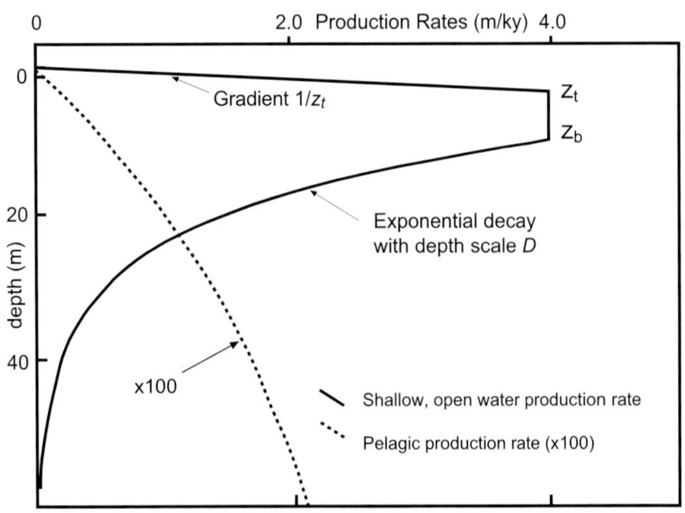

Fig. B.4.— Typical production function of the shallow-water, open-marine type (bold), and the pelagic type (stippled). Production expressed as meters of vertical accumulation per ky. Optimal open-marine production is limited at the upper end by depth zt above which wave energy reduces production. Depth zb at the lower end of the optimal zone marks the onset of exponential decay of production caused by decrease of light. After Warrlich et al. (2002) , modified.

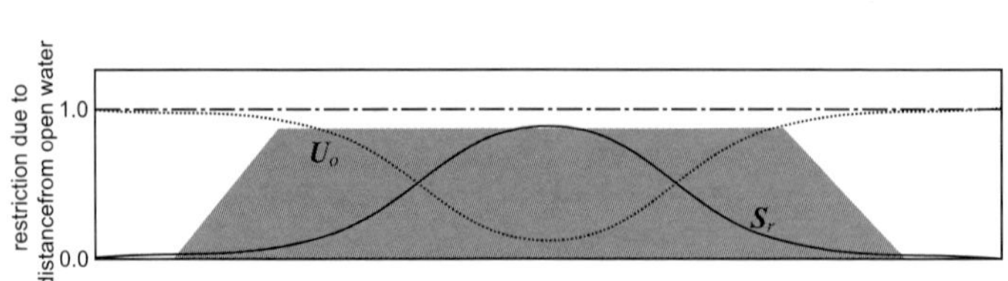

Fig. B.5.— Modeling the effects of restriction on an atoll (shaded). Open-marine production is proportional to U_0 and therefore decreases towards the center of the atoll because of stress related to restriction. On the other hand, S_r, the function that controls production by the restricted community, increases towards the center because the decrease of water energy reduces stress on the restricted community. Both curves are calculated from the smoothed depth profile of the atoll. Vertical axis shows the fraction of optimal production of open-marine and restricted type respectively. Total production is the sum of open-marine and restricted production. As optimal open-marine production significantly exceeds optimal restricted production, total production across the atoll is highest at the margins and lowest at the center. After Warrlich et al. (2002), modified.

water depth, z, restriction, U_0, and sediment load, L. Assuming that the various stress factors act independently, the stress function can be written as

$$S(z, L, U_0) = D_0(z) \cdot U_0(x, y) \cdot L_0(L)$$

with D_0, U_0 and L_0 representing the stresses by water depth, restriction and sediment load respectively. The functions D_0 and U_0 have been discussed under items 1 and 2 above. L_0 is explained in the context of sediment dispersal. It should be noted that $S = 0$ indicates maximum stress, whereas $S = 1$ indicates no stress and maximum production.

Sediment dispersal

Sediment is dispersed by separately modeling sediment generation (volume and proportion of sand and mud), entrainment, transport and deposition. For the last three steps, the total shear stress acting on the grains, τ_{total} , is calculated as the sum of stresses from currents, waves and topographic slope. The sediment load is entrained if τ_{total} exceeds the critical value for entrainment; it is moved in the shear-stress vector field until τ_{total} falls below a critical value for deposition. Fig. B.6 shows an example of a stress vector field and sediment transport paths over a platform and an offshore high. Transport paths of coarse sediment are strongly influenced by slope gradient, transport paths of fine material are mainly governed by the direction of currents.

Once deposition commences, its rate is governed by two conditions. (1) Sedimentation rate, ds/dt, is proportional to sediment load, L,

$$ds/dt = kL$$

where k is the constant of proportionality. If k is large, sedimentation is rapid, the load is reduced quickly, and transport distances are short. The program defines a characteristic transport distance,X, whereby $X = 1/k$. X is small for sand and large for mud. (2) The continuity equation holds, i.e. sediment mass is conserved in all steps of deposition, erosion and sediment transport. This can be stated as

$$ds/dt = -dL/dS.$$

The equation implies that the sedimentation rate is equal to the change in sediment load along the path of sediment transport, S. (Note that S in the above equation is unrelated to the stress function S introduced in the section on sediment production).

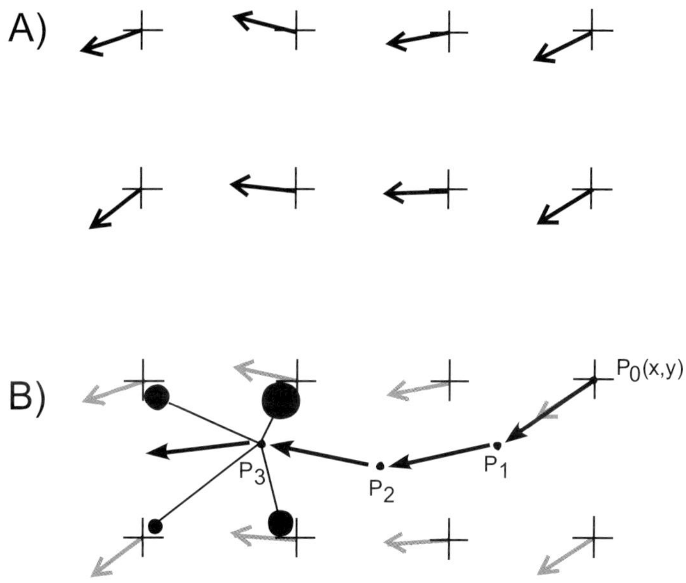

Fig. B.6.— Scheme of sediment entrainment, transport and deposition in CARBONATE 3D. A) Surface of model is gridded and sediment is entrained and deposited only at grid nodes but may be transported anywhere. Three cells are shown with their nodes marked by crosses. Arrows represent sediment transport vectors of unit length. Transport vectors are calculated from shear stress caused by waves and topographic slope. B) Sediment transport and deposition in vector field of A. Sediment is entrained at upper right node, i.e. at point P_0 (x,y). The sediment is moved through points P_1 and P_2 without deposition because shear stress is sufficiently high. Direction and velocity of sediment movement at P_1 and P_2 are interpolated from the stress vectors at the four surrounding nodes. At P_3 shear stress has fallen below the critical level and deposition ensues. The sediment is partitioned among the surrounding nodes such that the load received by each node (black circles) is inversely proportional to the distance of the node from P3. Based on G. Warrlich (written communication).

Additional options

CARBONATE 3D includes a number of options not discussed here. Siliciclastic and carbonate material can be input at the margins of the modeling space. Erosion and grain diminution of material can be calculated from sediment properties and the current vector field. Furthermore, certain facies attributes, such as sand/mud ratios, can be calculated.

APPENDIX C

Principles of reflection seismology

Reflection seismology has become one of the most important disciplines for sedimentologists. It is a highly sophisticated technique whose data may be superior to the data from the best outcrops in certain respects. In addition, the seismic tool has "seen" much more of the Earth's sediment mass than the eyes of field geologists and core loggers together. Sequence stratigraphy, in particular, relies heavily on seismic data. There is one catch, however: unlike photographs, the evidence in seismic data is not immediately obvious. The data result from very delicate measurements and subsequent processing and some basic understanding of the seismic reflection process is essential for geologists dealing with seismic data. This is the reason for the brief summary of principles and peculiarities of reflection seismics given below. (See Sheriff and Geldart, 1995, for in-depth treatment of the subject).

ORIGIN OF SEISMIC REFLECTIONS

The basis of reflection seismology are man-made elastic waves that travel in the subsurface and get reflected or refracted at material boundaries. The relevant rock properties governing the behavior of these waves are rock density, ρ, and sonic velocity, V, that represents the speed at which seismic waves travel in the rock. Reflection seismology is particularly well suited to image sedimentary rocks because almost without exception, ρ and V change very gradually parallel to bedding but abruptly perpendicular to bedding, i.e. at bedding planes. Consequently, the bedding patterns are particularly well imaged by the seismic technique. Fig. C.1 shows the behavior of seismic waves at a material boundary, such as a bedding plane. The incident wave is split into a reflected part that travels upward and a refracted part that continues to propagate downward albeit in a slightly different direction. Part of the refracted wave may be reflected at a deeper material boundary. In this way, many superimposed layers can be "seen" by the seismic tool. For seismic waves, bedded sedimentary rocks are like many layers of tinted glass - they reflect enough energy to be visible from above but they also let sufficient energy pass to illuminate deeper layers. The amplitude of the reflected wave, a measure of the "reflectivity" of the boundary, is called the reflection coefficient, RC. It is given by

$$RC = (\rho_2 V_2 - \rho_1 V_1)/(\rho_2 V_2 + \rho_1 V_1)$$

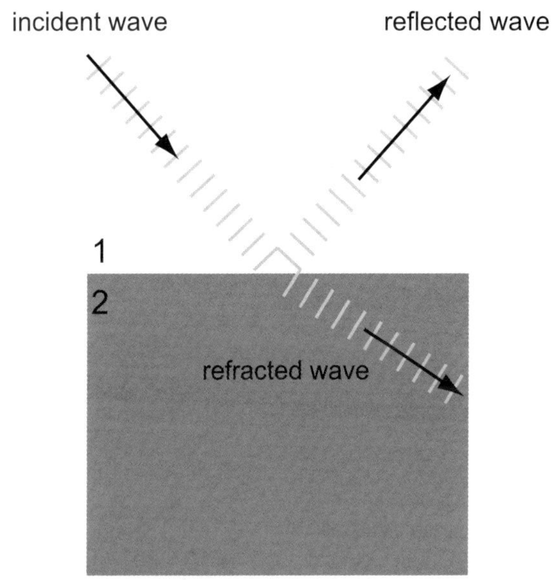

Fig. C.1.— Seismic waves, indicated as wave fronts, at the lithologic boundary of layer 1 and layer 2 (shaded). Downgoing wave is split into a refracted part that continues downward, and a reflected part that travels back to the surface. Partitioning of energy between reflected and refracted waves depends on the difference of acoustic impedance between the two layers and is calculated by the reflection coefficient, RC. Big impedance differences at the boundary result in a large fraction of reflected energy – the layer is a good "mirror". If the impedance difference is zero, the boundary is seismically invisible.

The products ρV represent the acoustic impedance of the layers 1 and 2 respectively. Seismic surveys usually proceed by making "bangs" at the surface and listening to the echo with geophones. The process is repeated at equally spaced locations along surface profiles and produces a seismic cross section of the subsurface. Such seismic sections resemble geologic cross sections in many ways but two peculiarities set them apart.

1) Reflections recorded at a shot point are assumed to have originated vertically below this shot point and are plotted there. This assumption may be considerably wrong for dipping reflectors (Fig. C.2), for reflecting point sources, such as the abrupt end of a reflector (Fig. C.3) and for reflections that originated not in the plane of section but

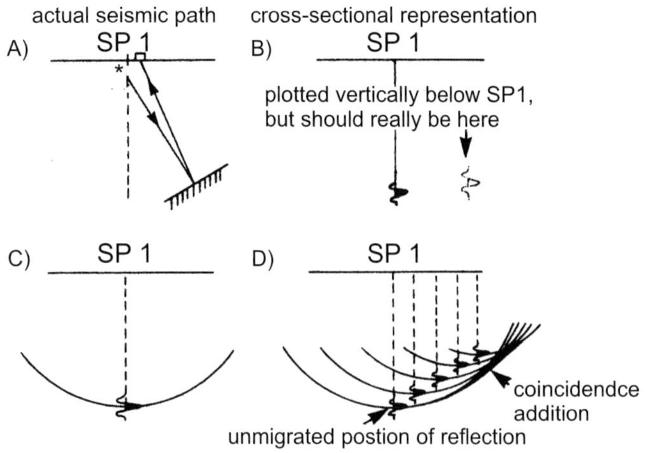

Fig. C.2.— In constructing a seismic section, the reflections at first are plotted vertically below the shot point. This is correct for horizontal reflectors but incorrect for the dipping reflector shown here. The error is corrected by the numerical process of "migration", illustrated schematically in Figs. C – D. C) shows all the possible sources that might have produced the reflection observed at shot point 1 – they lie approximately on a semi-circle around the shot point. D) shows this process repeated for subsequent shot points. The circles coincide in a zone on the right. This zone of coincidence of circles approximately represents the true position of the reflector. After Anstey (1982), modified.

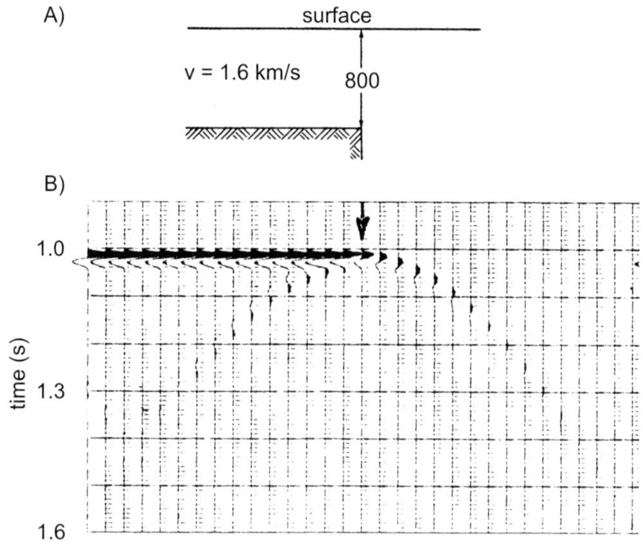

Fig. C.3.— Sharp edges or reflecting points in the subsurface diffract rather than reflect and refract seismic waves. The seismic record of these point sources consists of reflections that describe one branch of a hyperbola with the true position of the reflector at the apex of the hyperbola. Migration removes diffraction hyperbolas and preserves only the reflection at the apex. After Trorey (1970).

sideways of it ("side echoes").
2) vertical scale is travel time, not depth. Conversion to depth is not straightforward because usually the sonic velocity changes from layer to layer.

The peculiarities of raw seismic data can be removed by data processing. Peculiarity 1 can be eliminated by "migrating" the reflections in the time domain. The principle of time-migrating reflections from dipping layers is illustrated in Fig. C.2. Each diffraction hyperbola is collapsed into its apex as the place where the physical reflection is generated (Fig. C.3). Correction of side echoes requires tracing reflections in a grid of lines. To eliminate peculiarity 2 one needs to calculate the interval velocities of the various layers (Fig. C.4) and use them for "depth migration" of the reflections. Time migration and depth migration improve significantly if the survey grid gets denser. Migration is particularly successful in "3D seismic data" where the spacing of the survey lines is almost as close as the spacing of shot points on the lines. In this way, the entire rock is insonified and powerful computers can display and process the data as a 3D volume of data.

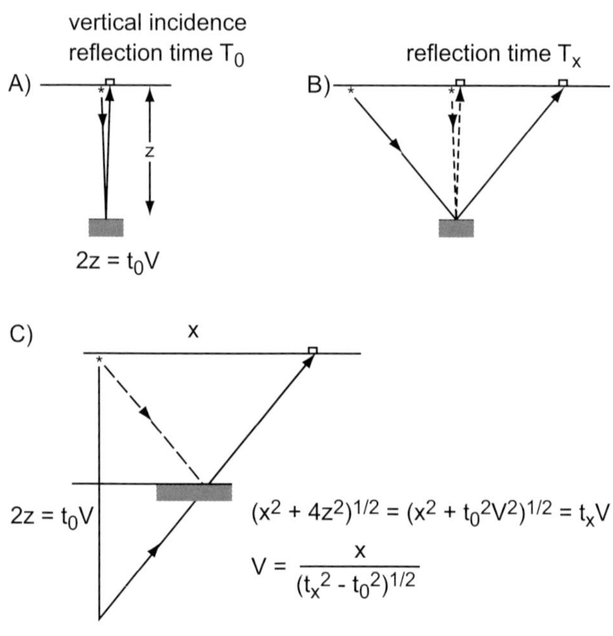

Fig. C.4.— The velocity in a layer between the surface and a subsurface reflection, the interval velocity, V, can be determined by comparing seismic travel times to a reflector along different ray paths. In simple cases, two measurements (shown in A and B) and the theorem of Pythagoras (shown in C) suffice to calculate V. After Anstey (1982), modified.

SYNTHETIC SEISMIC TRACES AND SEISMIC MODELS OF OUTCROPS

The oil industry routinely measures the seismically relevant rock properties in boreholes and these boreholes then serve as points of "ground truth" for seismic interpretation.

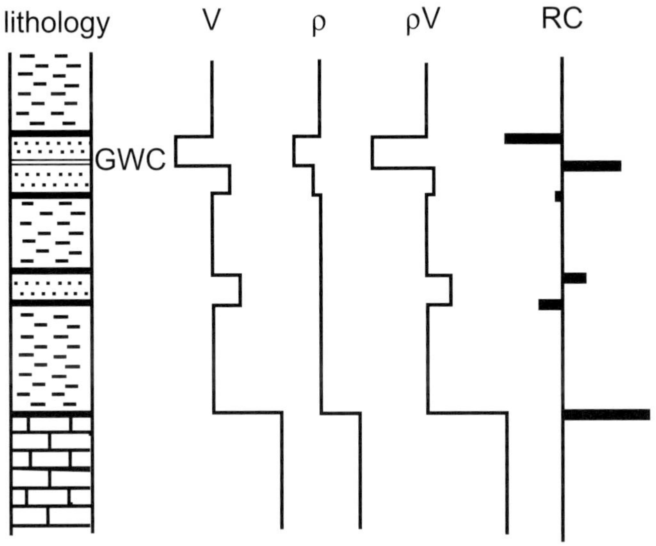

Fig. C.5.— Derivation of the reflection coefficient from the rock column in a borehole. From left to right: Lithologic column consisting of layers of shale, sandstone and limestone and a gas-water contact (GWC) in the uppermost sandstone; log of sonic velocity, V; log of rock density, ρ, and acoustic impedance, ρV; finally, on far right, the reflection coefficients (RC) calculated from the changes in impedance. The reflection coefficient is positive if a hard layer is overlain by a soft layer and negative if a soft layer is overlain by a hard one. After Anstey (1982), modified.

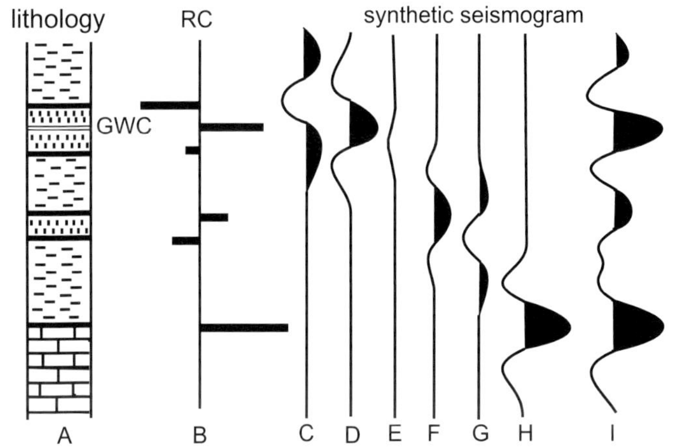

Fig. C.6.— Construction of synthetic seismogram of the lithologic column in Fig. C.5. The RC log is convolved in a computer with a seismic wavelet. This operation produces a seismic reflection at each of the bars in the RC log . The wavelet consists of a central peak accompanied by two flanking peaks of opposite sign and much lower amplitude. Traces C – H are the reflections generated at each bar in the RC log – reflections at positive RC's have black centers, at negative RC's white centers. Column I represents the complete synthetic trace, consisting of the arithmetic sum of traces C – H. After Anstey (1982), modified.

To optimize the ground truthing of seismic surveys, synthetic seismic traces are constructed at the location of the borehole and directly compared with the real seismic traces at the location. The construction of synthetic traces is illustrated in Figs. C.5, C.6. Very similar procedures were used to build the seismic models of outcrops shown in this book. The main difference is that in outcrops the acoustic impedance function has to be constructed from measurements of spot samples and lithologic observations for want of wireline logs of physical properties. Most outcrops may be viewed as two-dimensional sections and therefore 2D models of impedance distribution are constructed from observed bedding, distribution of lithologies and measured petrophysical properties (e.g. Rudolph et al., 1989; Stafleu et al., 1994; Anselmetti et al., 1997; Bracco-Gartner and Schlager, 1997). Most outcrop models are vertical incidence models. This means that the impedance model of the outcrop is examined at closely spaced vertical sections and synthetic seismic traces are constructed analogous to the procedure for boreholes in Figs. C.5, C.6. The vertical-incidence models are equivalent to perfectly migrated sections.

SEISMIC RESOLUTION

Geologists who observe rocks directly can easily switch to another tool for higher spatial resolution. There exists a continuous range of tools from the unarmed eye to the electron microscope. Moreover, the resolution of chemical analytical tools has grown in step with microscopy. The world of seismology is different. Usually, there is only one data set available and its fundamental properties are largely determined by the acquisition technique. Subsequent processing can improve resolution but within rather narrow limits. Therefore, the physical principles of seismic resolution are a matter of prime importance in seismic interpretation. *Vertical resolution* increases with wave frequency. Frequency is expressed as number of cycles per second (1 cycle/s = 1 Hertz). Higher frequency means sharper pulses and therefore better resolution (Fig. C.7). As a rule of thumb, two reflectors can be recognized as separate events if the distance is at least 1/4 of the wave length. Wave length, λ, is given by

$$\lambda = V/f$$

where V is the sonic velocity and f the frequency. The frequency of waves transmitted into the ground depends on the nature of the "bang" created by the sound source during acquisition. Once seismic waves are formed, however, their frequency changes as they travel in the subsurface. High-frequency waves are attenuated faster than low-frequency ones. Therefore, the frequency content of seismic data decreases with increasing penetration. The downward decrease in frequency and resolution is obvious in most vertical seismic sections. *Horizontal resolution* is also limited and decreases with increasing depth of penetration. The seismic "illumination" of the subsurface resembles the beam of a flashlight – it broadens and its intensity decreases with increasing distance. In seismology, horizontal resolution is indicated by the width of the Fresnel zone. Fig. C.8 shows

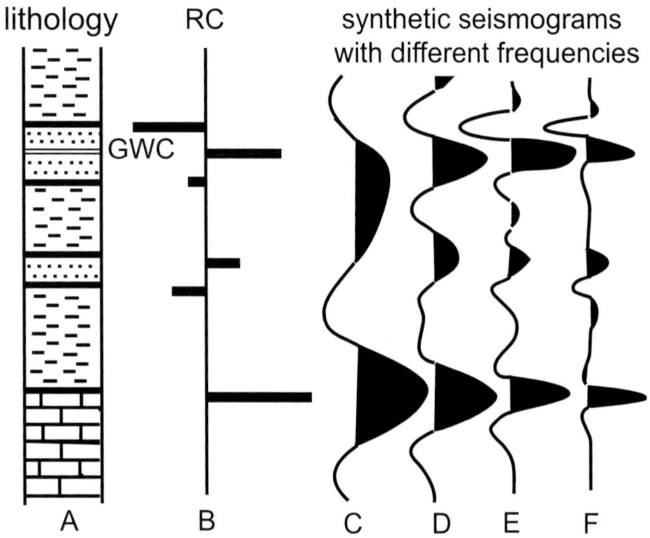

Fig. C.7.— The effect of pulse breadth (seismic frequency) on seismic resolution illustrated by synthetic seismograms. Frequency in (F) is about four times higher than in (C). After Anstey (1982), modified.

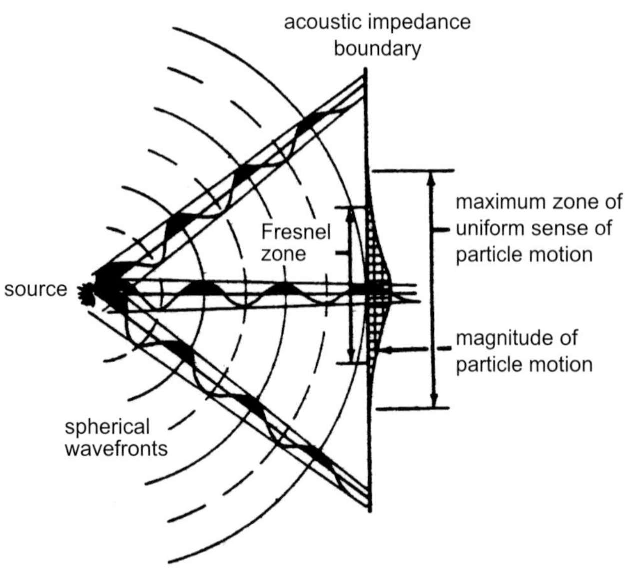

Fig. C.8.— Spherical wave fronts impinge on a flat reflector. Only the cross-hatched part of the reflector contributes to the reflected wave. The Fresnel zone is even narrower because the contribution of the distal parts of the cross-hatched zone is insignificant. After Neidell (1979). (Reprinted by permission of the AAPG whose permission is required for further use).

spherical wave fronts hitting a flat reflector. The Fresnel zone is the region of the reflector where particle motion is in the same direction and therefore contributes positively to the reflected wave. The radius of the Fresnel zone, r_f, is given by

$$r_f = (V/4)(t/f_c)^{1/2}$$

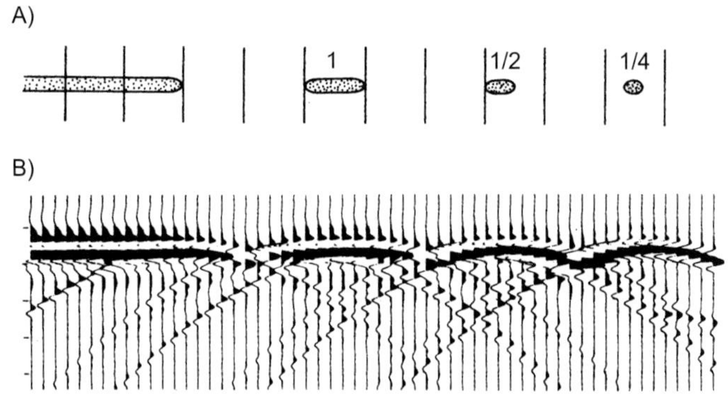

Fig. C.9.— Fresnel zone and lateral seismic resolution. A) Cross sections of layers with width expressed as fraction of Fresnel radius. B) Seismic images of layers in A. Layers smaller than 1 Fresnel radius no longer appear flat. Rather, they are imaged as reflecting point sources that create a diffraction hyperbola. After Neidell and Poggialiolmi (1977). (Reprinted by permission of the AAPG whose permission is required for further use).

where V is the average sonic velocity of the overburden, t is two-way travel time, and f_c is the frequency of the seismic waves. Thus, the Fresnel zone increases (and resolution decreases) with increasing velocity, increasing travel time and decreasing frequency.

SEISMIC ATTRIBUTES

From the basic measurements of travel time, amplitude and frequency of seismic traces, attributes can be calculated to better examine structural and stratigraphic patterns . Particularly in 3D data, attributes may be powerful tools to visualize subsurface geology (e.g. Brown, 1996). It must be kept in mind, though, that despite their vastly different appearance, attributes of the same data set all are derived from the same basic measurements. Therefore, they are not independent of each other. Seismic attributes derived from time are particularly useful for illustrating structure. Attributes related to amplitude and frequency are particularly useful in stratigraphic and sedimentologic work. A general word of caution is in order: algorithms for calculating seismic attributes seem to appear on the market much faster than they can be tested and calibrated on geologic objects. Thus, the geologic meaning of many attributes is not well constrained.

References

ABREU, V. S., HARDENBOL, J., HADDAD, G. A., BAUM, G. R., DROXLER, A. W., AND VAIL, P. R., 1998, Oxygen isotope synthesis: a Cretaceous ice-house? , in P. C. De Graciansky, J. Hardenbol, T. Jacquin, and P. R. Vail, eds., Mesozoic and Cenozoic sequence stratigraphy of European basins: SEPM Special Publication, v. 60, p. 75-81.

ADAMS, E. W., MORSILLI, M., SCHLAGER, W., KEIM, L., AND VAN, H. T., 2002, Quantifying the geometry and sediment fabric of linear slopes; examples from the Tertiary of Italy (Southern Alps and Gargano Promontory): Sedimentary Geology, v. 154, p. 11-30.

ADAMS, E. W., AND SCHLAGER, W., 2000, Basic types of submarine slope curvature: Journal of Sedimentary Research, v. 70, p. 814-828.

ADAMS, E. W., SCHRÖDER, S., GROTZINGER, J. P., AND MCCORMICK, D. S., 2004, Digital reconstruction and stratigraphic evolution of a microbial-dominated, isolated carbonate platform (terminal Proterozoic, Nama Group, Namibia): Journal of Sedimentary Research, v. 74, p. 479-497.

ADEY, W. H., MACINTYRE, I. G., STUCKENRATH, R., AND DILL, R. F., 1977, Relict barrier reef system off St. Croix: its implications with respect to late Cenozoic coral reef development in the western Atlantic: Proceedings of the Third International Coral Reef Symposium, Miami, p. 15-21.

AHR, W. M., 1973, The carbonate ramp: an alternative to the shelf model: Transactions of the Gulf Coast Association of Geologists, v. 23, p. 221-225.

AIGNER, T., 1985, Storm Depositional Systems: Lecture Notes in Earth Sciences: Berlin, Springer, v. 3, 174 p.

ALLAN, J. R., AND MATTHEWS, R. K., 1977, Carbon and oxygen isotopes as diagenetic and stratigraphic tools: surface and subsurface data, Barbados, West Indies: Geology, v. 5, p. 16-20.

ALLEN, J. R. L., 1982, Sedimentary structures: their character and physical basis: Developments in Sedimentology: Amsterdam, Elsevier, v. 30, 593 p.

ALLEN, J. R. L., 1985, Principles of Physical Sedimentology: London, Allen Unwin, 272 p.

ANDERSON, J., RODRIGUEZ, A., ABDULAH, K., FILLON, R. H., BANFIELD, L., MCKEOWN, H., AND WELLNER, J., 2004, Late Quaternary stratigraphic evolution of the northern Gulf of Mexico margin: A synthesis, in J. B. Anderson, and R. H. Fillon, eds., Late Quaternary Stratigraphic Evolution of the Northern Gulf of Mexico Margin: SEPM Special Publication, v. 79, p. 1-24.

ANDERSON, J. B., AND FILLON, R. H., EDS. 2004, Late Quaternary stratigraphic evolution of the northern Gulf of Mexico margin: SEPM Special Publication, v. 79, 311 p.

ANDRESEN, N., REIJMER, J. J. G., AND DROXLER, A. W., 2003, Timing and distribution of calciturbidites around a deeply submerged carbonate platform in a seismically active setting (Pedro Bank, northern Nicaragua Rise, Caribbean Sea): International Journal of Earth Sciences, v. 92, p. 573-592.

ANSELMETTI, F., EBERLI, G. P., AND ZAN-DONG, D., 2000, From the Great Bahama Bank into the Straits of Florida: a margin architecture controlled by sea-level fluctuations and ocean currents: Geological Society of America Bulletin, v. 112, p. 829-844.

ANSELMETTI, F. S., EBERLI, G. P., AND BERNOULLI, D., 1997, Seismic modeling of a carbonate platform margin (Montagna della Maiella, Italy): variations in seismic facies and implications for sequence stratigraphy, in I. Palaz, and K. J. Marfurt, eds., Carbonate seismology: Geophysical Developments: Tulsa, Society of Exploration Geophysicists, v. 6, p. 373-406.

ANSTEY, N. A., 1982, Simple Seismics: Boston, International Human Research Development Corporation, 168 p.

AUSTIN, J. A., SCHLAGER, W., AND PARTY, 1986, Proceedings Ocean Drilling Program Initial Reports: College Station, Ocean Drilling Program, v. 101, 501 p.

BAK, P., TANG, C., AND WIESENFELD, K., 1987, Self-organized criticality. An explanation of 1/f noise.: Physical Review Letters, v. 59, p. 381-384.

BARRELL, J., 1912, Criteria for the recognition of ancient delta deposits: Geological Society of America Bulletin, v. 23, p. 377-446.

BATES, AND JACKSON, 1987, Glossary of Geology: Alexandria, American Geological Institute, 788 p.

BATHURST, R. G. C., 1971, Carbonate sediments and their diagenesis: Developments in Sedimentology: Amsterdam, Elsevier, v. 12, 620 p.

BELOPOLSKY, A. V., AND DROXLER, A. W., 2004, Seismic expressions and interpretation of carbonate sequences: the Maldives platform, equatorial Indian Ocean: American Association of Petroleum Geologists Studies in Geology, v. 49, 46 p.

BERGER, W. H., 1989, Global maps of ocean productivity, in W. H. Berger, V. S. Smetacek, and G. Wefer, eds., Productivity of the ocean: present and past: Chichester, Wiley, p. 429-455.

BERGER, W. H., AND WINTERER, E. L., 1974, Plate stratigraphy and the fluctuating carbonate line, in K. J. Hsu, and H. C. Jenkyns, eds., Pelagic Sediments: on Land and Under the Sea: International Association of Sedimentologists Special Publications, v. 1, p. 11-48.

BERNOULLI, D., 2001, Mesozoic-Tertiary carbonate platforms, slopes and basins of the external Apennines and Sicily, in G. B. Vai, and I. P. Martini, eds., Anatomy of an orogen: the Appenines and adjacent Mediterranean Basins.: Amsterdam, Kluwer, p. 307-326.

BETZLER, C., 1997, Ecological controls on geometries of carbonate platforms: Miocene/Pliocene shallow-water microfaunas and carbonate biofacies from the Queensland Plateau (NE Australia): Facies, v. 37, p. 147-166.

BETZLER, C., BRACHERT, T. C., AND NEBELSICK, J., 1997, The warm temperate carbonate province - a review of the facies, zonations, and delimitations: Courier Forschungs-Institut Senckenberg, v. 201, p. 83-99.

BIDDLE, K. T., SCHLAGER, W., RUDOLPH, K. W., AND BUSH, T. L., 1992, Seismic model of a progradational carbonate platform, Picco di Vallandro, the Dolomites, northern Italy: American Association of Petroleum Geologists Bulletin, v. 76, p. 14-30.

BILLUPS, K., AND SCHRAG, D. P., 2002, Paleotemperatures and ice volume of the past 27 Myr revisited with paired Mg/Ca and 18O/16O measurements on benthic foraminifera: Paleoceanography, v. 17, p. 3-11.

BIRKELAND, P. W., 1999, Soils and Geomorphology: New York, Oxford University Press, 430 p.

BLENDINGER, W., 1994, The carbonate factory of Middle Triassic buildups in the Dolomites, Italy: a quantitative analysis: Sedimentology, v. 41, p. 1147-1159.

BLENDINGER, W., 2001, Triassic carbonate buildup flanks in the Dolomites, northern Italy: breccias, boulder fabric and the importance of early diagenesis: Sedimentology, v. 48, p. 919-933.

BLENDINGER, W., BOWLIN, B., ZIJP, F. R., DARKE, G., AND EKROLL, M., 1997, Carbonate buildup flank deposits: an example from the Permian (Barents Sea, Northern Norway) challenges classical facies models: Sedimentary Geology, v. 112, p. 89-103.

BOSELLINI, A., 1984, Progradational geometries of carbonate platforms: examples from the Triassic of the Dolomites, northern Italy: Sedimentology, v. 31, p. 1-24.

BOSELLINI, A., 1988, Outcrop models for seismic stratigraphy: examples from the Triassic of the Dolomites, in A. W. Bally, ed., Atlas of Seismic Stratigraphy: American Association of Petroleum Geologists Studies in Geology, v. 27-2, p. 194-205.

BOSELLINI, A., MORSILLI, M., AND NERI, C., 1999, Long-term event stratigraphy of the Apulia platform margin (Upper Jurassic to Eocene, Gargano, southern Italy): Journal of Sedimentary Research, v. 69, p. 1241-1252.

BOSENCE, D., AND WALTHAM, D., 1990, Computer modeling the internal architecture of carbonate platforms: Geology, v. 18, p. 26-30.

BOSSCHER, H., AND SCHLAGER, W., 1992, Computer simulation of reef growth: Sedimentology, v. 39, p. 503-512.

BOSSCHER, H., AND SCHLAGER, W., 1993, Accumulation rates of carbonate platforms: Journal of Geology, v. 101, p. 345-355.

BOULVAIN, F., 2001, Facies architecture and diagenesis of Belgian Late Frasnian carbonate mounds: Sedimentary Geology, v. 145, p. 269-294.

BRACCO GARTNER, G. L., AND SCHLAGER, W., 1999, Discrimination between onlap and lithologic interfingering in seismic models of outcrops: American Association of Petroleum Geologists Bulletin, v. 83, p. 952-971.

BRACHERT, T. C., FORST, M. H., PAIS, J. J., LEGOINHA, P., AND REIJMER, J. J. G., 2003, Lowstand carbonates, highstand sandstones? : Sedimentary Geology, v. 155, p. 1-12.

BRACK, P., MUNDIL, R., OBERLI, F., MEIER, M., AND RIEBER, H., 1996, Biostratigraphic and radiometric age data question the Milankovitch characteristics of the Latemar cycles (Southern Alps, Italy): Geology, v. 24, p. 371-375.

BRANDLEY, R. T., AND KRAUSE, F. F., 1997, Upwelling, thermoclines and wave-sweeping on an equatorial carbonate ramp: Lower Carboniferous strata of western Canada, in N. P. James, and A. D. Clarke, eds., Cool-water carbonates: SEPM Special Publication, v. 56, p. 365-390.

BRANDNER, R., FLÜGEL, E., KOCH, R., AND YOSE, L. A., 1991, The northern margin of the Schlern/Sciliar-Rosengarten/Catinaccio Platform: Field Trip Guidebook, Dolomieu Conference on Carbonate Platforms and Dolomitization: Ortisei, Tourist Office Ortisei, v. Excursion A, 61 p.

BROECKER, W. S., AND PENG, T. H., 1982, Tracers in the Sea: New York, Eldigio Press, 690 p.

BROWN, A., 1996, Interpretation of Three-Dimensional Seismic Data: American Association of Petroleum Geologists Memoir, v. 42, 424 p.

BUCHBINDER, B., BENJAMINI, C., AND LIPSON-BENITAH, S., 2000, Sequence development of late Cenomanian-Turonian carbonate ramps, platforms and basins in Israel: Cretaceous Research, v. 21, p. 813-843.

BUFFLER, R. T., 1991, Chapter 13: Seismic stratigraphy of the deep Gulf of Mexico basin and adjacent margins, The Geology of North America, Geological Society of America, v. J, p. 353-387.

BURCHETTE, T. P., AND WRIGHT, V. P., 1992, Carbonate ramp depositional systems: Sedimentary Geology, v. 79, p. 3-57.

CALDER, W. A., 1984, Size, Function, and Life History: Cambridge, Harvard University Press, 431 p.

CALVET, F., AND TUCKER, M. E., 1995, Mud-mounds with reefal caps in the upper Muschelkalk (Triassic), eastern Spain, in C. L. V. Monty, D. W. J. Bosence, P. H. Bridges, and B. R. Pratt, eds., Carbonate Mud-Mounds - Their Origin and Evolution: International Association of Sedimentologists Special Publication, v. 23, p. 311-333.

CAMOIN, G. F., GAUTRET, P., MONTAGGIONI, L. F., AND GABIOCH, G., 1999, Nature and environmental significance of microbialites in Quaternary reefs: the Tahiti paradox: Sedimentary Geology, v. 126, p. 271-304.

CARTER, R. M., ABBOTT, S. T., FULTHORPE, C. S., AND HAYWICK, D. W., 1991, Application of global sea-level and sequence-stratigraphic models in Southern Hemisphere Neogene strata from New Zealand, in D. I. M. MacDonald, ed., Sedimentation, Tectonics and Eustasy: Sea-level Changes at Active Margins: International Association of Sedimentologists Special Publication, v. 12, p. 41-65.

CATHRO, D. L., AUSTIN, J. A., AND MOSS, G. D., 2003, Progradation along a deeply submerged Oligocene-Miocene heterozoan carbonate shelf: how sensitive are clinoforms to sea level variations? : American Association of Petroleum Geologists Bulletin, v. 87, p. 1547-1574.

CHERNS, L., AND WRIGHT, V. P., 2000, Missing molluscs as evidence of large-scale, early skeletal aragonite dissolution in a Silurian sea: Geology, v. 28, p. 791-794.

CHOQUETTE, P. W., AND PRAY, L. C., 1970, Geologic nomenclature and classification of porosity in sedimentary carbonates: American Association of Petroleum Geologists Bulletin, v. 54, p. 20-250.

CHOW, N., AND GEORGE, A. D., 2004, Tepee-shaped agglutinated microbialites: an example from a Famennian carbonate platform on the Lennard Shelf, northern Canning Basin, Western Australia: Sedimentology, v. 51, p. 253-265.

CHRISTIE-BLICK, N., MOUNTAIN, G. S., AND MILLER, K. G., 1990, Seismic stratigraphic record of sea-level change, in R. R. Revelle, ed., Sea-level change: Studies in Geophysics: Washington D.C., National Academy Press, p. 116-140.

CLOETINGH, S. A. P. L., 1988, Intraplate stresses: a tectonic cause for third-order cycles in apparent sea level, in C. K. Wilgus, B. S. Hastings, C. G. S. C. Kendall, H. W. Posamentier, C. A. Ross, and J. C. Van Wagoner, eds., Sea-level Changes - an Integrated Approach: SEPM Special Publication, v. 42, p. 19-29.

COLLINS, L. B., 1988, Sediments and history of the Rottnest Shelf, southwest Australia: a swell-dominated, non-tropical carbonate margin: Sedimentary Geology, v. 60, p. 15-49.

COLLINS, L. B., FRANCE, R. E., AND ZHU, Z. R., 1997, Warm-water platform and cool-water shelf carbonates of the Abrolhos Shelf, southwest Australia, in N. P. James, and J. A. D. Clarke, eds., Cool-water Carbonates: SEPM Special Publications, v. 56, p. 23-36.

CORTÉS, J., MACINTYRE, I. G., AND GLYNN, P. W., 1994, Holocene growth history of an eastern Pacific fringing reef, Punta Islotes, Costa Rica.: Coral Reefs, v. 13, p. 65-73.

CREVELLO, P. D., AND SCHLAGER, W., 1980, Carbonate debris sheets and turbidites, Exuma Sound, Bahamas: Journal of Sedimentary Petrology, v. 50, p. 1121-1148.

CROWELL, J. C., 2000, Pre-Mesozoic ice ages: their bearing on understanding the climate system: Geological Society of America Memoirs, v. 192, 106 p.

CURRAY, J. R., 1964, Transgressions and regressions, in R. L. Miller, ed., Papers in Marine Geology: New York, MacMillan, p. 175-203.

D'ARGENIO, B., FERRERI, V., AMODIO, S., AND PELOSI, N., 1997, Hierarchy of high-frequency orbital cycles in Cretaceous carbonate platform strata: Sedimentary Geology, v. 113, p. 169-193.

D'ARGENIO, B., FERRERI, V., RASPINI, A., AMODIO, S., AND BUONOCUNTO, F. P., 1999, Cyclostratigraphy of a carbonate platform as a tool for high-precision correlation: Tectonophysics, v. 315, p. 357-385.

DARWIN, C. R., 1842, The Structure and Distribution of Coral Reefs (reprint edition): Tucson, University of Arizona Press, 214 p.

DAVIES, P. J., BUBELA, B., AND FERGUSON, J., 1978, The formation of ooids: Sedimentology, v. 25, p. 703-729.

DAVIES, P. J., SYMONDS, P. A., FEARY, D. A., AND PIGRAM, C. J., 1989, The evolution of the carbonate platforms of northeast Australia, in P. D. Crevello, J. L. Wilson, J. F. Sarg, and J. F. Read, eds., Controls on Carbonate Platform and Basin Development: SEPM Special Publication, v. 44, p. 233-258.

DE MOL, B., VAN RENSBERGEN, P., PILLEN, S., VAN HERREWEGHE, K., VAN ROOIJ, D., MCDONNELL, A., HUVENNE, V., IVANOV, M., AND SWENNEN, R., 2002, Large deep-water coral banks in the Porcupine Basin, southwest of Ireland: Marine Geology, v. 188, p. 193-231.

DELLA PORTA, G., 2003, Depositional anatomy of a Carboniferous high-rising carbonate platform (Cantabrian Mountains, NW Spain): Amsterdam, Vrije Universiteit, 250 p.

DEMICCO, R. V., 1998, Cycopath 2D - a two-dimensional, forward model of cyclic sedimentation on carbonate platforms: Computers and Geosciences, v. 24, p. 405-423.

DICKSON, J. A. D., 2004, Echinoderm skeletal preservation: calcite-aragonite seas and the Mg/Ca ratio of Phanerozoic oceans: Journal of Sedimentary Research, v. 74, p. 355-365.

DODD, J. R., AND STANTON, R. J., 1981, Paleoecology - Concepts and Applications: New York, Wiley, 559 p.

DOLAN, J. F., 1989, Eustatic and tectonic controls on deposition of hybrid siliciclastic/carbonate basinal cycles: discussion with examples: American Association of Petroleum Geologists Bulletin, v. 73, p. 1233-1246.

DREYBRODT, W., 1988, Processes in Karst Systems: Physics, Chemistry, and Geology: Berlin, Springer, 290 p.

DRISCOLL, N. W., WEISSEL, J. K., KARNER, G. D., AND MOUNTAIN, G. S., 1991, Stratigraphic response of a carbonate platform to relative sea level changes: Broken Ridge, southeast Indian Ocean: American Association of Petroleum Geologists Bulletin, v. 75, p. 808-831.

DROXLER, A. W., BRUCE, C. H., SAGER, W. W., AND HAWKINS, D. H., 1988, Pliocene-Pleistocene variations in aragonite content and planktonic oxygen-isotope record in Bahamian periplatform ooze, Hole 633A, in A. Austin James, Jr., W. Schlager, and A. Palmer, eds., Proceedings of the Ocean Drilling Program Scientific Results: College Station, Ocean Drilling Program, v. 101, p. 221-244.

DROXLER, A. W., HADDAD, G. A., MUCCIARONI, D. A., AND CULLEN, J. L., 1990, Pliocene-Pleistocene aragonite cyclic variations in Ocean Drilling Program holes 714A and 716A (the Maldives) compared to hole 633A (the Bahamas): records of climate-induced CaCO3 preservation at intermediate water depths, in R. A. Duncan, J. Backman, and L. C. Peterson, eds., Proceedings of the Ocean Drilling Program Scientific Results: College Station, Ocean Drilling Program, v. 115, p. 539-577.

DROXLER, A. W., AND SCHLAGER, W., 1985, Glacial versus interglacial sedimentation rates and turbidite frequency in the Bahamas: Geology, v. 13, p. 799-802.

DROXLER, A. W., SCHLAGER, W., AND WHALLON, C. C., 1983, Quaternary aragonite cycles and oxygen-isotope record in Bahamian carbonate ooze: Geology, v. 11, p. 235-239.

DRUMMOND, C. N., AND WILKINSON, B. H., 1993, Aperiodic accumulation of cyclic peritidal carbonate: Geology, v. 21, p. 1023-1026.

DULLO, W. C., CAMOIN, G. F., BLOMEIER, D., COLONNA, M., EISENHAUER, A., FAURE, G., CASANOVA, J., AND THOMASSIN, B. A., 1998, Morphology and sediments of the fore-slopes of Mayotte, Comoro Islands: direct observations from a submersible., in G. F. Camoin, and P. J. Davies, eds., Reefs and Carbonate Platforms in the Pacific and the Indian Oceans: International Association of Sedimentologists Special Publication, v. 25, p. 219-236.

DUNHAM, R. J., 1962, Classification of carbonate rocks according to depositional texture, in W. E. Ham, ed., Classification of Carbonate Rocks: American Association of Petroleum Geologists Memoir, v. 1, p. 108-121.

DUNHAM, R. J., 1970, Stratigraphic reefs versus ecologic reefs: American Association of Petroleum Geologists Bulletin, v. 54, p. 1931-1932.

DUVAL, B., CRAMEZ, C., AND VAIL, P. R., 1998, Stratigraphic cycles and major marine source rocks, in P. C. De Graciansky, J. Hardenbol, T. Jacquin, and P. R. Vail, eds., Mesozoic and Cenozoic Sequence Stratigraphy of European Basins: SEPM Special Publication, v. 60, p. 43-51.

EBERLI, G. P., 2000, The record of Neogene sea-level changes in the prograding carbonates along the Bahamas transect - Leg 166 synthesis, in P. K. Swart, G. P. Eberli, M. J. Malone, and J. F. Sarg, eds., Proceedings Ocean Drilling Program, Scientific Results: College Station, Ocean Drilling Program, v. 166, p. 167-178.

EBERLI, G. P., ANSELMETTI, F. S., KENTER, J. A. M., MCNEILL, D. F., AND MELIM, L. A., 2001, Calibration of seismic sequence stratigraphy with cores and logs, in R. N. Ginsburg, ed., Subsurface Geology of a Prograding Carbonate Platform Margin, Great Bahama Bank: Results of the Bahamas Drilling Project: SEPM Special Publications, v. 70, p. 241-265.

EBERLI, G. P., ANSELMETTI, F. S., KROON, D., SATO, T., AND WRIGHT, J. D., 2002, The chronostratigraphic significance of seismic reflections along the Bahamas transect: Marine Geology, v. 185, p. 1-17.

EBERLI, G. P., AND GINSBURG, R. N., 1988, Aggrading and prograding Cenozoic seaways, northwest Great Bahama Bank, in A. W. Bally, ed., Atlas of Seismic Stratigraphy: American Association of Petroleum Geologists Studies in Geology, v. 27-2, p. 97-103.

EBERLI, G. P., SWART, P. K., MALONE, M. J., AND PARTY, 1997, Proceedings Ocean Drilling Program Initial Reports: College Station, Ocean Drilling Program, v. 166, 850 p.

EGENHOFF, S. O., PETERHÄNSEL, A., BECHSTÄDT, T., ZÜHLKE, R., AND GRÖTSCH, J., 1999, Facies architecture of an isolated carbonate platform: tracing the cycles of the Latemar (Middle Triassic, northern Italy): Sedimentology, v. 46, p. 893-912.

ELDREDGE, N., AND GOULD, S. J., 1972, Punctuatad equilibria: an alternative to phyletic gradualism., in T. J. M. Schopf, ed., Models in Paleobiology: San Francisco, Freeman, p. 88-115.

ELVEBAKK, G., HUNT, W., AND STEMMERIK, L., 2002, From isolated buildups to buildup mosaics: 3D seismic sheds new light on upper Carboniferous - Permian fault controlled carbonate buildups, Norwegian Barents Sea: Sedimentary Geology, v. 152, p. 7-17.

EMBRY, A. F., 1993, Transgressive-regressive (T-R) sequence analysis of the Jurassic succession of the Sverdrup Basin, Canadian Arctic Archipelago: Canadian Journal of Earth Sciences, v. 30, p. 301-320.

EMBRY, A. F., AND KLOVAN, J. E., 1971, A Late Devonian reef tract on northeastern Banks Island, Northwest Territories: Canadian Petroleum Geology Bulletin, v. 19, p. 730-781.

EMERY, D., MYERS, K., BERTRAM, G., GRIFFITHS, C., MILTON, N., REYNOLDS, T., RICHARDS, M., AND STURROCK, S., 1996, Sequence Stratigraphy.: Oxford, Blackwell, 297 p.

EMILIANI, C., 1992, Planet Earth: Cambridge, Cambridge University Press, 715 p.

ENOS, P., 1977, Carbonate sediment accumulations of the South Florida Shelf Margin, in P. Enos, and R. D. Perkins, eds., Quaternary depositional framework of South Florida: Geological Society of America Memoir, v. 147, p. 1-130.

ENOS, P., AND SAMANKASSOU, E., 1998, Lofer cyclothems revisited (Late Triassic, Northern Alps, Austria): Facies, v. 38, p. 207 - 228.

ERLICH, R. N., BARRETT, S. F., AND GUO, J. B., 1990, Seismic and geologic characteristics of drowning events on carbonate platforms: American Association of Petroleum Geologists Bulletin, v. 74, p. 1523-1537.

ESTEBAN, M., AND KLAPPA, C. F., 1983, Subaerial exposure, in P. A. Scholle, D. G. Bebout, and C. H. Moore, eds., Carbonate Depositional Environments: American Association of Petroleum Geologists Memoirs, v. 33, p. 1-95.

EVERTS, A. J. W., 1991, Interpreting compositional variations of calciturbidites in relation to platform-stratigraphy: an example from the Paleogene of SE Spain: Sedimentary Geology, v. 71, p. 231-242.

FALCONER, K., 1990, Fractal Geometry: Chichester, Wiley, 288 p.

FEARY, D. A., AND JAMES, N. P., 1995, Cenozoic biogenic mounds and buried Miocene (?) barrier reef on a predominantly cool-water carbonate continental margin, Eucla Basin, western Great Australian Bight: Geology, v. 23, p. 427-430.

FEARY, D. A., AND JAMES, N. P., 1998, Seismic stratigraphy and geological evolution of the Cenozoic, cool-water, Eucla Platform, Great Australian Bight: American Association of Petroleum Geologists Bulletin, v. 82, p. 792-816.

FEDER, J., 1988, Fractals: New York, Plenum Press, 283 p.

FISCHER, A. G., 1964, The Lofer cyclothems of the Alpine Triassic: Kansas State Geological Survey Bulletin, v. 169, p. 107-150.

FISCHER, A. G., 1982, Long-term climatic oscillations recorded in stratigraphy, in W. H. Berger, and J. C. Crowell, eds., Climate in Earth History: National Academy of Sciences, Studies in Geophysics: Washington, D.C., National Academy Press, p. 97-104.

FISHER, W. L., AND MCGOWEN, J. H., 1967, Depositional systems in the Wilcox Group of Texas and their relationship to occurrence of oil and gas: Transactions of the Gulf Coast Association of Geological Societies, v. 17, p. 105-125.

FLEMINGS, P. B., AND GROTZINGER, J. P., 1996, STRATA: Freeware for analyzing classic stratigraphic problems: GSA Today, v. 6, p. 1-7.

FLUEGEMAN, R. H., JR., AND SNOW, R. S., 1989, Fractal analysis of long-range paleoclimatic data: oxygen isotope record of Pacific core V28-239: Pure and Applied Geophysics, v. 131, p. 307-313.

FLÜGEL, E., 1982, Microfacies Analysis of Limestones: Berlin, Springer, 633 p.

FLÜGEL, E., 2004, Microfacies of Carbonate Rocks, Springer, 976 p.

FLÜGEL, E., AND KIESSLING, W., 2002, A new look at ancient reefs, in W. Kiessling, E. Flügel, and J. Golonka, eds., Phanerozoic Reef Patterns: SEPM Special Publications, v. 72, p. 3-10.

FÖLLMI, K. B., WEISSERT, H., BISPING, M., AND FUNK, H. P., 1994, Phosphogenesis, carbon-isotope stratigraphy, and carbonate-platform evolution along the Lower Cretaceous northern Tethyan margin: Geological Society of America Bulletin, v. 106, p. 729-746.

FOUKE, B. W., ZWART, E. W., EVERTS, A. J. W., AND SCHLAGER, W., 1995, Carbonate platform stratal geometries and the question of subaerial exposure: Sedimentary Geology, v. 97, p. 9-19.

FRAKES, L. A., FRANCIS, J. E., AND SYKTUS, J. I., 1992, Climate Mode of the Phanerozoic: The History of the Earth's Climate over the Past 600 Million Years: Cambridge, Cambridge University Press, 274 p.

GARDNER, T. W., JORGENSEN, D. W., SHUMAN, C., AND LEMIEUX, C. R., 1987, Geomorphic and tectonic process rates: effects of measured time interval: Geology, v. 15, p. 259-261.

GEBELEIN, C. D., STEINEN, R. P., GARRETT, P., HOFFMAN, E. J., QUEEN, J. M., AND PLUMMER, L. N., 1980, Subsurface dolomitization beneath the tidal flats of central west Andros Island, in D. H. Zenger, J. B. Dunham, and R. L. Ethington, eds., Concepts and Models of Dolomitization: SEPM Special Publications, v. 28, p. 31-49.

GENIN, A., NOBLE, M., AND LONSDALE, P. F., 1989, Tidal currents and anticyclonic motions on two North Pacific seamounts: Deep-Sea Research, v. 36, p. 1803-1815.

GIBBS, M. T., BICE, K. L., BARRON, E. J., AND KUMP, L. R., 2000, Glaciation in the early Paleozoic "greenhouse": the roles of paleogeography and atmospheric CO2, in B. T. Huber, K. G. MacLeod, and S. L. Wing, eds., Warm Climates in Earth History.: Cambridge, Cambridge University Press, p. 386-422.

GILLESPIE, J. L., AND NELSON, C. S., 1997, Mixed siliciclastic-skeletal carbonate facies on Wanganui shelf, New Zealand: a contribution to the temperate carbonate model, in N. P. James, and J. A. D. Clarke, eds., Cool-water Carbonates: SEPM Special Publications, v. 56, p. 127-140.

GINSBURG, R. N., 1971, Landward movement of carbonate mud: new model for regressive cycles in carbonates (abs.): American Association of Petroleum Geologists Bulletin, v. 55, p. 340.

GINSBURG, R. N., ED.,. 2001, Subsurface Geology of a Prograding Carbonate Platform Margin, Great Bahama Bank: Results of the Bahamas Drilling Project: SEPM Special Publications, v. 70, 271 p.

GISCHLER, E., AND LOMANDO, A., 1999, Recent sedimentary facies of isolated carbonate platforms, Belize-Yucatan system, Central America: Journal of Sedimentary Research, v. 69, p. 747-763.

GOLDHAMMER, R. K., DUNN, P. A., AND HARDIE, L. A., 1990, Depositional cycles, composite sea-level changes, cycle stacking patterns, and the hierarchy of stratigraphic forcing: examples from Alpine Triassic platform carbonates: Geological Society of America Bulletin, v. 102, p. 535-562.

GOLDHAMMER, R. K., HARRIS, M. T., DUNN, P. A., AND HARDIE, L. A., 1993, Sequence stratigraphy and systems tract development of the Latemar platform, Middle Triassic of the Dolomites (Northern Italy): outcrop calibration keyed by cycle stacking patterns, in R. G. Loucks, and J. F. Sarg, eds., Carbonate Sequence Stratigraphy: American Association of Petroleum Geologists Memoirs, v. 57, p. 353-387.

GOLDHAMMER, R. K., OSWALD, E. J., AND DUNN, P. A., 1994, High-frequencey, glacio-eustatic cyclicity in the Middle Pennsylvanian of the Paradox Basin: an evaluation of Milankovitch forcing, in P. L. De Boer, and D. G. Smith, eds., Orbital Forcing and Cyclic Sequences: International Association of Sedimentologists, Special Publications, v. 19, p. 243-283.

GOLDSTEIN, R. H., AND FRANSEEN , E. K., 1995, Pinning points: a method providing quantitative constraints on relative sea-level history: Sedimentary Geology, v. 95, p. 1-10.

GOLDSTEIN, R. H., FRANSEEN , E. K., AND MILLS, M. S., 1990, Diagenesis associated with subaerial exposure of Miocene strata, southeastern Spain: implications for sea-level change and preservation of low-temperature fluid inclusions in calcite cement: Geochimica Cosmochimica Acta, v. 54, p. 699-704.

GRABAU, A. W., 1924, Principles of Stratigraphy: New York, Seiler, 1185 p.

GRAMMER, G. M., GINSBURG, R. N., AND HARRIS, P. M., 1993, Timing of deposition, diagenesis, and failure of steep carbonate slopes in response to a high-amplitude/high-frequency fluctuation in sea level, Tongue of the Ocean, Bahamas, in R. G. Loucks, and J. F. Sarg, eds., Carbonate Sequence Stratigraphy: American Association of Petroleum Geologists Memoirs, v. 57, p. 107-132.

GREENLEE, S. M., 1988, Tertiary depositional sequences, offshore New Jersey and Alabama, in A. W. Bally, ed., Atlas of Seismic Stratigraphy: American Association of Petroleum Geologists Studies in Geology, v. 27-2, p. 67-80.

GRIGG, R. W., 1982, Darwin Point: a threshold for atoll formation: Coral Reefs, v. 1, p. 29-34.

GROTZINGER, J. P., AND JAMES, N. P., 2000, Precambrian carbonates: evolution of understanding, in J. P. Grotzinger, and N. P. James, eds., Carbonate Sedimentation and Diagenesis in the Evolving Precambrian World: SEPM Special Publications, v. 67, p. 3-20.

HAAK, A. B., AND SCHLAGER, W., 1989, Compositional variations in calciturbidites due to sea-level fluctuations, late Quaternary, Bahamas: Geologische Rundschau, v. 78, p. 477-486.

HALLAM, A., 1977, Secular changes in marine inundation of USSR and North America through the Phanerozoic: Nature, v. 269, p. 769-772.

HALLAM, A., 1998, Interpreting sea level, in P. Doyle, and M. R. Bennett, eds., Unlocking the Stratigraphical Record: Advances in Modern Stratigraphy: Chichester, Wiley, p. 421 - 439.

HALLEY, R. B., AND EVANS, C. C., 1983, The Miami limestone. A guide to selected outcrops and their interpretation (with a discussion of diagenesis in the formation): Miami, Miami Geological Society, 67 p.

HALLEY, R. B., AND HARRIS, P. M., 1979, Fresh water cementation of a 1,000 year-old oolite: Journal of Sedimentary Petrology, v. 49, p. 969-988.

HALLOCK, P., 1988, The role of nutrient availability on bioerosion: consequences to carbonate buildups: Palaeogeography, Palaeoclimatiology, Palaeoecology, v. 63, p. 275-291.

HALLOCK, P., 2001, Coral reefs, carbonate sediments, nutrient, and global change, in G. D. J. Stanley, ed., The History and Sedimentology of Ancient Reef Systems: New York, Kluwer/Plenum, p. 387-427.

HANDFORD, C. R., AND LOUCKS, R. G., 1993, Carbonate depositional sequences and systems tracts - responses of carbonate platforms to relative sea-level changes, in R. G. Loucks, and J. F. Sarg, eds., Carbonate Sequence Stratigraphy: American Association of Petroleum Geologists Memoirs, v. 57, p. 3-42.

HAQ, B. U., 1991, Sequence stratigraphy, sea-level change, and significance for the deep sea, in D. I. M. Macdonald, ed., Sedimentation, Tectonics and Eustasy: Sea-Level Changes at Active Margins: International Association of Sedimentologists Special Publications, v. 12, p. 3-39.

HAQ, B. U., HARDENBOL, J., AND VAIL, P. R., 1987, Chronology of fluctuating sea levels since the Triassic (250 million years ago to present): Science, v. 235, p. 1156-1167.

HARDENBOL, J., THIERRY, J., FARLEY, M. B., JACQUIN, T., DE GRACIANSKY, P. C., AND VAIL, P. R., 1998, Mesozoic and Cenozoic sequence chronostratigraphic framework of European basins, in P. C. De Graciansky, J. Hardenbol, T. Jacquin, and P. R. Vail, eds., Mesozoic and Cenozoic Sequence Stratigraphy of European Basins: SEPM Special Publications, v. 60, p. 3-14.

HARDIE, L. A., 1996, Secular variations in seawater chemistry: An explanation for the coupled secular variation in the mineralogies of marine limestones and potash evaporites over the past 600 m.y.: Geology, v. 24, p. 279-283.

HARDIE, L. A., AND SHINN, E. A., 1986, Carbonate depositional environments, modern and ancient. Part 3 - Tidal flats: Colorado School of Mines Quarterly, v. 80, 74 p.

HARRIS, P. M., 1979, Facies Anatomy and Diagenesis of a Bahamian Ooid Shoal: Sedimenta, University of Miami, Comparative Sedimentology Laboratory, v. 7, 163 p.

HARRIS, P. M., AND KOWALIK, W. S., 1994, Satellite images of carbonate depositional settings: American association of Petroleum Geologists Methods in Exploration, v. 11, 147 p.

HARRIS, P. M., AND SALLER, A. H., 1999, Subsurface expression of the Capitan depositional system and implications for hydrocarbon reservoirs, northeastern Delaware Basin, in A. H. Saller, P. M. Harris, B. L. Kirkland, and S. J. Mazzullo, eds., Geologic Framework of the Capitan Reef.: SEPM Special Publications, v. 65, p. 37-49.

HARRIS, P. T., HEAP, A. D., WASSENBERG, T., AND PASSLOW, V., 2004, Submerged coral reefs in the Gulf of Carpentaria, Australia: Marine Geology, v. 207, p. 185-191.

HARRISON, C. G. A., 1990, Long-term eustasy and epeirogeny in continents, in R. R. Revelle, ed., Sea-Level Change: Washington, D.C., National Academy Press, p. 141-158.

HARRISON, C. G. A., 2002, Power spectrum of sea level change over 15 decades of frequency: Geochemistry Geophysics Geosystems, v. 3/8, p. 1-17.

HARWOOD, J. M., AND TOWERS, P. A., 1988, Seismic sedimentologic interpretation of a carbonate slope, north margin of Little Bahama Bank, in J. A. Austin, Jr., W. Schlager, and A. Palmer, eds., Proceedings Ocean Drilling Program, Initial Reports: College Station, Ocean Drilling Program, v. 101, p. 263-277.

HAY, W. W., 1985, Potential errors in estimates of carbonate rock accumulating through geologic time, in E. T. Sundquist, and W. S. Broecker, eds., The Carbon Cycle and Atmospheric CO2: Natural Variations, Archaean to Present: American Geophysical Union Geophysical Monographs, v. 32, p. 573-583.

HAY, W. W., 1988, Paleoceanography: A review for the GSA Centennial: Geological Society of America Bulletin, v. 100, p. 1934-1956.

HAY, W. W., FLÖGEL, S., AND SÖDING, E., 2004, Is the initiation of glaciation on Antarctica related to a change in the structure of the ocean? : Global and Planetary Change, p. doi:10.1016/j.gloplacha.2004.09.005.

HEEZEN, B. C., THARP, M., AND EWING, M., 1959, The Floors of the Ocean. 1. The North Atlantic: Geological Society of America Special Paper, v. 65, 122 p.

HEIM, A., 1934, Stratigraphische Kondensation: Eclogae Geologicae Helvetiae, v. 27, p. 372-383.

HELLAND-HANSEN, W., HELLE, H. B., AND SUNDE, K., 1994, Seismic modelling of Tertiary sandstone clinothems, Spitsbergen: Basin Research, v. 6, p. 181-191.

HENRICH, R., FREIWALD, A., BICKERT, T., AND SCHÄFER, P., 1997, Evolution of an Arctic open-shelf carbonate platform, Spitsbergen Bank, in N. P. James, and J. A. D. Clarke, eds., Cool-water Carbonates: SEPM Special Publications, v. 56, p. 163-184.

HERGARTEN, S., 2002, Self-Organized Criticality in Earth Systems: Berlin, Springer, 272 p.

HILLGÄRTNER, H., 1999, The Evolution of the French Jura Platform During the Late Berriasian to Early Valanginian: Controlling Factors and Timing: Geofocus: Fribourg, University of Fribourg, v. 1, 201 p.

HILLGÄRTNER, H., AND STRASSER, A., 2003, Quantification of high-frequency sea-level fluctuations in shallow-water carbonates: an example from the Berriasian - Valanginian (French Jura): Palaeogeography, Palaeoclimatology, Palaeoecology, v. 200, p. 43-63.

HINE, A. C., LOCKER, S. D., TEDESCO, L. P., MULLINS, H. T., HALLOCK, P., BELKNAP, D. F., GONZALES, J. L., NEUMANN, A. C., AND SNYDER, S. W., 1992, Megabreccia shedding from modern low-relief carbonate platforms, Nicaraguan Rise: Geological Society of America Bulletin, v. 104, p. 928-943.

HINE, A. C., AND MULLINS, H. T., 1983, Modern carbonate shelf-slope breaks, in D. J. Stanley, and G. T. Moore, eds., The Shelfbreak: Critical Interface on Continental Margins: SEPM Special Publications, v. 33, p. 169-188.

HINE, A. C., AND NEUMANN, A. C., 1977, Shallow carbonate-bank margin growth and structure, Little Bahama Bank, Bahamas: American Association of Petroleum Geologists Bulletin, v. 61, p. 376-406.

HINNOV, L. A., 2000, New perspectives on orbitally forced stratigraphy: Annual Reviews of Earth and Planetary Sciences, v. 28, p. 419-475.

HOFFMAN, P., 1974, Shallow and deepwater stromatolites in Lower Proterozoic platform-basin facies change, Great Slave Lake, Canada: American Association of Petroleum Geologists Bulletin, v. 58, p. 856-867.

HOMEWOOD, P. W., 1996, The carbonate feedback system: interaction between stratigraphic accommodation, ecological succession and the carbonate factory: Bulletin de la Societe geologique de France, v. 167, p. 701-715.

HOMEWOOD, P. W., AND EBERLI, G. P., EDS. 2000, Genetic stratigraphy on the exploration and production scales, Case studies from the Upper Devonian of Alberta and the Pennsylvanian of the Paradox Basin: Elf Exploration Production Editions Memoirs, v. 24, 290 p.

HOTTINGER, L., 1989, Conditions for generating carbonate platforms: Memorie della Società Geologica Italiana, v. 40, p. 265-271.

HOUSE, M. R., 1985, A new approach to an absolute time scale from measurements of orbital cycles and sedimentary microrhythms: Nature, v. 316, p. 721-725.

HSUI, A. T., RUST, K. A., AND KLEIN, G. D., 1993, A fractal analysis of Quaternary, Cenozoic-Mesozoic, and Late Pennsylvanian sea-level changes: Journal of Geophysical Research, v. 98, p. 21963-21967.

HUNT, D., AND TUCKER, M. E., 1992, Stranded parasequences and the forced regressive wedge systems tract: deposition during base-level fall: Sedimentary Geology, v. 81, p. 1-9.

HUNT, D., AND TUCKER, M. E., 1993, Sequence stratigraphy of carbonate shelves with an example from the mid-Cretaceous (Urgonian) of southeast France, in H. W. Posamentier, C. P. Summerhayes, B. U. Haq, and G. P. Allen, eds., Sequence Stratigraphy and Facies Associations: International Association of Sedimentologists Special Publication, Blackwell, v. 18, p. 307-341.

IMBRIE, J., AND IMBRIE, K. P., 1979, Ice Ages: Solving the Mystery: New York, Macmillan, 224 p.

IMMENHAUSER, A., CREUSEN, A., ESTEBAN, M., AND VONHOF, H. B., 2000, Recognition and interpretation of polygenic discontinuity surfaces in the Middle Cretaceous Shuaiba, Nahr Umr, and Natih Formations of northern Oman: GeoArabia, v. 5, p. 299-322.

IMMENHAUSER, A., HILLGÄRTNER, H., SATTLER, U., BERTOTTI, G., SCHOEPFER, P., HOMEWOOD, P. W., VAHRENKAMP, V., STEUBER, T., MASSE, J. P., DROSTE, H., TAAL-VAN KOPPEN, J., VAN DER KOOIJ, B., VAN BENTUM, E., VERWER, K., HOOGERDUIJN-STRATING, E., SWINKELS, W., PETERS, J., IMMENHAUSER-POTTHAST, I., AND AL MASKERY, S., 2004, Barremian-lower Aptian Qishn Formation, Haushi-Huqf area, Oman: a new outcrop analogue for the Kharaib/Shuaiba reservoirs: GeoArabia, v. 9, p. 153-194.

IMMENHAUSER, A., KENTER, J. A. M., GANSSEN, G., BAHAMONDE, J. R., VAN VLIET, A., AND SAHER, M. H., 2002, Origin and significance of isotope shifts in Pennsylvanian carbonates (Asturias, NW Spain): Journal of Sedimentary Research, v. 72, p. 82-94.

IMMENHAUSER, A., SCHLAGER, W., BURNS, S. J., SCOTT, R. W., GEEL, T., LEHMANN, J., VAN DER GAAST, S., AND BOLDER-SCHRIJVER, L. J. A., 1999, Late Aptian to late Albian sea-level fluctuations constrained by geochemical and biological evidence (Nahr Umr Formation, Oman): Journal of Sedimentary Research, v. 69, p. 434-446.

IMMENHAUSER, A., VAN DER KOOIJ, B., VAN VLIET, A., SCHLAGER, W., AND SCOTT, R. W., 2001, An ocean-facing Aptian-Albian carbonate margin, Oman: Sedimentology, v. 48, p. 1187-1207.

INESON, J. R., AND SURLYK, F., 2000, Carbonate megabreccias in a sequence stratigraphic context; evidence from the Cambrian of North Greenland, in D. Hunt, and R. L. Gawthorpe, eds., Sedimentary Responses to Forced Regressions: Geological Society London Special Publications, v. 172, p. 47-68.

IRWIN, M. L., 1965, General theory of epeiric clear water sedimentation: American Association of Petroleum Geologists Bulletin, v. 49, p. 445-459.

ISERN, A., ANSELMETTI, F., BLUM, P., AND PARTY, S., 2002, Leg 194 - Marion Plateau, northeast Australia: Ocean Drilling Program Initial Reports: College Station, Ocean Drilling Program, v. 194, 116 p.

JACQUIN, T., ARNAUD-VANNEAU, A., ARNAUD, H., RAVENNE, C., AND VAIL, P. R., 1991, Systems tracts and depositional sequences in a carbonate setting: a study of continuous outcrops from platform to basin at the scale of seismic lines: Marine and Petroleum Geology, v. 8, p. 122-139.

JAMES, N. P., 1997, The cool-water carbonate depositional realm, in N. P. James, and J. A. D. Clarke, eds., Cool-water Carbonates: SEPM Special Publications, v. 56, p. 1-20.

JAMES, N. P., AND BONE, Y., 1989, Petrogenesis of Cenozoic, temperate water calcarenites, south Australia: a model for meteoric/shallow burial diagenesis of shallow water calcite sediments: Journal of Sedimentary Research, v. 59, p. 191-203.

JAMES, N. P., BONE, Y., COLLINS, L. B., AND KYSER, T. K., 2001, Surficial sediments of the Great Australian Bight: facies dynamics and oceanography on a vast cool-water carbonate shelf: Journal of Sedimentary Research, v. 71, p. 549-567.

JAMES, N. P., AND BOURQUE, P. A., 1992, Reefs and mounds, in R. G. Walker, and N. P. James, eds., Facies models: St. Johns, Geological Association of Canada, p. 323-345.

JAMES, N. P., COLLINS, L. B., BONE, Y., AND HALLOCK, P., 1999, Subtropical carbonates in a temperate realm: modern sediments on the southwest Australian shelf: Journal of Sedimentary Research, v. 69, p. 1297-1321.

JAMES, N. P., FEARY, D. A., SURLYK, F., SIMO, J. A., BETZLER, C., HOLBOURN, A. E., LI, Q., MATSUDA, H., MACHIYAMA, H., BROOKS, G. R., ANDRES, M. W., HINE, A. C., MALONE, M. J., AND PARTY, 2000, Quaternary bryzoan reef mounds in cool-water, upper slope environments: Great Australian Bight: Geology, v. 28, p. 647 - 650.

JAMES, N. P., AND KENDALL, A. C., 1992, Introduction to carbonate and evaporite facies models, in R. G. Walker, and N. P. James, eds., Facies Models: St.Johns, Geological Association of Canada, p. 265-275.

JAMES, N. P., AND MACINTYRE, I. G., 1985, Carbonate depositional environments, modern and ancient. Part 1 - Reefs: Zonation, depositional facies, diagenesis: Colorado School of Mines Quarterly, v. 80, 70 p.

JENKYNS, H. C., AND WILSON, P. A., 1999, Stratigraphy, paleoceanography, and evolution of Cretaceous Pacific guyots; relics from a greenhouse Earth: American Journal of Science, v. 299, p. 341-392.

JERLOV, N. G., 1976, Marine Optics: Elsevier Oceanography Series: Amsterdam, Elsevier, v. 14, 231 p.

JERVEY, M. T., 1988, Quantitative geological modeling of siliciclastic rock sequences and their seismic expression, in C. K. Wilgus, B. S. Hastings, C. G. S. C. Kendall, H. W. Posamentier, C. A. Ross, and J. C. Van Wagoner, eds., Sea-level changes: an integrated approach: SEPM Special Publication, v. 42, p. 47-69.

JOHNSON, M. A., KENYON, N. H., BELDERSON, R. H., AND STRIDE, A. H., 1982, Sand transport, in A. H. Stride, ed., Offshore Tidal Sands: Processes and Deposits.: London, Chapman Hall, p. 58-94.

KEIM, L., AND SCHLAGER, W., 1999, Automicrite facies on steep slopes (Triassic, Dolomites, Italy): Facies, v. 41, p. 15-26.

KEIM, L., AND SCHLAGER, W., 2001, Quantitative compositional analyses of a Triassic carbonate platform (Southern Alps, Italy): Sedimentary Geology, v. 139, p. 261-283.

KENDALL, C. G. S. C., AND LERCHE, I., 1988, The rise and fall of eustasy, in C. K. Wilgus, B. S. Hastings, H. Posamentier, J. Van Wagoner, C. A. Ross, and C. G. S. C. Kendall, eds., Sea-level changes: an integrated approach: SEPM Special Publications, v. 42, p. 3-18.

KENNARD, J. M., SOUTHGATE, P. N., JACKSON, M. J., O'BRIEN, P. E., CHRISTIE-BLICK, N., HOLMES, A. E., AND SARG, J. F., 1992, New sequence perspective on the Devonian reef complex and the Frasnian-Famennian boundary, Canning Basin, Australia: Geology, v. 20, p. 1135-1138.

KENTER, J. A. M., 1990, Carbonate platform flanks: slope angle and sediment fabric: Sedimentology, v. 37, p. 777-794.

KENTER, J. A. M., BRACCO GARTNER, G. L., AND SCHLAGER, W., 2001, Seismic models of a mixed carbonate-siliciclastic shelf margin: Permian upper San Andres Formation, Last Chance Canyon, New Mexico: Geophysics, v. 66, p. 1744-1748.

KENTER, J. A. M., GINSBURG, R. N., AND TROELSTRA, S. R., 2001, Sea-level-driven sedimentation patterns on the slope and margin, in R. N. Ginsburg, ed., Subsurface Geology of a Prograding Carbonate Platform Margin, Great Bahama Bank: Results of the Bahamas Drilling Project: SEPM Special Publications, v. 70, p. 61-100.

KENYON, P. M., AND TURCOTTE, D. L., 1985, Morphology of a delta prograding by bulk sediment transport: Geological Society of America, Bulletin, v. 96, p. 1457-1465.

KERANS, C., AND TINKER, S., 1999, Extrinsic stratigraphic controls on development of the Capitan reef complex, in A. H. Saller, P. M. Harris, B. L. Kirkland, and S. J. Mazzullo, eds., Geologic framework of the Capitan Reef.: SEPM Special Publication, v. 65, p. 15-36.

KIER, J. S., AND PILKEY, O. H., 1971, The influence of sea level changes on sediment carbonate mineralogy, Tongue of the Ocean, Bahamas: Marine Geology, v. 11, p. 189-200.

KIESSLING, W., 2002, Secular variations in the Phanerozoic reef ecosystem, in W. Kiessling, E. Flügel, and J. Golonka, eds., Phanerozoic Reef Patterns: SEPM Special Publications, v. 72, p. 625-690.

KIEVMAN, C. M., 1998, Match between late Pleistocene Great Bahama Bank and deep-sea oxygen isotope records of sea level: Geology, v. 26, p. 635-638.

KIRKBY, M. J., 1987, General models of long-term slope evolution through mass movement, in M. G. Anderson, and K. S. Richards, eds., Slope Stability: Chichester, Wiley, p. 359-380.

KNOERICH, A., AND MUTTI, M., 2003, Controls of facies and sediment composition on the diagenetic pathway of shallow-water Heterozoan carbonates: the Oligocene of the Maltese Islands: International Journal of Earth Sciences, v. 92, p. 494-510.

KOZIR, A., 2004, Microcodium revisited; root calcification products of terrestrial plants on carbonate-rich substrates: Journal of Sedimentary Research, v. 74, p. 845-857.

LADD, H. S., 1973, Bikini and Eniwetok atolls, Marshall Islands, in O. A. Jones, and R. Endean, eds., Biology and Geology of Coral Reefs: New York, Academic Press, v. 1, p. 93-111.

LADD, H. S., AND HOFFMEISTER, J. E., 1945, Geology of Lau, Fiji: Bernice P. Bishop Museum Bulletin, v. 181, p. 1-190.

LASKAR, J., JOUTEL, F., AND BOUDIN, F., 1993, Orbital, precessional, and insolation quantities for the Earth from -20Myr to +10Myr: Astronomy and Astrophysics, v. 270, p. 522-533.

LEEDER, M., 1999, Sedimentology and Sedimentary Basins: Oxford, Blackwell, 608 p.

LEES, A., 1975, Possible influence of salinity and temperature on modern shelf carbonate sedimentation: Marine Geology, v. 19, p. 159-198.

LEES, A., AND BULLER, A. T., 1972, Modern temperate-water and warm-water shelf carbonate sediments contrasted: Marine Geology, v. 13, p. M67-M73.

LEES, A., AND MILLER, J., 1985, Facies variation in Waulsortian buildups. Part 2. Mid-Dinantian buildups from Europe and North America: Geological Journal, v. 20, p. 159-180.

LEES, A., AND MILLER, J., 1995, Waulsortian banks, in C. L. V. Monty, D. W. J. Bosence, P. H. Bridges, and B. R. Pratt, eds., Carbonate Mud Mounds: International Association of Sedimentologists Special Publication, v. 23, p. 191-271.

LEINFELDER, R. R., NOSE, M., SCHMID, D. U., AND WERNER, W., 1993, Microbial crusts of the Late Jurassic: composition, palaeoecological significance and importance in reef construction: Facies, v. 29, p. 195-230.

LIDZ, B., REICH, C. D., AND SHINN, E. A., 2003, Regional Quaternary submarine geomorphology in the Florida Keys: Geological Society of America Bulletin, v. 115, p. 845-866.

LIGHTY, R. G., MACINTYRE, I. G., AND STUCKENRATH, R., 1978, Submerged early Holocene barrier reef southeast Florida shelf: Nature, v. 275, p. 59-60.

LOURENS, L. J., AND HILGEN, F. J., 1997, Long-period variations in the Earth's obliquity and their relation to third-order eustatic cycles and late Neogene glaciations.: Quaternary International, v. 40, p. 43-52.

LOUTIT, T. S., HARDENBOL, J., VAIL, P. R., AND BAUM, G. R., 1988, Condensed sections: the key to age dating and correlation of continental margin sequences, in C. K. Wilgus, B. S. Hastings, C. G. S. C. Kendall, H. W. Posamentier, C. A. Ross, and J. C. Van Wagoner, eds., Sea-Level Changes: An Integrated Approach: SEPM Special Publications, v. 42, p. 183-213.

LOWENSTAM, H. A., AND WEINER, S., 1989, On Biomineralization: New York, Oxford University Press, 324 p.

LUCIA, F. J., 1995, Rock-fabric/petrophysical classification of carbonate pore space for reservoir characterization: American Association of Petroleum Geologists Bulletin, v. 79, p. 1275-1300.

LUDWIG, K. R., HALLEY, R. B., SIMMONS, K. R., AND PETERMAN, Z. B., 1988, Strontium-isotope stratigraphy of Enewetak Atoll: Geology, v. 16, p. 173-177.

LYNTS, G. W., JUDD, J. W., AND STEHMANN, C. F., 1973, Late Pleistocene history of Tongue of the Ocean, Bahamas: Geological Society of America Bulletin, v. 84, p. 2605-2684.

MACARTHUR, R. H., AND WILSON, E. O., 1967, Theory of Island Biogeography: Princeton, Princeton University Press, 203 p.

MACHEL, H. G., 2004, Concepts and models of dolomitization: a critical reappraisal, in C. J. R. Braithwaite, G. Rizzi, and G. Darke, eds., The Geometry and Petrogenesis of Dolomite Hydrocarbon Reservoirs: Geological Society London, Special Publications, v. 235, p. 7-63.

MACINTYRE, I. G., RÜTZLER, K., NORRIS, J. N., SMITH, K. P., CAIRNS, S. D., BUCHER, K. E., AND STENECK, R. S., 1991, An early Holocene reef in the Western Atlantic - submersible investigations of a deep relict reef off the west coast of Barbados, WI: Coral Reefs, v. 10, p. 167-174.

MACKENZIE, F. T., BISCHOFF, W. D., BISHOP, F. C., LOIJENS, M., SCHOONMAKER, J., AND WOLLAST, R., 1983, Magnesian calcites: low-temperature occurrence, solubility and solid solution behaviour, in R. J. Reeder, ed., Carbonates: Mineralogy and Chemistry: Mineralogical Society of America, Reviews in Mineralogy, v. 11, p. 97-144.

MACNEIL, F. S., 1954, The shape of atolls; an inheritance from subaerial erosion forms: American Journal of Science, v. 252, p. 402-427.

MALLON, A. J., SWARBICK, R. E., AND KATSUBE, T. J., 2005, Permeability of fine-grained rocks: New evidence from chalks: Geology, v. 33, p. 21-24.

MANDELBROT, B. B., 1967, How long is the coast of Britain? Statistical self-similarity and fractional dimension: Science, v. 156, p. 636-638.

MCNEILL, D. F., EBERLI, G. P., LIDZ, B. H., SWART, P. K., AND KENTER, J. A. M., 2001, Chronostratigraphy of a prograded carbonate platform margin: a record of dynamic slope sedimentation, western Great Bahama Bank, in R. N. Ginsburg, ed., Subsurface Geology of a Prograding Carbonate Platform Margin, Great Bahama Bank: Results of the Bahamas Drilling Project: SEPM Special Publications, v. 70, p. 101-135.

MCNEILL, D. F., GINSBURG, R. N., CHANG, S. B. R., AND KIRSCHVINK, J. L., 1988, Magnetostratigraphic dating of shallow-water carbonates from San Salvador, the Bahamas: Geology, v. 16, p. 8-12.

MCNEILL, D. F., GRAMMER, G. M., AND WILLIAMS, S. C., 1998, A 5My chronolgy of carbonate platform margin aggradation, southwestern Little Bahama Bank, Bahamas: Journal of Sedimentary Research, v. 68, p. 603-614.

MELIM, L. A., ANSELMETTI, F. S., AND EBERLI GREGOR, P., 2001, The importance of pore type on permeability of Neogene carbonates, Great Bahama Bank, in R. N. Ginsburg, ed., Subsurface Geology of a Prograding Carbonate Platform Margin, Great Bahama Bank: Results of the Bahamas Drilling Project: SEPM Special Publications, v. 70, p. 217-238.

MENARD, H. W., 1986, Islands: New York, Scientific American Books, 230 p.

MEYER, F. O., 1989, Siliciclastic influence on Mesozoic platform development: Baltimore Canyon Trough, Western Atlantic, in P. D. Crevello, J. L. Wilson, J. F. Sarg, and J. F. Read, eds., Controls on Carbonate Platform and Basin Development: SEPM Special Publication, v. 44, p. 213-232.

MIALL, A. D., 1992, Exxon global cycle chart: an event for every occasion? : Geology, v. 20, p. 787-790.

MIALL, A. D., 1997, The Geology of Stratigraphic Sequences: Berlin, Springer, 433 p.

MIDDLETON, M. F., 1987, Seismic stratigraphy of Devonian reef complexes, northern Canning basin, Western Australia: American Association of Petroleum Geologists Bulletin, v. 71, p. 1488-1498.

MILLIMAN, J. D., 1974, Marine Carbonates. Recent Sedimentary Carbonataes, Part I: Berlin, Springer, 375 p.

MINERO, C. J., 1988, Sedimentation and diagenesis along an island-sheltered platform margin, El Abra Formation, Cretaceous of Mexico: Paleokarst, v. 18, p. 385-405.

MITCHUM, R. M., AND VAN WAGONER, J. C., 1991, High-Frequency Sequences and Their Stacking Patterns - Sequence-Stratigraphic Evidence of High-Frequency Eustatic Cycles: Sedimentary Geology, v. 70, p. 131-160.

MOLDOVANYI, E. P., WAAL, F. M., AND YAN, Z. J., 1995, Regional exposure events and platform evolution of Zhujiang Formation carbonates. Pearl River Mouth Basin: evidence from primary and diagenetic seismic facies, in D. A. Budd, A. H. Saller, and P. M. Harris, eds., Unconformities and Porosity in Carbonate Strata: American Association of Petroleum Geologists Memoirs, v. 63, p. 125-140.

MONTAGGIONI, L. F., 1985, Makatea Island, Tuamotu archipelago.: 5th International Coral Reef Congress Proceedings v.1, Tahiti, p. 103-158.

MONTAGGIONI, L. F., 2000, Postglacial reef growth: Earth Planetary Science, v. 331, p. 319-330.

MONTANEZ, I. P., AND OSLEGER, D. A., 1993, Parasequence stacking patterns, third-order accommodation events and sequence stratigraphy of Middle to Upper Cambrian platform carbonates, Bonanza King Formation, southern Great Basin, in R. G. Loucks, and J. F. Sarg, eds., Carbonate Sequence Stratigraphy: American Association of Petroleum Geologists Memoirs, v. 57, p. 305-326.

MONTY, C. L. V., 1995, The rise and nature of carbonate mud-mounds: an introductory actualistic approach, in C. L. V. Monty, D. W. J. Bosence, P. H. Bridges, and B. R. Pratt, eds., Carbonate Mud-Mounds - their Origin and Evolution: International Association of Sedimentologists Special Publications, v. 23, p. 11-48.

MONTY, C. L. V., BOSENCE, D. W. J., BRIDGES, P. H., AND PRATT, B. R., EDS. 1995, Carbonate Mud-mounds: Their Origin and Evolution: International Association of Sedimentologists Special Publication, v. 23, 537 p.

MOORE, C. H., 2001, Carbonate Reservoirs: Amsterdam, Elsevier, 444 p.

MORSE, J. W., 2004, Formation and diagenesis of carbonate sediments, in F. T. Mackenzie, ed., Sediments, Diagenesis, and Sedimentary Rocks: Treatise on Geochemistry: Amsterdam, Elsevier, v. 7, p. 67-85.

MORSE, J. W., AND MACKENZIE, F. T., 1990, Geochemistry of Sedimentary Carbonates: Developments in Sedimentology: Amsterdam, Elsevier, v. 48, 707 p.

MULLINS, H. T., 1983, Comment on "Eustatic control of turbidites and winnowed turbidites": Geology, v. 11, p. 57 - 58.

MULLINS, H. T., HEATH, K. C., VAN BUREN, H. M., AND NEWTON, C. R., 1984, Anatomy of modern deep-ocean carbonate slope: northern Little Bahama Bank: Sedimentology, v. 31, p. 141-168.

MUTTI, M., AND HALLOCK, P., 2003, Carbonate systems along nutrient and temperature gradients: some sedimentological and geochemical constraints: International Journal of Earth Sciences, v. 92, p. 463-475.

NAISH, T., AND KAMP, P. J. J., 1997, Sequence stratigraphy of sixth-order (41 k.y.) Pliocene-Pleistocene cyclothems, Wanganui basin, New Zealand: A case for the regressive systems tract: Geological Society of America Bulletin, v. 109, p. 978-999.

NEIDELL, N. S., 1979, Stratigraphic modeling and interpretation; geophysical principles and techniques: American Association of Petroleum Geologists Education Course Notes, v. 13, 141 p.

NEIDELL, N. S., AND POGGLIAGLIOLMI, F., 1977, Stratigraphic modeling and interpretation - geophysical principles and techniques, in C. E. Payton, ed., Seismic stratigraphy - applications to hydrocarbon exploration: American Association of Petroleum Geologists Memoirs, v. 26, p. 389 - 416.

NELSON, C. S., 1982, Compendium of sample data for temperate carbonate sediments, Three Kings region, Northern New Zealand, Occasional Report, University of Waikato, Department of Earth Sciences, Hamilton, University of Waikato, p. 95.

NELSON, C. S., 1988, An introductory perspective on non-tropical shelf carbonates: Sedimentary Geology, v. 60, p. 3-12.

NELSON, C. S., HANCOCK, G. E., AND KAMP, P. J. J., 1982, Shelf to basin, temperate skeletal carbonate sediments, Three Kings Plateau, New Zealand: Journal of Sedimentary Petrology, v. 52, p. 717-732.

NELSON, C. S., AND JAMES, N. P., 2000, Pliocene Te Aute limestones, New Zealand: expanding concepts for cool-water shelf carbonates: New Zealand Journal of Geology and Geophysics, v. 46, p. 407-424.

NEUHAUS, D., BORGOMANO, J., JAUFFRED, J., MERCADIER, C., OLOTU, S., AND GRÖTSCH, J., 2004, Quantitative seismic reservoir characterization of an Oligocene-Miocene Carbonate Buildup: Malampaya Field, Philippines., in G. P. Eberli, J. L. Massaferro, and J. F. Sarg, eds., Seismic Imaging of Carbonate Reservoirs and Systems: American Association of Petroleum Geologists Memoirs, v. 81, p. 169-183.

NEUMANN, A. C., AND HEARTY, P. J., 1996, Rapid sea-level changes at the close of the last interglacial (substage 5e) recorded in Bahamian island geology: Geology, v. 24, p. 775-778.

NEUMANN, A. C., AND LAND, L. S., 1975, Lime mud deposition and calcareous algae in the Bight of Abaco, Bahamas: a budget: Journal of Sedimentary Petrology, v. 45, p. 763-768.

NEUMANN, A. C., AND MACINTYRE, I., 1985, Reef response to sea level rise: keep-up, catch-up or give-up: 5th International Coral Reef Congress Tahiti, Proceedings v.3, p. 105-110.

NEUWEILER, 1995, Dynamische Sedimentationsvorgaenge, Diagenese und Biofazies unterkretazischer Plattformraender (Apt/Alb; Soba-Region, Prov. Cantabria, N-Spanien): Berliner geowissenschaftliche Abhandlungen: Berlin, v. E17, 235 p.

NEUWEILER, F., D'ORAZIO, V., IMMENHAUSER, A., GEIPEL, G., HEISE, K.-H., COCOZZA, C., AND MIANO, T. M., 2003, Fulvic acid-like organic compounds control nucleation of marine calcite under suboxic conditions: Geology, v. 31, p. 681-684.

NEUWEILER, F., GAUTRET, P., THIEL, V., LANGE, R., MICHAELIS, W., AND REITNER, J., 1999, Petrology of lower Cretaceous carbonate mud mounds (Albian, N. Spain): insight into organomineralic deposits of the geological record: Sedimentology, v. 46, p. 837-859.

NEUWEILER, F., MEHDI, M., AND WILMSEN, M., 2001, Facies of Liassic sponge mounds, Central High Atlas, Morocco: Facies, v. 44, p. 243-264.

NEUWEILER, F., REITNER, J., AND MONTY, C., 1997, Biosedimentology of microbial buildups: Facies, v. 36, p. 195-284.

NEUWEILER, F., RUTSCH, M., GEIPEL, G., REIMER, A., AND HEISE, K. H., 2000, Soluble humic substances from in situ precipitated microcrystalline calcium carbonate, internal sediment, and spar cement in a Cretaceous carbonate mud-mound: Geology, v. 28, p. 851-854.

NUMMEDAL, D., GUPTA, S., PLINT, A. G., AND COLE, R. D., 1995, The falling stage systems tract: definition, character and expression in several examples from the Cretaceous from the U.S. Western Interior: Sedimentary responses to forced regressions, p. 45-48.

NUMMEDAL, D., RILEY, G. W., AND TEMPLET, P. L., 1993, High-resolution sequence architecture: a chronostratigraphic model based on equilibrium profile studies, in H. E. Posamentier, C. P. Summerhayes, B. U. Haq, and G. P. Allen, eds., Sequence Stratigraphy and Facies Associations: International Association of Sedimentologists Special Publications, v. 18, p. 55-68.

OPDYKE, B. N., AND WILKINSON, B. H., 1990, Paleolatitude distribution of Phanerozoic marine ooids and cements: Palaeogeography, Palaeoclimatology, Palaeoecology, v. 78, p. 135-148.

OPEN UNIVERSITY, O. C. T., 1989a, Seawater: its composition, properties and behavior: Oxford, Pergamon Press, 186 p.

OPEN UNIVERSITY, O. C. T., 1989b, Ocean circulation: Oxford, Pergamon Press, 236 p.

PASSLOW, V., 1997, Slope sedimentation and shelf to basin sediment transfer: a cool-water carbonate example from the Otway margin, southeastern Australia., in N. P. James, and J. A. D. Clarke, eds., Cool-water carbonates: SEPM Special Publication, SEPM Special Publication, v. 56, p. 107-125.

PETERHÄNSEL, A., AND PRATT, B. R., 2001, Nutrient-triggered bioerosion on a giant carbonate platform masking the postextinction Famennian benthic community: Geology, v. 29, p. 1079-1082.

PINET, P. R., AND POPENOE, P., 1985, Shallow seismic stratigraphy and post-Albian geologic history of the northern and central Blake Plateau: Geological Society of America Bulletin, v. 96, p. 627-638.

PIPKIN, B. W., GORSLINE, D. S., CASEY, R. E., AND HAMMOND, D. E., 1987, Laboratory Exercises in Oceanography: New York, Freeman, 257 p.

PITMAN, W. C., 1978, Relationship between eustasy and stratigraphic sequences of passive margins: Geological Society of America Bulletin, v. 89, p. 1389-1403.

PLAYFORD, P. E., 1980, Devonian "Great Barrier Reef" of Canning Basin, Western Australia: American Association of Petroleum Geologists Bulletin, v. 64, p. 814-840.

PLAYFORD, P. E., 2002, Palaeokarst, pseudokarst, and sequence stratigraphy in Devonian reef complexes of the Canning Basin, Western Australia, in M. Keep, and S. J. Moss, eds., The Sedimentary Basins of Western Australia: Exploration Society of Australia Symposium Proceedings: Perth, v. 3, p. 763-793.

PLAYFORD, P. E., COCKBAIN, A. E., HOCKING, R. M., AND WALLACE, M. W., 2001, Novel paleoecology of a postextinction reef: Famennian (Late Devonian) of the Canning basin, northwest Australia: comment: Geology, v. 29, p. 1155-1156.

PLAYFORD, P. E., HURLEY, N. F., KERANS, C., AND MIDDLETON, M. F., 1989, Reefal platform development, Devonian of the Canning Basin, Western Australia, in P. D. Crevello, J. L. Wilson, J. F. Sarg, and J. F. Read, eds., Controls on Carbonate Platform and Basin Development: SEPM Special Publication, v. 44, p. 187-202.

PLINT, A. G., AND NUMMEDAL, D., 2000, The falling stage systems tract: recognition and importance in sequence stratigraphic analysis, in D. Hunt, and R. L. Gawthorpe, eds., Sedimentary Responses to Forced Regressions: Geological Society London Special Publications, v. 172, p. 1-17.

PLOTNICK, J. E., 1986, A fractal model for the distribution of stratigraphic hiatuses: Journal of Geology, v. 94, p. 885-890.

POMAR, L., 1991, Reef geometries, erosion surfaces and high-frequency sea-level changes, upper Miocene reef complex, Mallorca, Spain: Sedimentology, v. 38, p. 243-270.

POMAR, L., 1993, High-resolution sequence stratigraphy in prograding Miocene carbonates: application to seismic interpretation, in R. G. Loucks, and J. F. Sarg, eds., Carbonate sequence stratigraphy: American Association of Petroleum Geologists Memoir, p. 389-408.

POMAR, L., 2000, Excursion Guide, Field Visits Mallorca, Palma de Mallorca, University of the Balearic Islands, Departament de Ciencias de la Terra, p. 85.

POMAR, L., AND WARD, W. C., 1995, Sea-level changes, carbonate production and platform architecture: the Llucmajor platform, Mallorca, Spain, in B. U. Haq, ed., Sequence Stratigraphy and Depositional Response to Eustatic, Tectonic and Climatic Forcing: Coastal Systems and Continental Margins: Dordrecht, Kluwer, v. 1, p. 87-112.

POMAR, L., AND WARD, W. C., 1999, Reservoir-scale heterogeneity in depositional packages and diagenetic patterns on a reef-rimmed platform, Upper Miocene, Mallorca, Spain: American Association of Petroleum Geologists Bulletin, v. 83, p. 1759-1773.

POMAR, L., WARD, W. C., AND GREEN, D. G., 1996, Upper Miocene reef complex of the Llucmajor area, Mallorca, Spain, in E. K. Franseen, M. Esteban, W. C. Ward, and J. M. Rouchy, eds., Models for Carbonate Stratigraphy from Miocene Reef Complexes of the Mediterranean Regions: SEPM Concepts in Sedimentology and Paleontology, v. 5, p. 191-225.

POPE, M. C., AND READ, J. F., 1997, High-resolution stratigraphy of the Lexington Limestone (late Middle Ordovician), Kentucky, U.S.A.: a cool-water carbonate-clastic ramp in a tectonically active foreland basin, in N. P. James, and J. A. D. Clarke, eds., Cool-Water Carbonates: SEPM Special Publications, v. 56, p. 411-430.

POSAMENTIER, H. W., AND ALLEN, G. P., 1999, Siliciclastic sequence stratigraphy - concepts and applications: SEPM Concepts in Sedimentology and Paleontology: Tulsa, Society for Sedimentary Geology, v. 7, 209 p.

POSAMENTIER, H. W., ALLEN, G. P., AND JAMES, D. P., 1992a, High resolution sequence stratigraphy - The East Coulee delta, Alberta: Journal of Sedimentary Petrology, v. 62, p. 310-317.

POSAMENTIER, H. W., ALLEN, G. P., JAMES, D. P., AND TESSON, M., 1992b, Forced regressions in a sequence stratigraphic framework: concepts, examples, and exploration significance: American Association of Petroleum Geologists Bulletin, v. 76, p. 1687-1709.

POSAMENTIER, H. W., JERVEY, M. T., AND VAIL, P. R., 1988, Eustatic controls on clastic deposition I - conceptual framework, in C. K. Wilgus, B. S. Hastings, H. Posamentier, J. Van Wagoner, and C. G. S. C. Kendall, eds., Sea-level Changes: an Integrated Approach: SEPM Special Publications, v. 42, p. 109-124.

POSAMENTIER, H. W., AND VAIL, P. R., 1988, Eustatic controls on clastic deposition II - sequence and systems tract models, in C. K. Wilgus, B. S. Hastings, C. G. S. C. Kendall, H. W. Posamentier, C. A. Ross, and J. C. Van Wagoner, eds., Sea-level Changes: an Integrated Approach: SEPM Special Publication, v. 42, p. 125-154.

POTTER, P. E., 1959, Facies model conference: Science, v. 129, p. 1292-1294.

PRATT, B. R., 1995, The origin, biota and evolution of deep-water mud-mounds, in C. L. V. Monty, D. W. J. Bosence, P. H. Bridges, and B. R. Pratt, eds., Carbonate Mud-mounds: International Association of Sedimentologists Special Publications, v. 23, p. 49-125.

PRATT, B. R., AND JAMES, N. P., 1986, The St.George Group (Lower Ordovician) of western Newfoundland: tidal flat island model for carbonate sedimentation in shallow epeiric seas: Sedimentology, v. 33, p. 313-343.

PRETO, N., AND HINNOV, L. A., 2003, Unraveling the origin of carbonate platform cyclothems in the Upper Triassic Dürrenstein Formation (Dolomites, Italy): Journal of Sedimentary Research, v. 73, p. 774-789.

PRETO, N., HINNOV, L. A., HARDIE, L. A., AND DE ZANCHE, V., 2001, Middle Triassic orbital signature recorded in the shallow-marine Latemar carbonate buildup (Dolomites, Italy): Geology, v. 29, p. 1123-1126.

PURDY, E., 1974, Reef configurations: cause and effect, in L. F. Laporte, ed., Reefs in Time and Space: SEPM Special Publications, v. 18, p. 9-76.

PURDY, E. G., GISCHLER, E., AND LOMANDO, A. J., 2003, The Belize margin revisited. 2. Origin of Holocene antecedent topography: International Journal of Earth Sciences, v. 92, p. 552-572.

PURDY, E. G., AND WINTERER, E. L., 2001, Origin of atoll lagoons: Geological Society of America Bulletin, v. 113, p. 837-854.

PURSER, B. H., ED.,. 1973, The Persian Gulf. Holocene carbonate sedimentation and diagenesis in a shallow epicontinental sea: New York, Springer-Verlag, 471 p.

RANKEY, E. C., 2002, Spatial patterns of sediment accumulation on a Holocene carbonate tidal flat, northwest Andros Island, Bahamas: Journal of Sedimentary Research, v. 72, p. 591-601.

READ, J. F., 1982, Carbonate platforms of passive (extensional) continental margins: types, characteristics and evolution: Tectonophysics, v. 81, p. 195-212.

READ, J. F., 1985, Carbonate platform facies models: American Association of Petroleum Geologists Bulletin, v. 69, p. 1-21.

READ, J. F., 1989, Controls on evolution of Cambrian-Ordovician passive margin, U.S. Appalachians, in P. D. Crevello, J. L. Wilson, J. F. Sarg, and J. F. Read, eds., Controls on Carbonate Platform and Basin Development: SEPM Special Publications, v. 44, p. 146-165.

READ, J. F., AND GOLDHAMMER, R. K., 1988, Use of Fischer plots to define third-order sea-level curves in Ordovician peritidal cyclic carbonates, Appalachians: Geology, v. 16, p. 895-899.

READING, H. R., AND LEVELL, B. K., 1996, Controls on the sdimentary rock record, in H. R. Reading, ed., Sedimentary Environments: Oxford, Blackwell, p. 5-36.

REECKMANN, A., AND GILL, J. B., 1981, Rates of vadose diagenesis in Quaternary dune and shallow marine calcarenites, Warnambool, Vitoria, Australia: Sedimentary Geology, v. 30, p. 157-172.

REID, R. P., VISSCHER, P. T., DECHO, A. W., STOLZ, J. F., BEBOUT, B. M., DUPRAZ, C., MACINTYRE, I. G., PAERL, H. W., PINCKNEY, J. L., PRUFERT, B. L., STEPPE, T. F., AND DESMARAIS, D. J., 2000, The role of microbes in accretion, lamination and early lithification of modern marine stromatolites: Nature, v. 406, p. 989-992.

REID, S. K., AND DOROBEK, S. L., 1993, Sequence stratigraphy and evolution of a progradational, foreland carbonate ramp, Lower Mississippian Mission Canyon Formation and stratigraphic equivalents, Montana and Idaho, in R. G. Loucks, and J. F. Sarg, eds., Carbonate sequence stratigraphy; recent developments and applications: American Association of Petroleum Geologists Memoir, v. 57, p. 327-352.

REID, S. K., AND DOROBEK, S. L., 1993, Sequence stratigraphy and evolution of a progradational, foreland carbonate ramp, Lower Mississippian Mission Canyon Formation and stratigraphic equivalents, Montana and Idaho, in R. G. Loucks, and J. F. Sarg, eds., Carbonate Sequence Stratigraphy: American Association of Petroleum Geologists Memoirs, v. 57, p. 327-352.

REIJMER, J. J. G., SCHLAGER, W., AND DROXLER, A. W., 1988, Site 632: Pliocene-Pleistocene sedimentation in a Bahamian basin, in J. A. Austin, W. Schlager, and A. Palmer, eds., Proceedings Ocean Drilling Program, Scientific Results: College Station, Ocean Drilling Program, v. 101, p. 213-220.

REIJMER, J. J. G., SPRENGER, A., TEN KATE, W. G. H. Z., SCHLAGER, W., AND KRYSTYN, L., 1994, Periodicities in the composition of Late Triassic calciturbidites (Eastern Alps, Austria), in P. L. De Boer, and D. G. Smith, eds., Orbital Forcing and Cyclic Sequences: International Association of Sedimentologists Special Publications, v. 19, p. 323-343.

REIJMER, J. J. G., TEN KATE, W. G. H. Z., SPRENGER, A., AND SCHLAGER, W., 1991, Calciturbidite composition related to exposure and flooding of a carbonate platform (Triassic, Eastern Alps): Sedimentology, v. 38, p. 1059-1074.

REITNER, J., ARP, G., THIEL, V., GAUTRET, P., GALLING, U., AND MICHAELIS, W., 1997, Organic matter in Great Salt Lake ooids (Utah, USA) - first approach to a formation via organic matrices: Facies, v. 36, p. 210-219.

REITNER, J., GAUTRET, P., MARIN, F., AND NEUWEILER, F., 1995b, Automicrites in a modern marine microbialite. Formation model via organic matrices (Lizard Island, Great Barrier Reef, Australia): Bulletin Institut Oceanographique de Monaco, v. special issue 14, p. 1-26.

Reitner, J., Neuweiler, F., Flajs, G., Vigener, M., Keupp, H., Meischner, D., Neuweiler, F., Paul, J., Warnke, K., Weller, H., Dingle, P., Hensen, C., Schaefer, P., Gautret, P., Leinfelder, R. R., Huessner, H., and Kaufmann, B., 1995a, Mud mounds: a polygenetic spectrum of fine-grained carbonate buildups: Facies, v. 32, p. 1-70.

Reitner, J., Thiel, V., Zankl, H., Michaelis, W., Worheide, G., and Gautret, P., 2000, Organic and biochemical patterns in cryptic microbialites, in R. E. Riding, and S. M. Awramik, eds., Microbial Sediments: Berlin, Springer, p. 149-160.

Remane, A., and Schlieper, C., 1971, Biology of Brackish Water: Stuttgart, Schweizerbart, 372 p.

Rendle, R. H., and Reijmer, J. J. G., 2002, Quaternary slope development of the western, leeward margin of the Great Bahama Bank: Marine Geology, v. 185, p. 143-164.

Revelle, R. R., Barnett, T. P., Barron, E. J., Bloom, A. L., Christie-Blick, N., Harrison, C. G. A., Hay, W. W., Matthews, R. K., Meier, M. F., Munk, W. H., Peltier, R. W., Roemmich, D., Sturges, W., Sundquist, E. T., Thompson, K. R., and Thompson, S. L., 1990, Overview and recommendations, in G. S. Committee, ed., Sea-Level Change: Washington, National Academy Press, p. 3-34.

Rich, J. L., 1951, Three critical environments of deposition, and criteria for recognition of rocks deposited in each of them: Geological Society of America Bulletin, v. 62, p. 1-20.

Rine, J. M., and Ginsburg, R. N., 1985, Depositional facies of a mud shoreface in Suriname, South America - a mud analogue to sandy shallow-marine deposits: Journal of Sedimentary Petrology, v. 55, p. 633-652.

Robbin, D. M., and Stipp, J. J., 1979, Depositional rate of laminated soilstone crusts, Florida Keys: Journal of Sedimentary Petrology, v. 49, p. 175-178.

Roberts, H. H., Aharon, P., and Phipps, C. V., 1988, Morphology and sedimentology of Halimeda bioherms from the eastern Java Sea (Indonesia): Coral Reefs, v. 6, p. 161-172.

Roth, S., and Reijmer, J. J. G., 2004, Holocene Atlantic climate variations deduced from carbonate periplatform sediments (leeward margin, Great Baham Bank): Paleoceanography, v. 19, p. 1-14.

Royer, D. L., Berner, R. A., Montanez, I. P., Tabor, N. J., and Beerling, D. J., 2004, CO2 as a primary driver of Phanerozoic climate: GSA Today, v. 14, p. 4-10.

Rudolph, K. W., Schlager, W., and Biddle, K. T., 1989, Seismic models of a carbonate foreslope-to-basin transition, Picco di Vallandro, Dolomite Alps, northern Italy: Geology, v. 17, p. 453-456.

Russo, F., Neri, C., Mastandrea, A., and Baracca, A., 1997, The mud-mound nature of the Cassian platform margins of the Dolomites. A case history: the Cipit boulders from Punta Grohmann (Sasso Piatto Massif, northern Italy): Facies, v. 36, p. 25-36.

Sadler, P. M., 1981, Sediment accumulation rates and the completeness of stratigraphic sections: Journal of Geology, v. 89, p. 569-584.

Sadler, P. M., 1999, The influence of hiatuses on sediment accumulation rates: GeoResearch Forum, v. 5, p. 15-40.

Sadler, P. M., Osleger, D. A., and Montanez, I. P., 1993, On the labeling, length, and objective basis of Fischer plots: Journal of Sedimentary Petrology, v. 63, p. 360-368.

Saller, A., Armin, R., Ichram, L. W., and Glenn-Sullivan, C., 1993, Sequence stratigraphy of aggrading and backstepping carbonate shelves, Oligocene, Central Kalimantan, Indonesia, in R. G. Loucks, and J. F. Sarg, eds., Carbonate Sequence Stratigraphy: American Association of Petroleum Geologists Memoirs, v. 57, p. 267-290.

Sandberg, P. A., 1983, An oscillating trend in Phanerozoic non-skeletal carbonate mineralogy: Nature, v. 305, p. 19-22.

Sarg, J. F., 1988, Carbonate sequence stratigraphy, in C. K. Wilgus, B. S. Hastings, C. G. S. C. Kendall, H. W. Posamentier, C. A. Ross, and J. C. Van Wagoner, eds., Sea-Level Changes: An Integrated Approach: SEPM Special Publications, v. 42, p. 155-182.

Sarg, J. F., 1989, Middle-Late Permian depositional sequences, Permian Basin, west Texas-New Mexico, in A. W. Bally, ed., Atlas of Seismic Stratigraphy: American Association of Petroleum Geologists Studies in Geology, v. 27-3, p. 140-157.

Sarg, J. F., 2001, The sequence stratigraphy, sedimentology, and economic importance of evaporite-carbonate transitions: a review: Sedimentary Geology, v. 140, p. 9-42.

Sarg, J. F., Markello, J. R., and Weber, L. J., 1999, The second-order cycle, carbonate-platform growth, and reservoir, source, and trap prediction, in P. M. Harris, J. A. Simo, and A. H. Saller, eds., Advances in Carbonate Sequence Stratigraphy: Applications to Reservoirs, Outcrops and Models: SEPM Special Publications, v. 63, p. 1-24.

Sattler, U., 2005, Subaerial exposure and flooding surfaces of carbonate platforms: Amsterdam, Vrije Universiteit (published dissertation), 168 p.

Saxena, S., and Betzler, C., 2003, Genetic sequence stratigraphy of cool water slope carbonates (Pleistocene Eucla Shelf, southern Australia): International Journal of Earth Sciences, v. 92, p. 482-493.

Schlager, W., 1980, Mesozoic calciturbidites in DSDP hole 416A - petrographic recognition of a drowned carbonate platform, in Y. Lancelot, and E. L. Winterer, eds., Initial Reports of the Deep Sea Drilling Project: Washington D.C., US Government Printing Office, v. 50, p. 733-749.

Schlager, W., 1981, The paradox of drowned reefs and carbonate platforms: Geological Society of America

Bulletin, v. 92, p. 197-211.

SCHLAGER, W., 1989, Drowning unconformities on carbonate platforms, in P. D. Crevello, J. L. Wilson, J. F. Sarg, and J. F. Read, eds., Controls on Carbonate Platform and Basin Development: SEPM Special Publications, v. 44, p. 15-26.

SCHLAGER, W., 1991, Depositional bias and environmental change - important factors in sequence stratigraphy: Sedimentary Geology, v. 70, p. 109-130.

SCHLAGER, W., 1992, Sedimentology and Sequence Stratigraphy of Reefs and Carbonate Platforms: American Association of Petroleum Geologists Continuing Education Course Note Series, v. 34, 71 p.

SCHLAGER, W., 1993, Accommodation and supply - a dual control on stratigraphic sequences: Sedimentary Geology, v. 86, p. 111-136.

SCHLAGER, W., 1994, Reefs and carbonate platforms in sequence stratigraphy, in S. D. Johnson, ed., High Resolution sequence stratigraphy: innovations and applications: Liverpool, v. Abstract Volume, p. 143-149.

SCHLAGER, W., 1996, The mismatch between outcrop unconformity and seismic unconformity, in C. A. Caughey, D. C. Carter, J. Clure, M. J. Gresko, P. Lowry, R. K. Park, and A. Wonders, eds., Proceedings of the International Symposium on Sequence Stratigraphy in SE Asia: Jakarta, Indonesian Petroleum Association, p. 3-18.

SCHLAGER, W., 1998, Exposure, drowning and sequence boundaries on carbonate platforms, in G. Camoin, and P. Davies, eds., Reefs and Carbonate Platforms of the W Pacific and Indian Oceans: International Association of Sedimentologists Special Publications, v. 25, p. 3-21.

SCHLAGER, W., 1999a, Scaling of sedimentation rates and drowning of reefs and carbonate platforms: Geology, v. 27, p. 183-186.

SCHLAGER, W., 1999b, Type 3 sequence boundaries, in P. M. Harris, A. H. Saller, and J. A. Simo, eds., Advances in Carbonate Sequence Stratigraphy - Application to Reservoirs, Outcrops and Models: SEPM Special Publications, v. 62, p. 35-45.

SCHLAGER, W., 2000, Sedimentation rates and growth potential of tropical, cool-water and mud-mound carbonate factories, in E. Insalaco, P. W. Skelton, and T. J. Palmer, eds., Carbonate Platform Systems: Components and Interactions: Geological Society Special Publications, v. 178, p. 217-227.

SCHLAGER, W., 2002, Sedimentology and sequence stratigraphy of carbonate rocks: Amsterdam, Vrije Universiteit / Earth and Life Sciences, 146 p.

SCHLAGER, W., 2003, Benthic carbonate factories of the Phanerozoic: International Journal of Earth Sciences, v. 92, p. 445-464.

SCHLAGER, W., 2004, Fractal nature of stratigraphic sequences: Geology, v. 32, p. 185-188.

SCHLAGER, W., AND ADAMS, E. W., 2001, Model for the sigmoidal curvature of submarine slopes: Geology, v. 29, p. 883-886.

SCHLAGER, W., BIDDLE, K. T., AND STAFLEU, J., 1991, Picco di Vallandro (Duerrenstein) - a platform-basin transition in outcrop and seismic model: Field Trip Guidebook, Dolomieu Conference on Carbonate Platforms and Dolomitization: Ortisei, Tourist Office Ortisei, v. Excursion D, 22 p.

SCHLAGER, W., AND BOLZ, H., 1977, Clastic accumulation of sulphate evaporites in deep water: Journal of Sedimentary Petrology, v. 47, p. 600-609.

SCHLAGER, W., BUFFLER, W., AND PHAIR, R., 1984, Geologic history of the southeastern Gulf of Mexico, in R. T. Buffler, W. Schlager, and e. al., eds., Deep Sea Drilling Project Initial Reports: Washington D.C., v. 77, p. 715-738.

SCHLAGER, W., AND CAMBER, O., 1986, Submarine slope angles, drowning unconformities, and self-erosion of limestone escarpments: Geology, v. 14, p. 762-765.

SCHLAGER, W., AND CHERMAK, A., 1979, Sediment facies of platform-basin transition, Tongue of the Ocean, Bahamas, in L. J. Doyle, and O. H. Pilkey, eds., Geology of Continental Slopes: SEPM Special Publication, v. 27, p. 193-207.

SCHLAGER, W., AND JAMES, N. P., 1978, Low-magnesian calcite limestones forming at the deep-sea floor, Tongue of the Ocean, Bahamas: Sedimentology, v. 25, p. 675-702.

SCHLAGER, W., MARSAL, D., VAN DER GEEST, P. A. G., AND SPRENGER, A., 1998, Sedimentation rates, observation span, and the problem of spurious correlation: Mathematical Geology, v. 30, p. 547-556.

SCHLAGER, W., REIJMER, J. J. G., AND DROXLER, A. W., 1994, Highstand shedding of rimmed carbonate platforms - an overview: Journal of Sedimentary Research, v. B64.

SCHLANGER, S. O., 1981, Shallow-water limestones in oceanic basins as tectonic and paleoceanographic indicators, in J. E. Warme, R. G. Douglas, and E. L. Winterer, eds., The Deep Sea Drilling Project: A decade of progress: SEPM Special Publications, v. 32, p. 209-226.

SCHOLLE, P. A., AND EKDALE, A. A., 1983, Pelagic environment, in P. A. Scholle, D. G. Bebout, and C. H. Moore, eds., Carbonate Depositional Environments: American Association of Petroleum Geologists Memoirs, v. 33, p. 619-690.

SCHROEDER, M., 1991, Fractals, chaos, power laws: New York, Freeman, 429 p.

SCHWARZACHER, W., 1954, Die Grossrhythmik des Dachsteinkalkes von Lofer: Mineralogisch-petrographische Mitteilungen, v. 4, p. 44-54.

SHAVIV, N. J., AND VEIZER, J., 2003, Celestial driver of Phanerozoic climate? : GSA Today, v. 13, p. 4-10.

SHERIFF, R. E., 1988, Interpretation of West Africa, line C, in A. W. Bally, ed., Atlas of Seismic Stratigraphy: American Association of Petroleum Geologists Studies in Geology, v. 27-2, p. 37-44.

SHERIFF, R. E., AND GELDART, L. P., 1995, Exploration Seismology: Cambridge, Cambridge University Press, 592 p.

SHINN, E. A., 1968, Burrowing in recent lime sediments of Florida and the Bahamas: Journal of Paleontology, v. 42, p. 879-894.

SHINN, E. A., 1983, Recognition and economic significance of ancient carbonate tidal flats - a comparison of modern and ancient examples, in P. A. Scholle, ed., Recognition of Depositional Environments of Carbonate Rocks: American Association of Petroleum Geologists Memoirs, v. 33, p. 172-210.

SIMO, J. A., EMERSON, N. R., BYERRS, C. W., AND LUDVIGSON, G. A., 2003, Anatomy of an embayment in an Ordovician epeiric sea, Upper Mississppi Valley, USA: Geology, v. 31, p. 545-548.

SLOSS, L. L., 1963, Sequences in the cratonic interior of North America: Geological Society of America Bulletin, v. 74, p. 93-114.

SOREGHAN, G. S., AND DICKINSON, W. R., 1994, Generic types of stratigraphic cycles controlled by eustasy: Geology, v. 22, p. 759-761.

SOUTHGATE, P. N., KENNARD, J. M., JACKSON, M. J., O'BRIEN, P. E., AND SEXTON, M. J., 1993, Reciprocal lowstand clastic and highstand carbonate sedimentation, subsurface Devonian reef complex, Canning basin, Western Australia, in R. G. Loucks, and J. F. Sarg, eds., Carbonate Sequence Stratigraphy: American Association of Petroleum Geologists Memoirs, v. 57, p. 157-180.

SPENCE, G. H., AND TUCKER, M. E., 1997, Genesis of limestone megabreccias and their significance in carbonate sequence stratigraphic models: a review: Sedimentary Geology, v. 112, p. 163-193.

SPEZZAFERRI, S., MCKENZIE, J. A., AND ISERN, A., 2002, Linking the oxygen isotope record of late Neogene eustasy to sequence stratigraphic patterns along the Bahamas margin: results from a paleoceanographic study of ODP Leg 166, Site 1006 sediments: Marine Geology, v. 185, p. 95-120.

STAFLEU, J., 1994, Seismic Models of Outcrops as an Aid in Seismic Interpretation: Amsterdam, Vrije Universiteit (published dissertation), 223 p.

STAFLEU, J., EVERTS, A. J., AND KENTER, J. A. M., 1994, Seismic models of a prograding carbonate platform: Vercors, south-east France: Marine and Petroleum Geology, v. 11, p. 514-527.

STANLEY, D. J., ADDY, S. K., AND BEHRENS, E. W., 1983, The mudline: variability of its position relative to shelfbreak, in D. J. Stanley, and G. T. Moore, eds., The Shelfbreak: Critical Interface on Continental Margins: SEPM Special Publications, p. 279-298.

STANLEY, S. M., AND HARDIE, L. A., 1998, Secular oscillations in the carbonate mineralogy of reef-building and sediment-producing organisms driven by tectonically forced shifts in seawater chemistry: Palaeogeography, Palaeoclimatology, Palaeoecology, v. 144, p. 3-19.

STANTON, R. J., AND FLÜGEL, E., 1989, Problems with reef models: the Late Triassic Steinplatte "Reef" (Northern Alps, Salzburg/Tyrol, Austria): Facies, v. 20, p. 1-138.

STANTON, R. J., JEFFERY, D. L., AND GUILLEMETTE, R. N., 2000, Oxygen minimum zone and internal waves as potential controls on location and growth of Waulsortian Mounds (Mississippian, Sacramento Mountains, New Mexico): Facies, v. 42, p. 161-176.

STEPHENS, N. P., AND SUMNER, D. Y., 2003, Famennian microbial reef facies, Napier and Oscar Ranges, Canning Basin, western Australia: Sedimentology, v. 50, p. 1283-1302.

STRAHLER, A. N., 1971, The Earth Sciences: New York, Harper Row, 824 p.

STRASSER, A., AND HILLGÄRTNER, H., 1998, High-frequency sea-level fluctuations recorded on a shallow carbonate platform (Berriasian and Lower Valanginian of Mount Saleve, French Jura): Eclogae Geologicae Helvetiae, v. 91, p. 375-390.

STRASSER, A., PITTET, B., HILLGÄRTNER, H., AND PASQUIER, J.-B., 1999, Depositional sequences in shallow carbonate-dominated sedimentary systems: concepts for a high-resolution analysis: Sedimentary Geology, v. 128, p. 201-221.

SUESS, E., 1888, Das Antlitz der Erde: Leipzig, Freytag-Tempsky, v. 2.

SURLYK, F., 1997, A cool-water carbonate ramp with bryozoan mounds: Late Cretaceous-Danian of the Danish Basin, in N. P. James, and J. A. D. Clarke, eds., Cool-water Carbonates: SEPM Special Publications, v. 56, p. 293-308.

SWART, P. K., ELDERFIELD, E., AND OSTLUND, G., 2001, The geometry of pore fluids from bore holes in the Great Bahama Bank, in R. N. Ginsburg, ed., Subsurface Geology of a Prograding Carbonate Platform Margin, Great Bahama Bank: Results of the Bahamas Drilling Project: SEPM Special Publications, v. 70, p. 163-173.

TAKAHASHI, T., KOBA, M., AND KAN, H., 1988, Relationship between reef growth and sea level on the northwest coast of Kume island, the Ryukyus: data from drill holes on the Holocene coral reef: 6th International Coral Reef Congress Townsville Proceedings v.3, p. 491-496.

TEBBENS, S. F., BURROUGHHS, S. M., AND NELSON, E. E., 2002, Wavelet analysis of shoreline change on the outer banks of North Carolina: an example of complexity in the marine sciences: National Academy of Sciences Proceedings, v. 99, p. 2554-2560.

THIEDE, J., 1981, Reworked neritic fossils in upper Mesozoic and Cenozoic central Pacific deep-sea sediment monitor sea-level changes: Science, v. 211, p. 1422-1424.

THOMPSON, J. B., 2001, Microbial whitings, in R. E. Riding, and S. M. Awramik, eds., Microbial Sediments: Berlin, Springer, p. 250-269.

THORNE, J., 1995, On the scale independent shape of prograding stratigraphic units, in C. B. Barton, and P. R. La Pointe, eds., Fractals in Petroleum Geology and Earth Processes: New York, Plenum Press, p. 97-112.

TIPPER, J., 2000, Patterns of stratigraphic cyclicity: Journal of Sedimentary Research, v. 70, p. 1262-1279.

TIPPER, J. C., 1993, Do seismic reflections necessarily have chronostratigraphic significance? : Geological Magazine, v. 130, p. 47-55.

TOWNSEND, C. R., BEGON, M., AND HARPER, J. L., 2003, Essentials of Ecology: Malden, Blackwell, 530 p.

TRICHET, J., AND DEFARGE, C., 1995, Non-biologically supported organomineralization: Bulletin de I'Institut oceanographique Monaco, v. 2, p. 203-236.

TROREY, A. W., 1970, A simple theory for seismic diffractions: Geophysics, v. 42, p. 1177-1182.

TUBMAN, K. M., AND CRANE, S. D., 1995, Vertical versus horizontal well log variability and application to fractal reservoir modeling, in C. C. Barton, and P. R. La Pointe, eds., Fractals in Petroleum Geology and Earth Processes: New York, Plenum Press, p. 279 - 293.

TUCKER, M. E., AND WRIGHT, V. P., 1990, Carbonate Sedimentology: Oxford, Blackwell, 496 p.

TURCOTTE, D. L., 1997, Fractals and Chaos in Geology and Geophysics: Cambridge, Cambridge University Press, 396 p.

VAIL, P. R., 1987, Seismic stratigraphic interpretation using sequence stratigraphy. Part 1: Seismic stratigraphy interpretation procedure, in A. W. Bally, ed., Atlas of Seismic Stratigraphy: American Association of Petroleum Geologists Studies in Geology, v. 27-1, p. 1-10.

VAIL, P. R., AUDEMARD, F., BOWMAN, S. A., EISNER, P. N., AND PEREZ-CRUZ, G., 1991, The stratigraphic signature of tectonics, eustasy, and sedimentation, in G. Einsele, W. Ricken, and A. Seilacher, eds., Cycles and Events in Stratigraphy: Berlin, Springer, p. 617-659.

VAIL, P. R., MITCHUM, R. M., TODD, R. G., WIDMIER, J. M., THOMPSON, S., SANGREE, J. B., BUBB, J. N., AND HATLELID, W. G., 1977, Seismic stratigraphy and global changes of sea level, in C. E. Payton, ed., Seismic Stratigraphy - Applications to Hydrocarbon Exploration: American Association of Petroleum Geologists Memoirs, v. 26, p. 49-212.

VAIL, P. R., AND TODD, R. G., 1981, Northern North Sea Jurassic unconformities, chronostratigraphy and sea level changes from seismic stratigraphy: Petroleum Geology of the Continental Shelf of North-West Europe, Conference Proceedings, London Institute of Petroleum, p. 216-235.

VAN BUCHEM, F. S. P., CHAIX, M., EBERLI, G. P., WHALEN, M. T., MASSE, P., AND MOUNTJOY, E., 2000, Outcrop to subsurface correlation of the Upper Devonian (Frasnian) in the Altberta Basin (W Canada) based on the comparison of Miette and Redwater carbonate buildup margins, in P. W. Homewood, and G. P. Eberli, eds., Genetic Stratigraphy on the Exploration and the Production Scales: Centre de Recherches Exploration-Production Elf-Aquitaine Memoirs, v. 24, p. 225-267.

VAN BUCHEM, F. S. P., RAZIN, P., HOMEWOOD, P. W., OTERDOOM, W. H., AND PHILIP, J., 2002, Stratigraphic organization of carbonate ramps and organic-rich intrashelf basins: Natih Formation (middle Createaceous) of northern Oman: American Association of Petroleum Geologists Bulletin, v. 86, p. 21-53.

VAN BUCHEM, F. S. P., RAZIN, P., HOMEWOOD, P. W., PHILIP, J. M., EBERLI, G. P., PLATEL, J.-P., ROGER, J., ESCHARD, R., DESAUBLIAUX, G. M. J., BOISSEAU, T., LEDUC, J.-P., LABOURDETTE, R., AND CANTALOUBE, S., 1996, High-resolution sequence stratigraphy of the Natih Formation (Cenomanian/Turonian) in North Oman: distribution of source rocks and reservoir facies: GeoArabia, v. 1, p. 65-91.

VAN HINTE, J. E., 1982, Synthetic seismic sections from biostratigraphy, in J. S. Watkins, and C. L. Drake, eds., Studies in Continental Margin Geology: American Association of Petroleum Geologists Memoirs, p. 675-685.

VAN LOON, H., ED.,. 1984, Climates of the oceans: Amsterdam, Elsevier., 716 p.

VAN WAASBERGEN, R. J., AND WINTERER, E. L., 1993, Summit geomorhology of western Pacific guyots, in M. S. Pringle, W. W. Sager, W. V. Sliter, and S. Stein, eds., The Mesozoic Pacific: Geology, Tectonics, and Volcanism.: American Geophysical Union Geophysical Monographs, v. 77, p. 335-366.

VAN WAGONER, J. C., HOYAL, D. C. J. D., ADAIR, N. L., SUN, T., BEAUBOUEF, R. T., DEFFENBAUGH, M., DUNN, P. A., HUH, C., AND LI, D., 2003, Energy dissipation and the fundamental shape of siliciclastic sedimentary bodies: American Association of Petroleum Geologists Search and Discovery, Article #40080.

VAN WAGONER, J. C., MITCHUM, R. M., CAMPION, K. M., AND RAHMANIAN, V. D., 1990, Siliciclastic sequence stratigraphy in well logs, cores, and outcrops: concepts for high-resolution correlation of time and facies: American Association of Petroleum Geologists Methods in Exploration, v. 7, 55 p.

VAN WAGONER, J. C., MITCHUM, R. M., POSAMENTIER, H. W., AND VAIL, P. R., 1987, Seismic stratigraphy interpretation using sequence stratigraphy. Part 2: key definitions of sequence stratigraphy, in A. W. Bally, ed., Atlas of Seismic Stratigraphy: American Association of Petroleum Geologists Studies in Geology, v. 27-1, p. 11-13.

VAN WAGONER, J. C., POSAMENTIER, H. W., MITCHUM, R. M., VAIL, P. R., SARG, J. F., LOUTIT, T. S., AND HARDENBOL, J., 1988, An overview of the fundamentals of sequence stratigraphy and key definitions, in C. K. Wilgus, B. S. Hastings, C. G. S. C. Kendall, H. W. Posamentier, C. A. Ross, and J. C. Van Wagoner, eds., Sea-level Changes: an Integrated Approach: SEPM Special Publications, v. 42, p. 39-45.

VEIZER, J., GODDERIS, Y., AND FRANCOIS, L. M., 2000, Evidence for decoupling of atmospheric CO2 and global climate during the Phanerozoic eon: Nature, v. 408, p. 698-701.

VEIZER, J., AND MACKENZIE, F. T., 2004, Evolution of sedimentary rocks, in F. T. Mackenzie, ed., Sediments, Diagenesis, and Sedimentary Rocks: Treatise on Geochemistry: Amsterdam, Elsevier, v. 7, p. 369-407.

VOGEL, K., KIENE, W., GEKTIDIS, M., AND RADTKE, G., 1996, Scientific results from investigations of microbial borers and bioerosion in reef environments: Goettinger Arbeiten zur Geologie Palaeontologie, v. special volume 2, p. 139-143.

WALLACE, M. W., HOLDGATE, G. R., DANIELS, J., GALLAGHER, S. J., AND SMITH, A., 2002, Sonic velocity, submarine canyons and burial diagenesis in Oligocene-Holocene cool-water carbonates, Gippsland Basin, southeast Australia: American Association of Petroleum Geologists Bulletin, v. 86, p. 1593-1608.

WANLESS, H. R., AND DRAVIS, J. J., 1989, Carbonate Environments and Sequences of Caicos Platform. Field Trip Guidebook T374: 28th International Geological Congress, Washington, D.C., p. 75.

WARRLICH, G. M. D., WALTHAM, D. A., AND BOSENCE, D. W. J., 2002, Quantifying the sequence stratigraphy and drowning mechanisms of atolls using a new 3-D forward stratigraphic modelling program (CARBONATE 3D): Basin Research, v. 14, p. 379-400.

WATTS, A. B., 2001, Isostasy and Flexure of the Lithosphere: Cambridge, Cambridge University Press, 458 p.

WATTS, A. B., KARNER, G. D., AND STECKLER, M. S., 1982, Lithospheric flexure and the evolution of sedimentary basins: Philosophical Transactions Royal Society London, v. A305, p. 249-281.

WEBB, G. E., 1996, Was Phanerozoic reef history controlled by the distribution of non-enzymatically secreted reef carbonates (microbial carbonate and biologically induced cement)? : Sedimentology, v. 43, p. 947-971.

WEBB, G. E., 2001, Biologically induced carbonate precipitation in reefs through time, in G. D. Stanley, ed., The History and Sedimentology of Ancient Reef Systems: New York, Kluwer, p. 159-203.

WENDT, J., BELKA, Z., AND MOUSSINE-POUCHKINE, A., 1993, New architectures of deep-water carbonate buildups: evolution of mud mounds into mud ridges (Middle Devonian, Algerian Sahara): Geology, v. 21, p. 723-726.

WENDT, J., AND KAUFMANN, B., 1998, Mud buildups on a Middle Devonian carbonate ramp (Algerian Sahara), in V. P. Wright, and T. P. Burchette, eds., Carbonate Ramps: Geological Society London Special Publications, v. 149, p. 397-415.

WENDTE, J., AND MUIR, I., 1995, Recognition and significance of an intraformational unconformity in Late Devonian Swan Hills reef complexes, Alberta., in D. A. Budd, A. H. Saller, and P. M. Harris, eds., Unconformities and Porosity in Carbonate Strata: American Association of Petroleum Geologists Memoirs, v. 63, p. 259-278.

WENDTE, J. C., STOAKES, F. A., AND CAMPBELL, C. V., 1992, Devonian-Early Mississippian carbonates of the Western Canada sedimentary basin: a sequence stratigraphic framework: SEPM Short Course Notes, v. 28, 255 p.

WEST, B. W., AND BROWN, J. H., 2004, Life's universal scaling laws: Physics Today, v. 57, p. 36-42.

WEST, G. B., AND BROWN, J. H., 2002, Allometric scaling of metabolic rate from molecule and mitochondria to cells and mammals: National Academy of Sciences Proceedings, v. 99, p. 2473-2478.

WEYL, 1970, Oceanography: New York, Wiley, 535 p.

WHITAKER, F. F., AND SMART, P. L., 1990, Active circulation of saline ground waters in carbonate platforms: evidence from the Great Bahama bank: Geology, v. 18, p. 200-203.

WHITE, W. B., 1984, Rate processes: chemical kinetics and karst landform developments, in R. G. LaFleur, ed., Groundwater as a Geomorphic Agent: Boston, Allen Unwin, p. 227-248.

WILLIAMS, D. F., 1988, Evidence for and against sea-level changes from the stable isotopic record of the Cenozoic, in C. K. Wilgus, B. S. Hastings, C. A. Ross, P. H., J. C. Van Wagoner, and C. G. S. C. Kendall, eds., Sea-level Changes - an Integrated Approach: SEPM Special Publications, p. 31-36.

WILLIAMS, T., KROON, D., AND SPEZZAFERRI, S., 2002, Middle and upper Miocene cyclostratigraphy of downhole logs and short- to long-term astronomical cycles in carbonate production of the Great Bahama Bank: Marine Geology, v. 185, p. 75-93.

WILLIS, B. J., AND GABEL, S. L., 2003, Formation of deep incisions into tide-dominated river deltas: implications for the stratigraphy of the Sego Sandstone, Book Cliffs, Utah, U.S.A.: Journal of Sedimentay Research, v. 73, p. 246-263.

WILSON, J. L., 1975, Carbonate Facies in Geologic History: New York, Springer, 471 p.

WINKER, C. D., AND BUFFLER, R. T., 1988, Paleogeographic evolution of early deep-water Gulf of Mexico and margins, Jurassic to Middle Cretaceous (Comanchean): American Ssociation of Petroleum Geologists Bulletin, v. 72, p. 318-346.

WINTERER, E. L., 1998, Cretaceous karst guyots: new evidence for inheritance of atoll morphology from subaerial erosional terrain: Geology, v. 26, p. 59-62.

WINTERER, E. L., VAN WAASBERGEN, R. J., MAMMERICKX, J., AND STUART, S., 1995, Karst morphology and diagenesisof top of Albian limestone platforms, Mid-Pacific Mountains, in E. L. Winterer, and W. W. Sager, eds., Proceedings of the Ocean Drilling Program, Scientific Results: College Station, Ocean Drilling Program, v. 143, p. 433-470.

WOLANSKI, E., 1993, Water circulation in the Gulf of Carpentaria: Journal of Marine Systems, v. 4, p. 401-420.

WOLF, K. H., 1965, Gradational sedimentary products of calcareous algae: Sedimentology, v. 5, p. 1-37.

WOOD, R. A., 1999, Paleoecology of the Capitan Reef, in A. H. Saller, P. M. Harris, B. L. Kirkland, and S. J. Mazzullo, eds., Geologic Framework of the Capitan Reef.: SEPM Special Publications, v. 65, p. 129-137.

WORTMANN, U. G., AND WEISSERT, H., 2001, Tying platform drowning to perturbations of the global carbon cycle with a delta 13C-curve from the Valanginian of DSDP Site 416: Terra Nova, p. 289-294.

WRIGHT, V. P., AND BURCHETTE, T. P., 1996, Shallow-water carbonate environments, in H. G. Reading, ed., Sedimentary Environments: Processes, Facies, Stratigraphy: Oxford, Blackwell, p. 325-394.

YANG, W., 2001, Estimation of duration of subaerial exposure in shallow-marine limestones- an isotopic approach: Journal of Sedimentary Research, v. 71, p. 778-789.

YATES, K. K., AND ROBBINS, L. L., 1999, Radioisotopic tracer studies of inorganic carbon and Ca in microbially derived CaCO3: Geochimica et Cosmochimica Acta, v. 63, p. 129-136.

ZAMPETTI, V., SCHLAGER, W., VAN KONIJNENBURG, J. H., AND EVERTS, A. J., 2004, Architecture and growth history of a Miocene carbonate platform from 3-D seismic reflection data, Luconia Province, Malaysia: Marine and Petroleum Geology, v. 21, p. 517-534.

ZANKL, H., 1969, Der Hohe Göll. Aufbau und Lebensbild eines Dachsteinkalk-Riffes in der Obertrias der nördlichen Kalkalpen.: Abhandlungen der Senckenbergischen Naturforschenden Gesellschaft: Frankfurt, Waldemar Kramer, v. 519, 122 p.

ZÜHLKE, R., BECHSTÄDT, T., AND MUNDIL, R., 2003, Sub-Milankovitch and Milankovitch forcing on a model Mesozoic carbonate platform - the Latemar (Middle Triassic, Italy): Terra Nova, v. 15, p. 69-80.

Index

First author index